MEDIA AND COMMUNICATIONS - TECHNOLOGIES, POLICIES AND CHALLENGES

KNOWLEDGE MANAGEMENT AND COMMUNICATION IN THE INFORMATION AGE

Media and Communications - Technologies, Policies and Challenges

Additional books in this series can be found on Nova's website under the Series tab.

Additional e-books in this series can be found on Nova's website under the e-book tab.

MEDIA AND COMMUNICATIONS - TECHNOLOGIES, POLICIES AND CHALLENGES

KNOWLEDGE MANAGEMENT AND COMMUNICATION IN THE INFORMATION AGE

RAFIQUL ISLAM
EDITOR

nova publishers
New York

Library of Congress Cataloging-in-Publication Data

ISBN: 978-1-62808-250-0

Published by Nova Science Publishers, Inc. † New York

CONTENTS

PREFACE

The information age came with the promise of transparency greater than anything witnessed heretofore by humanity. Of course, transparency is the essence of foresight and knowledge. The hope was for a greater accountability that would follow increased transparency, compelling policy makers to depend on knowledge and foresight rather than disinformation and hidden agendas. Recent events of the new millennium indicate that the increase has been in opacity and disinformation. The information age, often dubbed as the 'knowledge era', has become the antithesis of knowledge, however; even its most ardent proponents admit that. Despite globalization, the information age has failed to generate knowledge-based decision-making tools.

The failures of the information age to generate knowledge and remove or attenuate much of the basis of international conflicts are symptomatic of a deeper illness. Science has become the captive of the 'New Science' that focuses on short-term, tangible, and status-quo, and doesn't even allow the real science to face in a direction that would allow the knowledge-model to launch. This science is incompatible with nature, even in competition with nature, and is not even facing the same direction as the science that is needed to increase knowledge — the only process on which rational hopes of discovering the truth may be placed. Consequently, all the promises of the information age have been failing spectacularly. Today, we have medicines that don't cure, an economy that doesn't economize, education systems that don't educate, purification techniques that do not purify, and justice systems that award plaintiffs or deliver discriminatory treatment of an accused depending on ethnic origin, religious belief, or skin colour. Today, we do collect data at an unprecedented rate, and, yet, we do not process a single set of data with any technique that is truly natural. Today, in the name of simulation, airplanes do not emulate how birds fly, computers do not emulate how the brain computes, cameras do not emulate how the human visual cortex captures and fixes an image, electricity disdains to emulate lightning (insufficiently continuous) or the electric eel (too dependent on the material properties of the receiving medium), commercial ships forego natural sources of locomotion like wind, and submarines do not emulate how fish navigate beneath the water's surface. If nature is the truth, our New Science must be falsehood.

Chapter 1 - Following the Industrial Age, which lasted between the latter half of 18th century and the former half of 19th century, Information Age, also known as the Computer Age or Information Era, began to emerge at the end of 20th century. While Industrial Age could be seen as a shift from manual labor-based economy to machine-based economy,

Information Age could be regarded as a transformation from machine-based economy to knowledge-based economy. General Motors Cor. might be given as an example of the richest companies of the world during the Industrial Age while Google Inc. is among the richest companies of current age, namely Information Age.

Information Age has been characterized by a vast increase in the production and access to information and considerable advances in information technologies. These advances have affected many fields of our lives and brought various and obvious changes. Of these fields, one which is the main focus of this study is education. The advances in information technologies have increased the number of educational tools available to instructors and learners. Of these educational tools, one which is widely discussed in the last decade in literature is podcasting. However, the studies generally discuss the benefits and challenges or effectiveness of using podcasting either focusing on learner-produced podcasts or instructor-produced podcasts. They do not reflect an integrated approach to the discussion of the educational use of podcasting. This study is aimed at explaining the benefits and challenges of using podcasting as an educational tool with an integrated approach taking into account the different aspects of both learner-produced podcasts and instructor-produced podcasts. Three databases, Wiley Inter science, ERIC and ieeexplore were searched with the keywords 'education' and ‘podcasting’ to find peer-reviewed articles related to the use of podcasting in education published between the years 2000 and 2009 using Central Search interface. Numerous articles were considered but narrowed down to 15 articles that form integral part of this study. Articles were grouped according to their subjects and investigated to discuss the benefits and challenges of using podcasting as an educational tool. This study shows that Podcasting has various benefits in both classroom and distance education. Since podcast files are portable, they are time and place independent and, thereby, establish a continuous learning flow. Both producing and using podcast files are inexpensive and easy to operate. Due to its subscription-based nature, podcasting offers effective use of time. Instructor-produced podcasts are a preferred resource to revise classes for those attending the classes and to learn for those missing. The benefits of learner-produced podcasts outweigh the benefits of instructor-produced podcasts. The production of podcasts by learners leads them to develop a sense of ownership, spend time to make research and study in a planned and disciplined way. In addition, it allows for learner-learner and instructor-learner interaction requiring the active involvement of learners and collaboration among them. Podcasting helps learners to improve their oral skills and engage in knowledge creation. On the other hand, podcasting has some challenges such as requiring digital competency, technical infrastructure and time.

Chapter 2 - A variety of reasons exist for agencies to pursue managed ramps. Current funding constraints and difficulty in gaining environmental and public approval for large-scale construction projects has forced agencies to continue considering alternate solutions to roadway widening to mitigate congestion. Areas for potentially improving freeway performance are ramp management and location. The common motivations for ramp management explored in a recent study were to preserve quality of freeway flow, or speed or vehicle type restrictions due to horizontal and vertical limitations along the ramp. Applying managed lanes operational strategies to ramps could maximize existing capacity, manage demand, offer choices, improve safety and generate revenue.

A variety of freeway and ramp traffic mixes, number of freeway mainlanes, ramp spacing, and different ramp merge conditions were analyzed. The VISSIM microsimulation traffic software was used for modeling all scenarios. Performance measures used to compare outcomes from each combination of roadway environment factors were freeway mainlane and ramp weaving speeds.

Results confirm expectations for many scenarios in that higher freeway mainlane and ramp weaving speeds were recorded for longer managed lane ramp spacing from the freeway entrance ramp. Ramp merge conditions with a full auxiliary lane exhibited higher speed performance when compared against a forced merge and an acceleration lane ramp transition. The speed differential between freeway mainlane traffic and ramp weaving traffic was higher with a decreasing percentage of autos in freeway traffic mix.

Chapter 3 - A typical beam element in two dimension is chosen which incorporates the damage parameters (spread, location and intensity of the damage) in its stiffness matrix. A program in MATLAB (7.0) was generated to analyze the system using finite element technique. For that element, the intensity of damage is given as a main parameter, without quantifying or qualifying the type of damage. A general methodology is presented for identifying stiffness reduction by incorporating the results obtained from the finite element method with fuzzy logic.

Chapter 4 - Data storage and data communication currently pose a major problem for all parties involved with power quality and power systems monitoring. The problem arises from the tremendous amount of data involved. There is a common desire in the power industry to find new techniques for high accuracy data compression and data storage. This paper introduces the details of a novel lossless data compression technique that is very suitable for application to power system disturbance waveforms. The method is developed a new fast one dimensional (1-D) convolution decimation algorithm with fast wavelet packet. The algorithm is four times faster than standard convolution decimation. The Arithmetic coding technique is then used to reduce the amount of data more effectively, thereby lowering down the cost related with data storage and transmission. Full technique implementation details are given, and test cases based on power-quality waveforms are analyzed.

Chapter 5 - Faults Detection and Diagnosis (FDD) is a very important stake for industrial processes, since faults are often translated by unacceptable losses of productivity.

This article describes a new approach which allows detecting and diagnosing faults in a column of automated continuous distillation, which is a strongly unstable non-linear system.

It is based on the use of a RBF neuronal network (Radial Basis Function) as classifier of the various functioning modes (normal mode and abnormal mode). The determination of functions' centres of neurons' activation of the hidden layer is made by means of the K-means algorithm.

Principal Component Analysis (PCA) is employed for the graphic representation of the various functioning classes on a lower dimension space.

The studied distillation is a methylcyclohexane distillation from a toluene\ methylcyclohexane binary mixture.

Chapter 6 - Passenger and commercial vehicle crossings at the U.S.-Mexico border play an important role in the economies of the border region and both nations. The time it takes to cross the border from Mexico into the U.S. (and from the U.S. into Mexico), for both passenger and commercial vehicles, is an important element in making travel and business decisions. Unfortunately, there is no system in place to relay and archive crossing times of commercial and passenger vehicles. Public and Private stakeholders in the El Paso – Ciudad Juarez region have expressed a strong need for such a system.

This paper describes the development of a prototype Border Crossing Information System (BCIS). The objective of the prototype BCIS was to design and implement a centralized repository of archived border-crossing data and develop techniques to extract and relay traveler information. Since full-scale information system requires a considerable amount of time and resource, a prototype BCIS with limited functions was developed for this research. However, the overall goal of the prototype was to demonstrate how stakeholder needs could be fulfilled by - identifying components for a system, and integrating available technologies.

The prototype BCIS was developed and implemented using data warehouse concepts and principles, and a multi-tier architecture (client-server architecture). Using web-based tools, the prototype system provides archived border-crossing data and traveler information related to travel conditions at the border to stakeholders through a single window map-based website.

Chapter 7 - Artificial Intelligence (AI) and image processing can be utilized now in a robust control system. This paper addresses a scenario where AI and Image Processing are applied to a specific machine tool to achieve control of simple drilling process. Vision based machine control provides a broad functionality when compared to existing control systems (e.g. Numerical control (NC), Computer numerical control (CNC) etc). In this study AI and Image Processing are used by computer controlled drill machine to demonstrate the potential of an AI control system. Operational reliability & accuracy, reduced production cost and rapid recovery from malfunctions are the goals this paper seek. Neural network models were implemented to find formulae to describe the motions of the various axes in the drill model. A machine bed was constructed, requiring work positioning on three axes which are controlled indirectly by a computer running imaging software which will send work bed motion commands to the job positioning controller.

Chapter 8 - This paper provides an over view of superstrate loading effects on rectangular patch antennas with emphasis on theoretical design and practical techniques for last three decades. Several theoretical analysis methods have been summarized, including Transmission line, Method of Moments (MOM), Finite-Difference Time- Domain (FDTD) and Spectral Domain Immittance (SDI) as well as Modified Wolff Method (MWM). The characteristics of covered patch antenna have also been studied using genetic algorithm and neural network methods. In particular, effect of superstrate loading on co-polarisation and cross-polarisation as well as aperture coupled patch antenna with emphasis on radiation and directivity have also been studied. In addition to effect of dielectric loading on the patch performances, the various methods for their compensation have also been described. The paper aims those readers who desire to know about the latest developments in the field of superstrate loaded rectangular patch antennas.

Chapter 9 - One of the alternative technologies for communications that is widely used across the globe is Power Line Communications or Carrier “PLC” technology. Power Line carrier is a term used for the power cables as communication media. The broadband PLC is used for home automation and broadband data transfer i.e. internet in a low voltage distribution network in home electrical wring or between home and transformer station. The narrowband PLC run on overhead lines network between power plant and transformer station. This survey has been concentrated on over head network between substations.

The advantage of using power line carrier (PLC) for data transfer is that it does not require the installation of new cables. But the main disadvantage of PLC is a low transmission capacity.

In this work, the complex components of existing power-line carrier are explained and modeled with respect to frequency band up to 500 kHz. Also the exact teleprotection system (TPS) unit is designed as dynamic systems model, to enable the connection of four protection command on one unit.

In this work, the data rate is also developed for the data transfer from 1200 bps using BFSK to 2400 bps using QDPSK.

In order to achieve the authors' research goals a number of PLC and TPS systems studies were performed. Additionally, a set of tests were also performed to investigate the consistency of the PLC and TPS models. The results showed that the models responded satisfactorily according to the expected results of the tests.

Chapter 10 - A triangular fractal antenna with good radiation performances for multiband applications has been proposed in this paper, and found that increase in number iterations led to improvement is VSWR & return loss. It has also been observed that with increase in the number of iterations the bandwidth of the antenna increases and on second and third iterations, the antenna starts showing the multiband characteristics. The operating frequencies are found to be 8.9 and 9.6 GHz for II-iteration and 16.7, 18.3 19.6 & 22.2 GHz for III-iteration with acceptable bandwidth and S_{11}< -10 dB (VSWR < 2). The radiation characteristics, VSWR, directivity, gain and S_{11} of the proposed antenna are described and simulated using Ansoft HFSS and the obtained results have also been validated using HP810B network analyzer and found in good agreement.

Chapter 11 - All the crash data were based on the reported police crash report. Crash data were obtained from Nevada Department of Transportation (NDOT). Those data were collected for five years, from 1996 to 2000. The obtained data did not provide more information about psychological aspects of the individuals for traffic accidents. The fatality rate of older people exceeds than those of the younger people. Approximately 45 percent of fatal crashes were related to either alcohol or drugs. Fatal and total pedestrian crashes reduced from 1996 to 2000 by 35 percent and 10 percent, respectively. The age segments 64 years and younger comprised of approximately 3 fatalities per 100,000 population comparatively lower than the older population segments of 65 years and older, which contributed approximately 7 fatalities per 100,000 population. On the other hand, the injury crash rate is declined along with decrease in age of pedestrians. Fatal pedestrian crashes are greater in elder population groups than in younger population groups. Injury crashes are greater in younger population groups than elder population groups.

Chapter 12 - This paper provides a case study of superstrate loading effects on circular patch antennas with emphasis on theoretical and practical aspects for last three decades. Several theoretical analysis methods have been summarized, including Transmission line, Method of Moment (MOM), Finite Difference Time Domain (FDTD) as well as Modified Wolf Method (MWM). In particular, effect of superstrate loading on co-polarisation and cross-polarisation with emphasis on radiation and directivity has also been studied. The paper aims those readers who involve R & D in the field of superstrate loaded circular patch antennas.

Chapter 13 - The main objective of this paper is to show one way of optimization, analysis and design of microwave active circuits, based on the transformation of the reflection coefficient for an amplifier or a microwave oscillator. This improved method is based on a non linear optimization of the circuit. This way consists on introducing a transformation invariant which allows to test the appropriate transistor for a given frequency bandwidth.

The transformation invariant is called coefficient N, and it determines very quickly, the possible modifications of the active multiport network by external circuits connected to it. This procedure is applied by looking at the microwave transistor as a three- port or as an equivalent two-port. The present method appears to be accurate and fast in the design and optimization of a microwave oscillator or an amplifier based on H.F transistors.

Chapter 14 - Increasing demand for wireless data communication and services has introduced new technologies capable of handling higher data rate. 4th Generation wireless mobile systems are widely expected to cope with such demands. Orthogonal Frequency Division Multiplexing (OFDM) is proving to be as the multiple access technology and numerical modulation scheme in 4G. This paper investigates the Peak Average Power Ratio (PAPR) reduction in the OFDM-4G transmitter signal by using the pilot tones and unused carriers (PTUNC) algorithm. A dynamic adapting for a kind of 4G that can be selected depends on the chosen design parameters. The simulation results are shown for different OFDM-4G parameters.

Chapter 15 - This work, is placed within the framework of Detection by Pattern Recognition (DPR), the authors describe an approach with the detection of defects applied to a continuous distillation column which contains a binary mixture of toluene methylcyclohexane. Initially, by using the algorithm K-Nearest-Neighbors (KNN) and secondly, by using the Neural Networks of the type MLP (Multi Layer Perceptron) like classifiers of defects. The training base of the authors' classifiers consists of two modes: normal mode and abnormal mode which is caused by an increase in the distillation to 100% of its capacity. In continuation the authors work out a base of a test in order to evaluate the quality of the various classifiers. To this end, the authors proceed in two different ways: Firstly, the base of the test consists of samples in normal mode and samples in an abnormal mode which do not resemble the abnormal mode used during the training. The authors take for example the dysfunction of feed rate to 100% of its capacity. Secondly, the base of the test consists of samples in normal mode and samples in abnormal mode, but this time the abnormal mode is similar to the abnormal mode used during the phase of training.

Chapter 16 - In this paper, the authors study the multi – carrier direct – sequence code division multiple – access system when the received signal is divided into predetermined spreading sequence, these sub - bands modulates M carriers. Because of the no – knowledge of the phase of the received signal in each sub - band, the receiver cannot coherently combine outputs of the correlators. The authors consider in this paper two non - coherent combining schemes; the equal gain combining and the selecting combining. The calculation of the probabilities of detection and false alarm for both multi - carrier and single carrier systems in a Fading Rayleigh channel demonstrate that the performance of the multi - carrier system is better than the performance of the single carrier in equal gain combining, while the performance of the single- carrier system in the selecting combining is better than the performance of the multi- carrier. The MC-DS-CDMA system using the EGC is robust against the MAI interference in comparison with that using the selecting combining (SC).

Chapter 17 - A lean analysis for the production method of producing 5-burners gas oven at one of the biggest Jordanian companies hereinafter referred to as SABA is carried out. A new approach for process flow and activity relationship analysis is followed. Flow process chart, and Gantt chart are used as a lean tools for documenting the production process. The current method is questioned and brainstormed with the senior management of SABA, consequentlty lean performance measures like Tact time, utilization rate, queuing time etc. were developed, discussed, then used. Based on that, a new design for the production process that can meet the gradual increase of customer demand on 5-burners gas oven is introduced. The main advantages of the new designed production process is that it reduces the average production time by 46%, increases the average utilization of production resources by 36%, and increase production capacity of the production line by 100%. Another important advantage of the new design is that Work in Process (WIP) is decreases, bottleneck is diminished and waiting time of WIP between production stages is totally eliminated.

Chapter 18 - This paper presents two hybrid methods to solve a production planning problem. The problem is found in a producer of food-sweeteners, with several different kinds of products. The case study plant is found in the USA. The aim is to improve convergence for a big optimization problem. The first method uses a Genetic Algorithm in combination with an existing rigorous MILP model. The fitness value of each chromosome is evaluated by solving the MILP model with a highly reduced number of binary variables. The second method is a Tabu search that also uses the rigorous MILP to evaluate each node. The meta-heuristic models are presented and used to solve a few case problems and the results are discussed. Furthermore the two methods are compared in terms of solution times and convergence rate.

Chapter 19 - The sizes, complexity and power ratings of high voltage networks emphasize the need for highly reliable protection systems. Protection is based on information derived from the power system at one or more points, and fast selective protection applied to circuits with geographically separated terminals, such as cables and overhead lines, requires information interchange between these terminals. A secure and uninterrupted supply of electricity is only possible with the help of comprehensive protection and control functions, which ensure the reliable operation of the power system. As the complexity and ratings of electrical power systems increase, so do also the demands on the protective devices and systems which have to protect them from damage and to preserve the power system stability. Protection relays in conjunction with communication links provide the best possible means of selectively isolating faults on high voltage (HV) lines, transformers, reactors and other important items of electrical plants. To prevent the system from failure and damage, the tele protection system (TPS) has to selectively disconnect the faulty part within the shortest possible time.

Chapter 20 - This paper deals with an analysis of different scenarios of thermal runaway in a chemical reactor. The HAZOP method is chosen for the analysis of risk generated by a chemical reaction, while FMEA method is used for the risk analysis linked to the operation of the chemical reactor. It is indicated in this paper the existence of interactions between the chemical reaction and the behavior of the chemical reactor. However, this leads to develop a new analytical approach to this specific type of system, which is based on a combination of (FMEA, HAZOP, Karnaugh Table) methods. A simplified electrical schematic and a simple equation which describes the occurrence of different thermal runaway scenarios are obtained for this chemical reactor.

Chapter 21 - In this paper, a single-machine single-product manufacturing system is considered; the system is assumed to operate under constant production rate that exceeds customer demand rate, customer demand rate is constant and certain. The machine is not perfect and is prone to failure, upon which it needs repair. The machine is also not perfectly efficient and it manufactures non-confirming and imperfect items. The possibility of failure and imperfect production are combined and modeled into a single model. The machine times-to-failure, the repair time, and the item manufactured being non-confirming are random. When the machine is under repair, the demand is backordered till a maximum backorder level is reached and after that the demand is considered to be lost.

The state of the system changes in discrete time and the process is modeled as a Markov chain which is used to find the cost model. The Economic Manufacturing Quantity of a machine which is prone to failure and also may manufacture non-confirming items is derived. Modeling the process in a Markov chain is very beneficial in incorporating backorders and failures into the process.

The results clearly indicate that the total cost is convex with respect to the maximum inventory and the total cost increases when the probability for the machine not to fail decreases. The results also show that the repair success has an effect on the total cost. The total cost is higher when the probability of repair success is low than when the probability is high.

Chapter 22 - During the last 50 years, Artificial Intelligence (AI) made colossal progress. Beginning from the simple heuristic search in space now complicated methods of pattern recognition, planning, control and learning are developed. It seems that the authors are moving towards the creation of AI control system with a high degree of autonomy. Such systems are capable of efficient operation in unpredictable varying and hostile environment, including conflict situations. More and more practitioners in engineering will be involved in the process of AI control systems development. Unfortunately only the small part of up-to-date AI methods and algorithms are applicable to this task. The aim of this article is to list AI directions which, in author's opinion, are suitable for AI control system development. The authors will consider the control systems of unmanned aerial vehicles (UAVs). Such vehicles are characterized by the high dynamics and complicated system of restrictions on control parameters. Nevertheless, the main statements of this paper are applicable to control systems of other mobile objects based on AI principles.

Chapter 23 - This paper provides an over view of design, development and application of active integrated antennas. Active integrated antennas (AIA's) provide a new paradigm for designing modern microwave and millimeter-wave architecture with desirable features such as compactness, low cost, light weight, low profile, minimum power consumption, and multiple functionality. This paper reviews recent research related to this emerging technology with emphasis on its applications in high-efficiency radio-frequency (RF) front-end, filter, and phase arrays, as well as switching antenna. Optical controlling techniques for AIA's are also described.

Chapter 24 - Web page classification is a current research topic. The automated categorization (or classification) of web pages into topical categories has a long history, dating back at least in the early 1960. The Internet makes it possible to share and manipulate a vast quantity of information efficiently and effectively, but the rapid and chaotic growth experienced by the Net has generated a poorly organized environment that hinders the sharing and mining of useful data. The need for meaningful web-page classification techniques is therefore becoming an urgent issue. The main goal of this paper is to find whether Rules classifiers are helpful in web page classification and also the authors make a comparative study on the performance of these classifiers. In this paper the authors use Prism, ZeroR and OneR rule classifiers for the authors' classification purposes.

In: Knowledge Management and Communication …
Editor: Rafiqul Islam
ISBN: 978-1-62808-250-0

Chapter 1

KNOWLEDGE MANAGEMENT IN THE INFORMATION AGE: BENEFITS AND CHALLENGES OF USING PODCASTING AS AN EDUCATIONAL TOOL

***Fatih Gürsul*[1*] *and Sinem Canim*[2]**

[1]Istanbul University Informatics Department, Turkey
[2]Istanbul University Translation Studies Department, Turkey

ABSTRACT

Following the Industrial Age, which lasted between the latter half of 18th century and the former half of 19th century, Information Age, also known as the Computer Age or Information Era, began to emerge at the end of 20th century. While Industrial Age could be seen as a shift from manual labor-based economy to machine-based economy, Information Age could be regarded as a transformation from machine-based economy to knowledge-based economy. General Motors Cor. might be given as an example of the richest companies of the world during the Industrial Age while Google Inc. is among the richest companies of current age, namely Information Age.

Information Age has been characterized by a vast increase in the production and access to information and considerable advances in information technologies. These advances have affected many fields of our lives and brought various and obvious changes. Of these fields, one which is the main focus of this study is education. The advances in information technologies have increased the number of educational tools available to instructors and learners. Of these educational tools, one which is widely discussed in the last decade in literature is podcasting. However, the studies generally discuss the benefits and challenges or effectiveness of using podcasting either focusing on learner-produced podcasts or instructor-produced podcasts. They do not reflect an integrated approach to the discussion of the educational use of podcasting. This study is aimed at explaining the benefits and challenges of using podcasting as an educational tool with an integrated approach taking into account the different aspects of both learner-produced podcasts and instructor-produced podcasts. Three databases, Wiley Inter science, ERIC and ieeexplore were searched with the keywords 'education' and 'podcasting' to find peer-reviewed articles related to the use of podcasting in education

* Email: fatihg@istanbul.edu.tr.

published between the years 2000 and 2009 using Central Search interface. Numerous articles were considered but narrowed down to 15 articles that form integral part of this study. Articles were grouped according to their subjects and investigated to discuss the benefits and challenges of using podcasting as an educational tool. This study shows that Podcasting has various benefits in both classroom and distance education. Since podcast files are portable, they are time and place independent and, thereby, establish a continuous learning flow. Both producing and using podcast files are inexpensive and easy to operate. Due to its subscription-based nature, podcasting offers effective use of time. Instructor-produced podcasts are a preferred resource to revise classes for those attending the classes and to learn for those missing. The benefits of learner-produced podcasts outweigh the benefits of instructor-produced podcasts. The production of podcasts by learners leads them to develop a sense of ownership, spend time to make research and study in a planned and disciplined way. In addition, it allows for learner-learner and instructor-learner interaction requiring the active involvement of learners and collaboration among them. Podcasting helps learners to improve their oral skills and engage in knowledge creation. On the other hand, podcasting has some challenges such as requiring digital competency, technical infrastructure and time.

Keywords*:* Podcast, podcasting, M-learning, Higher Education

INTRODUCTION

Information Age is a period that began around 1970s but named in the late 1980s. On the other hand, the term became widespread around 1990s. At this age, also called Information Era or Compute Age, development is based on not only the production of knowledge but also the formation of knowledge patterns. For instance, Microsoft markets knowledge goods while Google aims at making the access of users simple, relevant and efficient through the knowledge patterns developed by itself using already existing knowledge. Therefore, although a world leader in software, Microsoft was established in 1975, Google was established in 1998 and became a competitor to Microsoft in a short period. Information Age came with developments in information technologies and brought significant changes to many fields of our daily life. One of these fields is education. By promoting the educational environment, information technologies have provided more effective use of existing educational approaches such as problem-based learning, case-based learning and project-based learning. On the other hand, the emerge of social networks due to the developments in information technologies of the digital age we live in such as Facebook, Twitter, Second Life, Blogs have triggered contemporary learning approaches such as Connectivism.

Connectivism is a theory which describes how learning happens in a digital age (Karacapilidis, 2009, p.113) and is becoming very popular since about 2005 (van der Zanden, 2009, p.73). Connectivism also addresses the challenges that many corporations face in knowledge management activities. Knowledge that resides in a database needs to be connected with the right people in the right context in order to be classified as learning (Siemens, 2005).

Connectivism is fed by educational tools such as podcasting. This study is aims to explain the benefits and challenges of using podcasting as an educational tool with an integrated approach taking into account the different aspects of both learner-produced podcasts and instructor-produced podcasts. Background

Podcasting as a concept first began to appear in the Fall of 2000. Ciconte and Jacob regarded podcasting as a tool for electronic communication in fundraising efforts of both small and large scale organizations. A non-profit organization, UNICEF began to use podcasting in early 2005 in order to inform its (prospective) donors of its program and funds. The early podcasts by UNICEF included only audio. Later on, in 2006 UNICEF started to produce video podcasts, also called vodcasts. According to Ciconte and Jacob, both audio and video versions of podcasts are an easy and inexpensive way of electronic communication to reach large numbers of people from various platforms (Ciconte & Jacob, 2008).

The term 'podcasting' started to be used in 2004 as a result of the growth in the popularity of RSS (Really Simple Syndication) technology for distributing audio content. RSS is a simple XML syntax for describing recent additions of content to a website (Barsky, 2006). With RSS technology, the user does not need to re-search for updates or refresh website but is automatically notified of recent additions to the website. Initially RSS could only be used for text files, but overtime it also started to be used for audio files including podcasts. Nowadays, user can subscribe to a web page containing RSS 2.0 tagged audio files on designated web pages and automatically download these files directly into their personal computer using an audio management program (Fernandez, Simo & Sallan, 2009). The audio management software used to subscribe to preferred content and transfer the updated content automatically to personal computers is called aggregator or podcatcher (Thomas, 2009: 351). iPodder, iTunes, Odeo, Dobbler are the commonly used examples of podcatchers. In addition, podcasts could also be downloaded from a website manually. The files downloaded to a personal computer could be played on the same medium, namely a personal computer using a media playing software such as Windows Media Player, iTunes, Winamp or transferred to a portable media playing device such as iPod, MP3, MP4 player and personal digital assistant in order to make use of its 'anytime, anywhere' feature.

The term 'podcasting' was created by former MTV VJ Adam Curry (Podcast Alley), and is a combination of 'broadcasting' and iPod, the popular MP3/MP4 player produced by Apple Computer (Lee, McLoughin & Chan, 2008; Lim, 2006; Donnelly & Berge, 2006; Matthews, 2006). The 'pod' in podcasting might give the misleading idea that podcast files could only played on iPod. However, any media playing device including personal computers, laptops, MP3 and MP4 players, mobile phones, personal digital assistants that are capable to playback podcast files could be used for listening/watching the files. Curry used AppleScript to develop a software application called iPodder which was open sourced and made available to other programmers and these programmers developed the software creating a complete software product (Matthews, 2006).

In its edition of 2005, the New Oxford American Dictionary defined the term 'podcasting' as 'a digital recording of a radio broadcast or similar program, made available on the internet for downloading to a personal audio player' (Oxford University Press, 2009). As the definition indicates, early podcasts only included audio. But today there are podcasts in video format which are also called a 'vodcast'. Although today podcast files are still generally in audio format, video podcasts are getting common with the launch of new media products supporting video format, the expansion of bandwidth capacity and the development of software used for creating and playing animations. It is obvious that with the arrival of 2006, there was an expansion in the definition of 'podcasting'. In 2006, Lim defined podcasting as the authoring of, and subscription to, audio and/or video files on the internet for downloading to the user's personal computer.

It is not with the arrival of podcasting that we started to access and play audio/video files on the internet. Video/audio streaming and downloading audio/video files from the internet have been common practices for several years. Those people who have little or no knowledge about podcasting generally end up with thinking that podcasting has no difference from streaming and designating the action with the term 'podcasting' is just a new commercial trend of IT market to attract the attention of people. Despite having some similar features such as ability to publish and access both through Internet, have feedback from the audience through comments, podcasting is a new type of content sharing. The section titled Special features of podcasting will give more idea about the difference of podcasting from media streaming.

Until here, we have mentioned of two types of podcasts: audio and video podcasts. However, there is one more type that also deserves attention. That is enhanced podcasts. As Thomas (2009) suggests, enhanced podcasts both include image and audio and are embedded with chapter markers that make it easy to go backward and forward through the file and saves time. Geoghegan and Klass (2007) state that enhanced podcasts might include chapter markers, hyperlinks and artworks, which make it more attractive and interesting for the listener in that listening is not static and linear but the listener could click on hyperlinks and markers to skip from one point to another and the artwork changes as the content of the show changes from one subject to another.

Podcasts consist of sequential files. Each of these files is called episode. These episodes vary in terms of duration according to the choice of the producer. For example, the episodes produced weekly by National Association of Photoshop Professionals (NAPP) to explain advanced Photoshop applications and published on the website 'Photoshop User Tv' (www.photoshopusertv.com) last for 20-25 minutes while the episodes published on the website of 'Grammar Girl' (http://grammar.quickanddirtytips.com/) and produced to share the answers for the questions about advanced English grammar asked by the visitors of the website last for approximately 5 minutes.

Special Features of Podcasting

Podcasting was selected as the Word of the Year for 2005 by the editors of the New Oxford American Dictionary and added to the following online update of the dictionary (Oxford University Press, 2009). Sahin (2006) stated that terms that are defined as special words with narrow and limited meaning and belonging to a special discipline and used to help those people related to that certain discipline communicate more easily transfer to the general language if the people using the general language are close to the discipline and know its terms. From this aspect, the presence of the term 'podcasting' in a general dictionary, New Oxford American Dictionary, might be seen as an indicator of common practice of podcasting.

Nowadays, people make use of the podcasting in many fields such as education, tourism, marketing, advertising, entertainment, sports and journalism. Of these fields, podcasting has received a spectacular increase in attention over the last few years (Fernandez, Simo & Sallan, 2008). This study will, therefore, focus on the use of podcasting in the field of education.

As an attempt to frame the concept of 'podcasting' more precisely, it would be beneficial to consider the following special features of podcasting:

- In podcasting, downloads are file-based (Dixon & Greeson, 2006). Each episode is downloaded entirely from Internet to a computer or a mobile device. Once download is complete, there is no need for Internet connection to play the downloaded episode. Podcasting, therefore, differentiates from media streaming which is playing digital media while it is being downloaded and, therefore, requires Internet connection during the playing time.
- Podcasting is subscription-based (Dixon and Greeson, 2006). It makes use of RSS technology. Once subscribed to an RSS feed, there is no need to visit the website to check if any new content has been published. RSS feed will periodically check if any updates have been published and automatically inform the subscriber of the new content. If the subscriber has a podcast aggregator, also known as 'podcatcher', installed on his/her computer, the new content will automatically be downloaded to the computer when Internet connection is available.
- The content could be consumed on a portable device (Dixon and Greeson, 2006) with the capability of playing media. iPod or other MP3, MP4 players and modern mobile phones and PDAs are among those devices that could play media.
- Podcast episodes could consist of audio or video or both. Episodes could also include text and sometimes could be supported by accompanying text files. Especially when episodes are produced for educational uses, the use of accompanying texts will enhance the power of podcasting. For example, in language teaching it is necessary for learner to learn both the pronunciation and spelling of words. Therefore, an episode teaching the pronunciation of words should be accompanied by a text showing the spelling of the words.

The use of podcasting is an opportunity for education. Several universities have set up podcasting. It was in 2002 that Georgia College and State University introduced podcasting in some courses. In 2004, at Duke University over 1600 first-year students were given MP3 players and used podcasting as an educational medium. Similarly, in 2005 the School of Education at Drexel provided students with MP3 players. These universities were followed by University of California, Berkeley, Princeton University, Stanford University and University of Canterbury. Although there are a good number of universities using podcasting for educational purposes only a few of which were mentioned above, the literature is still poor of examples, case studies and evaluations, most of them are merely surveys to investigate users' acceptance of podcasting and there is a lack of studies on the effect of podcasting on learning (Lazarri, 2009). This study explains the benefits and challenges of using podcasting as an educational tool with an integrated approach taking into account the different aspects both learner-produced podcasts and instructor-produced podcasts through data gathered from the articles on the use of podcasting in education included in three academic databases.

Table 1. List of articles reviewed

Year	Title of the study	Author(s)	Subject	Database
2009	Achieving Learning Goals with Student-Created Podcasts	Gary R. Armstrong; Joanne M. Tucker; Victor J. Massad	Benefits and challenges of podcasting in education	Wiley
2008	Diffusion of innovations: Smartphones and wireless anatomy learning resources	Robert B. Trelease	Possible uses of podcasting in different domains	Wiley
2008	Educational usages of podcasting	Howard Harris; Sungmin Park	Perspectives of educational use of podcasting	Wiley
2008	Talk the talk: Learner-generated podcasts as catalysts for knowledge creation	Mark J. W. Lee; Catherine McLoughlin; Anthony Chan	Benefits and challenges of using podcasting in education	Wiley
2007	Audio podcasting in a tablet PC-enhanced biochemistry course	Heather Lyles; Brian Robertson; Michael Mangino; James R. Cox	Effects of educational use of podcasting	Wiley
2007	Multimedia-casting syndication for educational purposes Considerations of a Podcast for use in higher education	Gregory O'Toole	Description of a podcasting implementation in education	Wiley
2007	The emerging Web 2.0 social software: an enabling suite of sociable technologies in health and health care education1	Maged N. Kamel Boulos; Steve Wheeler	Podcasting as a part of other collaborative educational web tools	Wiley
2009	Collaborative web learning tools: Wikis and blogs	Ramoón Rubio; Santiago Martín; Samuel MoráN	Podcasting as a part of other collaborative educational web tools	Wiley
2009	Creative Use of Podcasting in Higher Education and Its Effect on Competitive Agency	Marco Lazzari	Effects of educational use of podcasting	ERIC
2008	Beyond Technology for Technology's Sake: Adv-ancing Multi literacies in the Twenty-First Century	Carlin Borsheim; Kelly Merritt; Dawn Reed	Possible uses of podcasting and in different educational domains	ERIC
2008	Podcasts in Education: Let Their Voices Be Heard	Debra Sprague; Cynthia Pixley	Benefits and challenges of podcasting in education	ERIC
2008	Podcasting's Possibilities	Mimi Leonard	Possible uses of podcasting and in different educational domains	ERIC
2008	The Effectiveness of M-Learning in the Form of Podcast Revision Lectures in Higher Education	Chris Evans	Effectiveness of podcasting in education	ERIC
2006	History to Go: Why iTeach with iPods	Deborah L. Vess	Benefits and challenges of using podcasting in education	ERIC
2008	Distance Education: New Technologies and New Directions	Dale A. Harris; Chuck Krousgrill	Podcasting as a new technology	ieeexplore

THE INTEGRATED APPROACH

As stated before, the aim of this study is to discuss the benefits and challenges of using podcasting as an educational tool with an integrated approach. We will evaluate the aspects of both learner-produced and instructor-produced podcasts which intersect in some points and differentiate in others. With this aim in mind, we performed a search on three databases, Wiley Interscience, ERIC and ieeexplore. We used CentralSearch interface (http://ev4ww2km2g.cs.serialssolutions.com/) offered free by Istanbul University to its academics and students for conducting a simultaneous multiple query. The academics and students need to get a username and a password from the Computer Sciences Application and Research Center of Istanbul University to use the interface for free. We used the keywords 'education' and ‘podcasting’. We narrowed the search by selecting the ‘peer-reviewed’ option and determining the time period as the years between 2000 and 2009. The search in Wiley Interscience database provided 16 results. Eight results were discarded since one was a book review, one was an abstract of an article and one of a proceeding, two were news texts, one was a brochure and two were not directly related to the use of podcasting in education. The search in ERIC database provided six results. Finally, the search in ieeexplore database provided two results. One result could not be accessed because it was not included in the database any longer. In total, 15 articles were studied in-depth. They were grouped according to their subject of study in Table 1. Finally, the articles were investigated to discuss the benefits of and challenges of podcasting as an educational tool.

The articles are grouped under 9 subjects: benefits and challenges of podcasting in education (4), possible uses of podcasting in different domains (1), perspectives of educational use of podcasting (1), effects of educational use of podcasting (2), description of a podcasting implementation in education (1), podcasting as a part of other collaborative educational web tools (2), possible uses of podcasting and in different educational domains (2), effectiveness of podcasting in education (1), podcasting as a new technology (1).

Benefits of Using Podcasting as an Educational Tool

Podcasts are time and location independent digital files (Boulos & Wheeler, 2007; Evans, 2008; Sprague & Pixley, 2008). The capability of podcasting to transfer digital material to other portable devices provides an ‘anytime, anywhere’ media experience (Harris & Park, 2008; Lee et al., 2008; Wu; 2008). Therefore, it offers the opportunity to use time effectively. Learners can play podcast files while doing their daily activities such as walking, travelling, cooking, eating, doing sports. In other words, portability allows for more continuous engagement in course topics (Vess, 2006). According to Evans (2008) travelling is often a time when people engage in less cognitive activity; therefore, portability of podcasting might contribute to continuous engagement by allowing learner to continue the learning activities when it might not normally be possible. In addition, nowadays learners can transport their learning materials easily. In other words, by eliminating the requirement of access to a desktop computer and wired Internet access, podcasting goes one step beyond e-learning by making use of portability and can be defined as a mobile learning (m-learning) tool (Evans, 2008). Harris and Krousgrill (2008) stated that many employed learners wish to access course materials at their workplaces and podcasting could be a solution if there is firewall protection at the workplace that prevents Internet streaming. It could be deduced that podcasting extends learning activity outside classroom and class hours (Sprague & Pixley, 2008; Vess, 2006). The availability of podcasts for preparation of learners for classes before class hours leaves more time for discussion during the face-to-face class hours (Vess, 2006). This is another aspect of the effective use of time.

Another benefit of podcasting is that it is cheap and easy to use (Boulos & Wheeler, 2007; Vess, 2006). There are many free podcasting tools such as Audacity, Podomatic, Gcast, Chaptertoolme on the Internet. Users only need a computer and Internet access to start podcasting. In the case of both learner-produced podcasts and instructor-produced podcasts, learners do not need to spend any money to use podcast episodes while podcasts produced by others are either free or very cheap. In terms of ease of production, Sprague and Pixley (2008) stated that learning how to enhance and edit the audio to be used in podcasting might take time, but once learned, it is easy to do.

Podcasting makes use of RSS technology and is subscription-based. The subscription is made possible by a specific client application (Lazarri, 2008) such as iTunes or iPodder which are reported to be most common by Sprague and Pixley (2008). By using an aggregator or feed reader, learners do not need to check a web site periodically for new episodes and download manually if any available since feed readers access the podcasting server the learner has subscribed to and check the feed and download new episodes, if any, available directly to a computer (Lazzari, 2008; Harris & Krousgrill, 2008; Boulos & Wheeler, 2007; Sprague & Pixley, 2008; Lee et al., 2008; Evans, 2008).

Instructor-produced podcasts provide instructors with the opportunity to emphasize the information, which they feel to be critical for the learners, thus augmenting the teaching material (Harris & Park, 2008). Moreover, learners could easily review what they have already learned to get prepared for an exam (Sprague & Pixley, 2008). Evans (2008) suggested that since podcasts are self-contained broadcasts that immediately engage the learners with the outcomes of a given lecture, learners do not need to recontextualise themselves. However, in the case of notes taken by the learners' themselves or their peers in order to make revisions learners need to re-familiarize themselves with the material in the context in which they first encountered it and such a revision takes more time (ibid.). Moreover, when the learners miss any class hour, they could easily learn what they missed using podcasts (Sprague & Pixley, 2008; Lyles et al., 2007).

When it comes to learner-produced podcasts, it is seen that podcasting allows learners to develop a sense of ownership since they know that the podcasts developed by the learners themselves will be accessed by anybody, or at least by their peers, from anywhere and anytime; therefore, they try hard to produce a high-quality work (Sprague & Pixley, 2008; Harris & Krousgrill, 2008). This means that they will spend time to make research (Sprague & Pixley, 2008; Armstrong, Tucker and Massad, 2009; Lee et al., 2008; Evans, 2008), and to plan their work in a more disciplined way (Vess, 2006). In addition, after the learners present their podcasts, their peers will evaluate these works, which often may require learners' active participation (Lazarri, 2009) and promotes interaction (Vess, 2006; Harris & Krousgrill, 2008). Some might think that without face-to-face dialogue, podcasting could not promote interaction (Leonard, 2008); however, weblogs might be used as a space for instructor-learner and/or learner-learner interaction (Vess, 2006; Harris & Krousgrill, 2008).

In the case of learner-produced podcasts, before and during recordings, the learners will read the text of their podcast aloud many times. This will, firstly, help them to spot their mistakes in the written text since learners might not recognize their mistakes on the written text. Secondly, they will be able to improve their oral skills (Vess, 2006).

The production of podcasts by learner groups encourages collaborative learning (Lee et al., 2008; Vess, 2006; Lazarri, 2009; Harris & Krousgrill, 2008; Rubio, Martín & Morán, 2007) since learners of the team share the same goal and try to act as an orchestra to achieve it. Podcasting, thereby, also facilitates communication and social processes of teamwork because it requires open communication, mutual respect and collective problem solving to achieve a shared goal (Lee et al., 2008). Collaboration is not only facilitated in the design, recording and editing phases of production process but also in the post-production process of podcasting. Since podcasts are processable files, learners could re-edit their works according to the comments their instructors and/or peers. Therefore, it is evident that the end product will be an output of a collaboration of brains with various perspectives.

Lee et al. (2008) stated that during the course of producing podcasts, learners engage in knowledge-creation which involves creating innovative ideas. Therefore, podcasting could also be seen as a tool for motivating creative thinking. This also means that learning by producing podcasts should not be conceived of as a form of learning as acquisition in which learners passively receive knowledge, but as a process in which learners interacting with each other collaboratively engage in generating innovative ideas and solving problems to achieve a shared goal.

Challenges of Using Podcasting as an Educational Tool

Besides its numerous benefits mentioned above, the use of podcasting as an educational tool has some challenges. First of all, podcasting requires some level of digital competency (Vess, 2006; Sprague & Pixley, 2008). Learners need to learn how to record, edit audio/video and send podcast files to an instructor or upload to a server. Therefore, it is evident that instructors need to spend time to teach their students the basics of podcasting, help them to solve technical problems, and guide them in producing and uploading podcasts and creating RSS feeds. In addition, instructors may also need to advance in podcasting which also requires time.

Podcasting requires technical infrastructure. In rural areas, dial-up Internet connection is common, but in podcasting up-to-date hardware and strong Internet connection are recommended (Leonard, 2008). Therefore, in areas where the required technical infrastructure is not available, it would be considerably challenging to make use of podcasting technology. The solution might be using the laboratories offered by educational institutions. However, this might not be so easy due to the constraints related to working hours of laboratories and number of computers. In addition, during the recordings in laboratories the ambient noise would negatively affect the quality of podcasts (Leonard, 2008).

Another issue related to producing podcasts is that it takes time to produce podcasts. Producing a podcasts requires deciding on a topic, conducting the necessary research, writing out a script, practicing before recording, editing the recording and uploading to a server. Both learners and instructors might find these tasks challenging and abstain from producing podcasts; however, once they see the benefits of podcasting, time becomes a challenge to overcome (Sprague & Pixley, 2008).

CONCLUSION

Originating in 2004 as a result of the advance in RSS technology, podcasting is an innovative, easy-to-use tool for information sharing and communication with wider audiences at no or very low cost. Its use in the field of education gained popularity in the last decade. On one side, more universities are gradually adopting this new technology as a tool for mobile learning. On the other side, the number of scientific reports and papers discussing the impacts, benefits and challenges of using podcasting in the field of education is increasing. Though being important and valuable, these studies generally do not assess this new phenomenon with an integrated approach. They focus on either instructor-produced podcasts or learner produced podcasts. In this study, we tried to reflect the benefits and challenges of using podcasting as an educational tool with a new perspective which takes into account the different aspects of both learner-produced podcasts and instructor-produced podcasts.

In order to gather data, we reviewed 15 articles which we accessed through three databases, Wiley Interscience, ERIC and ieeexplore. The articles are peer-reviewed studies published between 2000 and 2009 about the use of podcasting as an educational tool. As the result of this literature review, we reached several conclusions. Firstly, podcasting has various benefits in face-to-face and distance education that call the attention of those learners and instructors who believe the power of information technologies in improving pedagogical methods and eager to update themselves. Since podcast files are portable, they are time and place independent and, thereby, establish a continuous learning flow. Both producing and using podcast files are cheap and easy. Due to its subscription-based nature, podcasting offers effective use of time. Instructor-produced podcasts are a preferred resource to revise classes for those attending the classes and to learn for those missing. The benefits of learner-produced podcasts outweigh the benefits of instructor-produced podcasts. The production of podcasts by learners leads them to develop a sense of ownership, spend time to make research and study in a planned and disciplined way. In addition, it allows for learner-learner and instructor-learner interaction requiring the active involvement of learners and collaboration among them. Podcasting helps learners to improve their oral skills and engage in knowledge creation. On the other hand, podcasting has some challenges such as requiring digital competence, technical infrastructure and time. In order to overcome the inadequate competence level of both instructors and learners, education institutions could give courses on podcasting. Instructors experienced in podcasting could take a step as volunteer podcasting instructors.

We think that the popularity of podcasting as an educational tool will increase due to its several and valuable benefits outweighing the challenges as the familiarity of both learners and instructors with the tool increases. The academia that could not avoid the reality of life will go on producing studies considering various aspects of podcasting as an educational tool and the discussions about the impact of this innovative technology will occupy much more place in the literature.

REFERENCES

[1] Armstrong, G. R., Tucker, J. M. and Massad, V. J. (2009). Achieving Learning Goals with Student-Created Podcasts, *Decision Sciences Journal of Innovative Education*, 7 (1), 149-154.

[2] Barsky, E. (2006). Introducing Web 2.0: RSS trends for health librarians. Retrieved May 20 2009 from http://article.pubs.nrc-cnrc.gc.ca/RPAS/rpv?hm=HInit&calyLang=eng&journal=jchla&volume=28&afpf=c07-014.pdf.

[3] Borsheim, C., Merritt, K.; Reed, D. (2008). Beyond Technology for Technology's Sake: Advancing Multiliteracies in the Twenty-First Century, *Clearing House: A Journal of Educational Strategies, Issues and Ideas*, 82 (2), 87-90.

[4] Boulos, M. N. K. and Wheeler, S. (2007). The emerging Web 2.0 social software: an enabling suite of sociable technologies in health and health care education1, *Health Information and Libraries Journal*, 24(1), 2-23.

[5] Ciconte, B. L. and Jacob, J. G. (2008). *Fundraising Basics: A Complete Guide,* London: Jones and Bartlett Publishers.

[6] Dixon, C. and Greeson, M. (2006). *Recasting the concept of podcasting: Part I.* Retrieved April 10 2009 from http://news.digitaltrends.com/talkback109.html.

[7] *Donnelly, K. M. and Berge, Z. L. (2006). Podcasting: Co-opting MP3 Players for Education and Training Purposes,* Online Journal of Distance Learning Administration, 9 (3).

[8] Evans, C. (2008). The Effectiveness of M-Learning in the Form of Podcast Revision Lectures in Higher Education, *Computers and Education*, 50 (2), 491-498.

[9] *Fernandez, V., Simo, P. and Sallan, J. M. (2009).* Podcasting: A new technological tool to facilitate good practice in higher education, *Computers and Education, 53* (2), 385-392.

[10] Geoghegan, M and Klass, D. (2007). *Podcast Solutions: the complete guide to audio and video podcasting,* New York: Apress.

[11] Harris, D. A. and Krousgrill, C. (2008). *Distance Education: New Technologies and New Directions,* Proceedings of the IEEE, 96 (6), 917.

[12] Harris, H. and Park S. (2008). Educational usages of podcasting, *British Journal of Educational Technology*, 39 (3), 548-551.

[13] Karacapilidis, N. (2009). *Solutions and Innovations in Web-Based Technologies for Augmented Learning.* Idea Group Inc Publishing, ISBN: 1605662380, 9781605662381 .

[14] Lazarri, M. (2009). Creative use of podcasting in higher education and its effect on competitive agency, *Computers and Education,* 52 (1), 27-34.

[15] Lee, M. J. W., McLoughin, C. and Chan, A. (2008). Talk the talk: Learner-generated podcasts as catalysts for knowledge creation, *British Journal of Educational Technology*, 39 (3), 501-521.

[16] Leonard, M. (2008). Podcasting's Possibilities, *Inquiry*, 13 (1), 20-25.

[17] Lim, K. Y. T. (2006). Now hear this: Exploring podcasting as a tool in geography education, In K. Purnell, J. Lidstone and S Hodgson (Eds.*), Changes in Geographic Education: Past, Present and Future.* Proceedings of the International Geographical Union Commission on Geographical Education Symposium. Brisbane.

[18] Lyles, H., Robertson, B., Mangino, M. and Cox, J. R. (2007). Audio podcasting in a tablet PC-enhanced biochemistry course, *Biochemistry and Molecular Biology Education*, 35 (6), 456-461.

[19] Matthews, K. (2006). *Research into Podcasting Technology Including Current and Possible Future Uses*. Retrieved May 15 2009 from http://mms.ecs.soton.ac.uk/ 2007/papers/32.pdf.

[20] O'Toole, G. (2007). Multimedia-casting syndication for educational purposes Considerations of a Podcast for use in higher education, *British Journal of Educational Technology*, 38 (5), 939-941.

[21] Oxford University Press (2009). *"Podcast" is the Word of the Year*. Retrieved April 15 2009 from http://www.us.oup.com/us/brochure/NOAD_podcast/.

[22] Rubio, R., Martín, S. and MoráN, S. (2007). Collaborative web learning tools: Wikis and blogs, *Computer Applications in Engineering Education (Early View)*, Retrieved April 10 2009 from Wiley InterScience Database.

[23] Sahin, H. (2006). Terimlerin Genel Dile Yansımasına Dair Bazı Gözlemler, *Erciyes Universitesi, Sosyal Bilimler Enstitusu Dergisi*, 20, 123-129.

[24] Sprague, D. and Pixley, C. (2008). Podcasts in Education: Let Their Voices Be Heard, *Computers in the Schools*, 25 (3-4), 226-234.

[25] Siemens, G. (2005). "Connectivism: A Learning Theory for the Digital Age", *International Journal Of Instructional Technology And Distance Learning*, 2(1), Retrieved February 3 2010 from . http://itdl.org/Journal/Jan_05/article01.htm

[26] Thomas, M. (2009). *Handbook of Research on Web 2.0 and Second Language Learning*, London: Idea Group Inc (IGI).

[27] Trelease, R. B. (2008). Diffusion of innovations: Smartphones and wireless anatomy learning resources, *Anatomical Sciences Education*, 1 (6), 233-239.

[28] Van der Zanden, P. (2009). The Facilitating University: *Positioning Next Generation Educational Technology*, Eburon Uitgeverij B.V. Publishing.ISBN 905972352X, 9789059723528

[29] Vess, D. L. (2006). History to Go: Why iTeach with iPods, *History Teacher*, 39 (4), 479-492.

In: Knowledge Management and Communication … ISBN: 978-1-62808-250-0
Editor: Rafiqul Islam

Chapter 2

ANALYSIS AND MODELING OF RAMPS USING VISSIM TO IMPROVE MANAGED LANE OPERATIONS

Ganesh J. Karkee*[*]*, Steven Venglar and Kwaku Obeng-Boampong
Texas Transportation Institute/Texas A&M University System, Texas, US

ABSTRACT

A variety of reasons exist for agencies to pursue managed ramps. Current funding constraints and difficulty in gaining environmental and public approval for large-scale construction projects has forced agencies to continue considering alternate solutions to roadway widening to mitigate congestion. Areas for potentially improving freeway performance are ramp management and location. The common motivations for ramp management explored in a recent study were to preserve quality of freeway flow, or speed or vehicle type restrictions due to horizontal and vertical limitations along the ramp. Applying managed lanes operational strategies to ramps could maximize existing capacity, manage demand, offer choices, improve safety and generate revenue.

A variety of freeway and ramp traffic mixes, number of freeway mainlanes, ramp spacing, and different ramp merge conditions were analyzed. The VISSIM microsimulation traffic software was used for modeling all scenarios. Performance measures used to compare outcomes from each combination of roadway environment factors were freeway mainlane and ramp weaving speeds.

Results confirm expectations for many scenarios in that higher freeway mainlane and ramp weaving speeds were recorded for longer managed lane ramp spacing from the freeway entrance ramp. Ramp merge conditions with a full auxiliary lane exhibited higher speed performance when compared against a forced merge and an acceleration lane ramp transition. The speed differential between freeway mainlane traffic and ramp weaving traffic was higher with a decreasing percentage of autos in freeway traffic mix.

Keywords: Managed lane, VISSIM, simulation, freeway, speed differential, forced merge, auxiliary lane, ramp spacing

[*] Corrersponding author: Texas Transportation Institute/Texas A&M University System 1100 NW Loop 410, Suite 400 San Antonio, Texas 78213 Email: g-karkee@tamu.edu.

INTRODUCTION

A variety of reasons exist for agencies to pursue managed ramps. Current funding constraints and difficulty in gaining environmental and public approval for large-scale construction projects has forced agencies to continue considering alternate solutions to roadway widening to mitigate congestion. Areas for potentially improving freeway performance are ramp management and location. The common motivations for ramp management are examined in this study and include ramp metering (and, in extreme conditions, closure) to preserve quality of freeway flow; speed or vehicle-type restrictions due to horizontal or vertical limitations along the ramp or in the vicinity of the ramp along the freeway; vehicle restrictions in support of mobility alternatives, such as bus-only and/or high occupancy vehicle-only ramps; and, hybrids or combinations of selected strategies. Strategy implementation may also vary by time of day, as in cases where ramp metering is utilized only during peak periods of demand.

Another motivation for managing ramps is in support of managed lanes that exist within a freeway corridor. Often the management of these ramps is less visible than directly-managed ramps and the strategies as subtle as ramp location to "feed" eligible vehicles to the managed portion of the facility.

Previous and related studies with similar objectives were performed in the areas of overall managed lanes design and ramp placement (Venglar, et al., 2002) and, to some extent, truck lane utilization (Middleton, et al., 2006). Simulation was used extensively during those investigations as a means of identifying the operational impacts of ramp placement design decisions and generating performance measures for various managed lane scenarios. Both to leverage that experience for managed ramps scenarios and to provide design guidance to the extent possible across the many possible combinations of freeway and ramp geometry, freeway and ramp traffic demand levels, freeway and ramp vehicle mixes and managed ramp strategies, researchers employed simulation in the current study. As in past studies, the VISSIM simulation model (VISSIM User Manual, 2006) was used in the analysis of study scenarios.

MANAGED RAMP SCENARIOS SUPPORTING MANAGED LANES

The research team faced competing concerns of desiring to provide utilitarian output for a broad range of possible managed ramp scenarios and keeping a practical limit on the number of scenarios analyzed. The fundamental decision was made to focus on only those scenarios involving freeway entrance ramps on the marginal, or right, side of the freeway whose vehicles would weave to the medial, or left, side of the freeway to access the facility's managed lane(s). The various factors researchers ultimately chose were based on their experience with previous, similar studies (Venglar, et al., 2002; Middleton, et al., 2006). It was assumed that conditions in the managed lane were such that no queue build up was existed on the entrance ramp to the managed lane. This eliminated the possibility of queue spill back from the managed lane entrance ramp into the freeway mainlanes.

Parameters used in the VISSIM simulation model are listed and briefly discussed below.

Number of Freeway Mainlanes

Simulation analyses were limited to scenarios that featured three or four freeway lanes across which managed ramp traffic would weave in order to reach a managed lane access point.

Freeway Vehicle Mix

Researchers chose a set of three vehicle mixes generally representative of urban freeway conditions. Since such facilities often include transit presence and trucks, a "normal mix" was chosen to include 90 percent automobiles, five percent trucks and five percent buses. The second mix was for freeways which either did not allow trucks or had limited or no truck use during peak periods. Finally, a "high truck" mix with 80 percent automobile, 15 percent truck and five percent bus composition was chosen. In all analyzed scenarios, the assumption was made that 95 percent of all freeway traffic used general-purpose freeway through lanes downstream of the entrance ramp and the remaining five percent of freeway traffic used the managed lanes. Traffic distribution from the freeway study section downstream to the managed lane and freeway general purpose lane is shown in Figure 1.

Freeway Volume

Two volume levels were chosen: moderately loaded and heavily loaded. A per lane flow rate of 1,000 vehicles per hour per lane (vphpl) was selected to represent moderate flow and a rate of 1,400 vphpl was selected for heavily loaded conditions (Kuhn, et al., 2008).

Ramp Merge Condition

Three of the most prevailing basic geometric configurations for the junction of an entrance ramp and a freeway's general purpose lanes were utilized: a forced, or direct, merge; acceleration lane; and a full auxiliary lane.

Ramp Vehicle Mix/User Group

Just as the freeway mainlanes had a selected set of generally-representative vehicle types, the managed ramp also has a set of vehicle types/user groups. To represent conditions where no heavy vehicles of any type are allowed to use a ramp as a result of one or more management strategies being applied, an "all automobile" group was used. Similar to the freeway, a "normal" mix of 90 percent automobiles, five percent buses and five percent trucks was developed. As ramps in some of the larger urban areas in Texas are used in support of high occupancy vehicle (HOV) operations, which can include high-occupancy toll (HOT) users and single-occupant toll users (SOT), an "HOV/HOT/SOT/bus" vehicle mix was developed. This mix included 85 percent automobiles and 15 percent buses (Kuhn, et al., 2008). Finally, recent Texas research on truck lanes (Middleton, et al., 2006) illustrated a need for a "truck only" mix that was composed of 100 percent heavy vehicles.

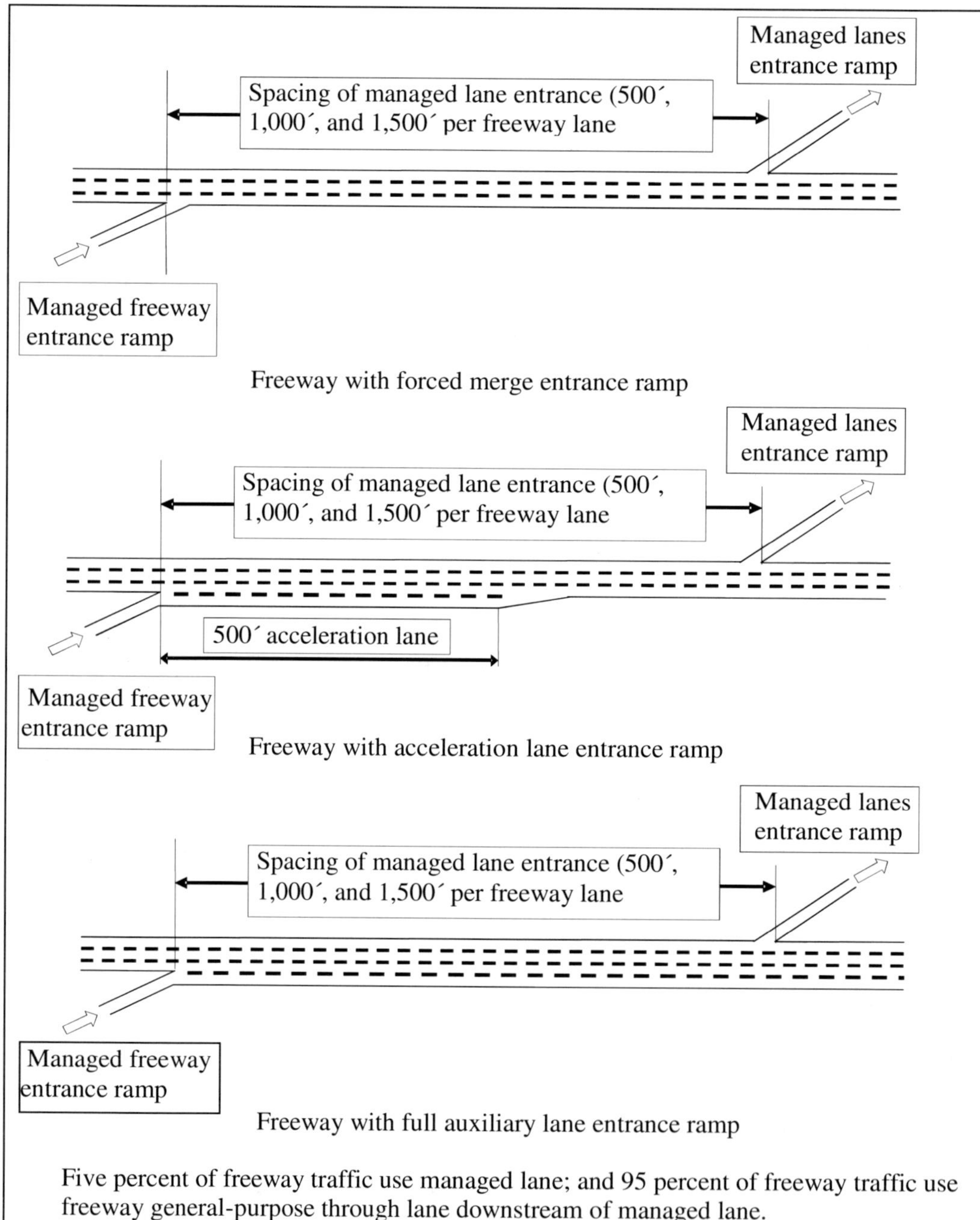

Figure 1. Freeway with forced merge, acceleration lane, and full auxiliary lane entrance ramps.

Ramp Volume and Proportion of Managed Ramp Traffic Weaving to Managed Lanes

Ramp volumes used in simulation were 375 vehicles per hour and 750 vehicles per hour (Kuhn, et al., 2008). The percentages of weaving traffic were selected as 25 percent and 50 percent of total ramp volumes.

Per Lane Spacing of Managed Lanes Entrance

Based on previous managed lanes research employing the same types of modeling tools (Venglar, et al., 2002), a three-value range was chosen: 500 feet (152.4 m), 1000 feet (304.8 m) and 1,500 feet (457.2 m) per lane of the freeway mainlanes/through lanes (Obeng-Boampong, et al., 2008).

Determination of Performance Measures

In the context of managed ramps which support managed lanes within a freeway corridor, the efficacy of ramp management is ultimately gauged by the relative success of the managed lanes themselves. Speed was chosen as the primary performance measure for this analysis as it provides the necessary information to analysts to directly determine freeway operations level and is readily understood across various potential stakeholder groups.

Speed

At the heart of any given managed lanes strategy is the explicit goal of maintaining high speed for the managed lanes (Kuhn, et al., 2005). In essence, while the goal of the overall managed lane is linked to community and stakeholder needs and objectives, the performance of the managed lane is judged by its ability to maintain mobility through higher-speed travel. As a result, the selected speed was the primary indicator of the level of performance for managed ramps scenarios supporting managed lanes.

Speed Differential and Safety

Traffic engineering research has shown that crash potential increases as the speed differential increases (Stover and Koepke, 2002). The safety impact of speed differentials is further documented in national practices and standards for roadway design (AASHTO, 2004). Lower speed differentials are associated with lower crash rates; a speed differential of approximately 10 mph (16 kph) is the transition point above which the crash rate or ratio begins to increase rapidly with increasing speed differential (Stover and Koepke, 2002; AASHTO, 2004).

Speed and speed differential are used in several ways in the course of analysis. Simulation output data were used from ramp modeling to support managed lanes. While all speed results were ultimately tabulated in the presentation of research findings, only those conditions with speed differentials of less than 10 mph (16 kph) were considered desirable and viable for design. Each set of simulation results was subjected to two tests: speed differential between approaching freeway traffic and traffic within the managed ramp weaving area, and speed differential between freeway through (i.e., non-weaving) and ramp-to-managed lanes (weaving) traffic. If either speed differential was observed to be in excess of 10 mph (16 kph), that scenario is not recommended as a potential managed ramp design condition.

VISSIM MODEL PREPARATION

Among the first model development tasks was configuring the VISSIM simulation model for the traffic environment typical of Texas freeway conditions. Previous research (Gomes, et al., 2003) led the analysts to immediately change the driving behavior parameter "waiting time before diffusion" in VISSIM from its default of 60 seconds to one second. If this change is not made, vehicles unable to maneuver to an exit ramp in a timely fashion will block freeway through lanes until the maximum wait time is reached and an unrealistic freeway bottleneck occurs. Analysts' experiences also led to a change in the exit ramp "look ahead" distance, or the distance when ramp-bound vehicles on the freeway begin to maneuver to the exit ramp lane. The default value of 656 feet (200 m) resulted in all vehicles traveling too close to ramps before maneuvering into the exit lane, again resulting in artificial congestion. A value of 1,968 feet (600 m) was determined to be more consistent with real-world exit ramp advanced notification signing and resulted in more representative "real-world" vehicle behavior.

Traffic Composition

Vehicle mixes and features used in the VISSIM model were representative of vehicles operating on Texas roadways. This task was relatively simple in terms of automobiles and buses, since automobile performance is common across many countries and bus performance also does not vary widely. The size and configuration of trucks, however, is much different in European countries than in the United States. Since VISSIM was developed in Germany, many of its truck and trailer size, axle configuration and weight characteristics are not well mated to heavy vehicles in the US. Previous research (Middleton, et al., 2006) was the guide in determining what types of trucks were typically found in Texas and what percentages of the truck traffic stream each comprised.

Model Coding

In all cases, the model was a relatively simple construction that included the freeway through lanes, the general purpose entrance ramp, and the downstream managed lanes entrance ramp. The length of each roadway segment and the number of lanes was fixed by the scenario. One important entry which had to be specified was the speed distribution on the links making up the roadway network being modeled. In order to replicate conditions on urban freeways in Texas, a free flow speed range of between 65.0 mph (104 kph) and 74.6 mph (119.4 kph) (VISSIM speed distribution 90) was chosen.

Figure 1 presents three merge conditions for the managed ramps - forced merge, merge with acceleration lane, and full auxiliary lane. The final step in preparation of performing the analytical simulation runs for all scenarios was the establishment of travel time detection zones. Knowing the travel time output for both weaving and non-weaving traffic and the length of the travel time detection zones, researchers were able to compute average space mean speed for both traffic streams.

RESULTS

Analysts created unique combinations of each individual roadway environment feature described earlier. Unique geometric files were created for each basic ramp and mainlane geometry condition and then search-and-replace methods were used to create variations on these files with different traffic mixes, volume levels and ramp weaving percentages. Custom programs were developed to run each file three times with a different random seed number. The freeway through lane speed and managed ramp weaving speed were extracted from each output file generated from VISSIM. A spreadsheet was developed to produce final average speeds for each scenario and generate graphs to interpret the output.

Proportion of Trucks on the Freeway

The impact of truck proportion in the mainlane traffic mix is shown in Figure 2. From Figure 2, it can be seen that, as the freeway truck percentage increased, freeway through lane speeds and ramp weaving speeds decreased. Freeway mainlane speeds and ramp weaving speeds were greater where the ramp merge features an auxiliary lane. Other areas investigated included:

- Proportion of automobiles on the ramp
- Freeway volume level
- Entry ramp flow level
- Proportion of ramp traffic weaving to managed lane

Figure 3 shows the speed differential between freeway mainlane and ramp weaving speed with various auto percentages in ramp traffic mix for the forced merge ramp condition. The speed differential is greater for a lower auto percentage in the ramp traffic mix. The speed differential is greater for a ramp spacing of 500 feet (152.4 m) per lane per freeway lane, followed by ramp spacing of 1,000 feet (304.8 m) and 1,500 feet (457.2 m) per freeway lane. Intuitively, gap acceptance opportunities for weaving maneuvers are greater as weaving length increases. Also consistent with real world scenarios is the fact that comparatively longer merging and weaving times and distances were required to accommodate heavy vehicle acceleration and deceleration compared with the light vehicles (autos).

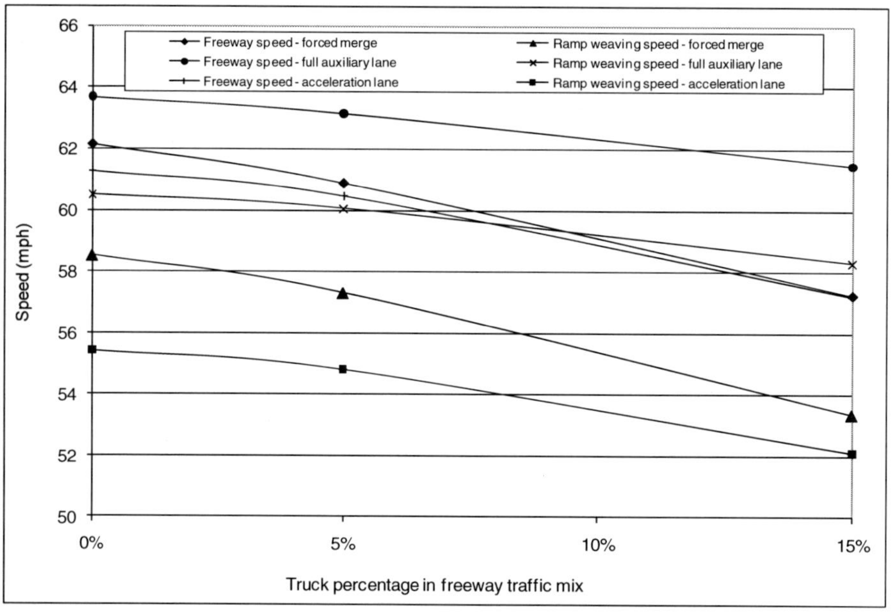

Figure 2. Freeway and ramp weaving speeds with different freeway truck mixes.

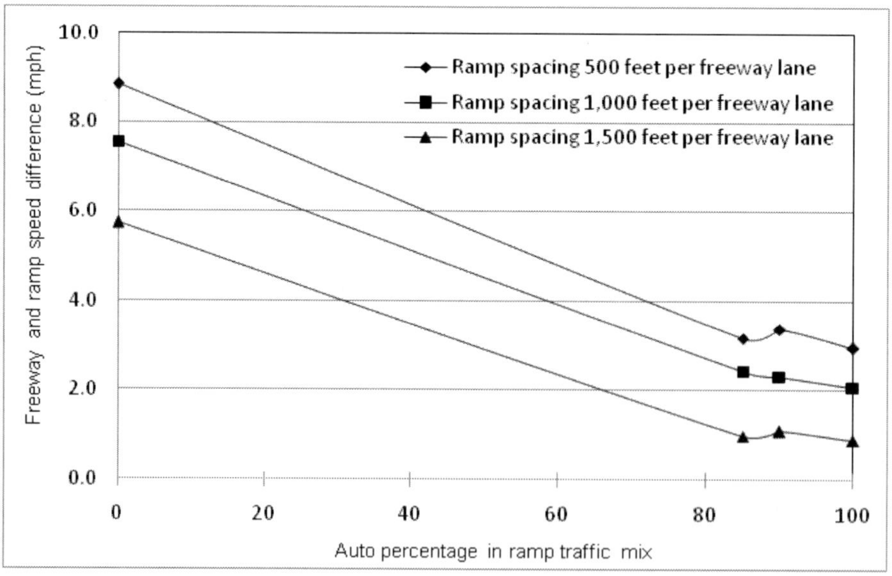

Figure 3. Freeway and ramp weaving speed difference with various auto percentages for forced merge ramp condition.

The speed differential between freeway mainlane and ramp weaving speed with various auto percentages in ramp traffic mix for the acceleration lane ramp condition is shown in Figure 4. Similar speed trends were observed for speed differential as in the forced merge condition, i.e., speed differential is higher for shorter ramp spacing and lower for longer ramp spacing. However, speed differential for acceleration lane ramp condition was slightly greater than the forced merge condition.

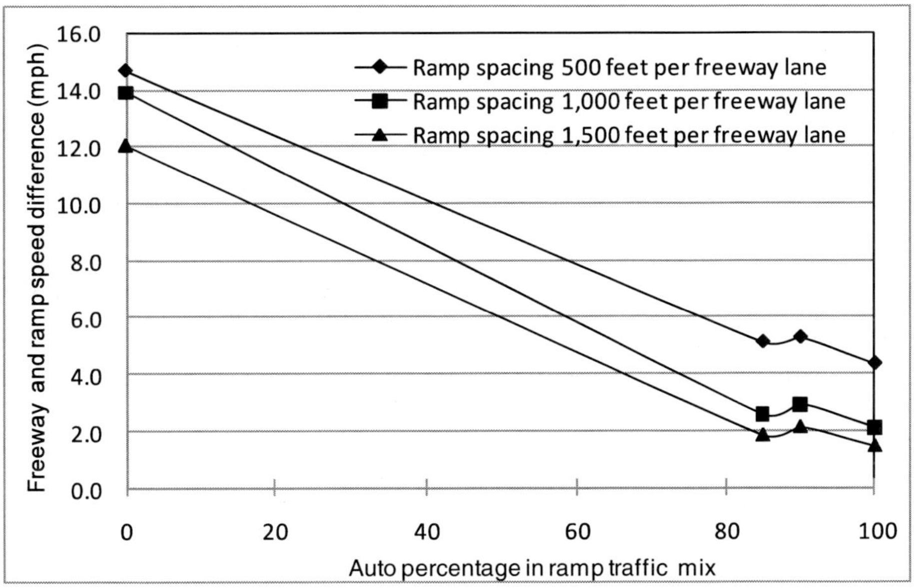

Figure 4. Freeway and ramp weaving speed difference with various auto percentages for acceleration lane ramp condition.

Figure 5 shows the speed differential between freeway main lane and ramp weaving speed for ramp merge conditions where an auxiliary lane is present, with various auto percentages in the ramp traffic mix. The speed differential is inversely proportional to the proportion of auto mix in ramp traffic mix, i.e., greater speed differential was observed for lower auto proportion in the ramp traffic mix. A similar speed differential trend was found when modeling forced merge ramp conditions. Speed differential was also observed to be lower for longer spacing/weaving distances, i.e., the speed differential was greater for 1,500 feet (457.2 m) per freeway mainlane ramp spacing than 500 feet (152.4 m) per freeway mainlane ramp spacing.

As discussed in the previous section, the composition of traffic mix includes trucks, buses, and HOV/HOT/SOT. For example, vehicle composition with 90 percent automobiles contains five percent buses and five percent trucks; another vehicle composition with 85 percent automobiles contains 15 percent buses. These variations in vehicle composition might have resulted in slight increase in speed differential corresponding to 90 percent and 85 percent automobiles in ramp traffic mix in Figures 3-5.

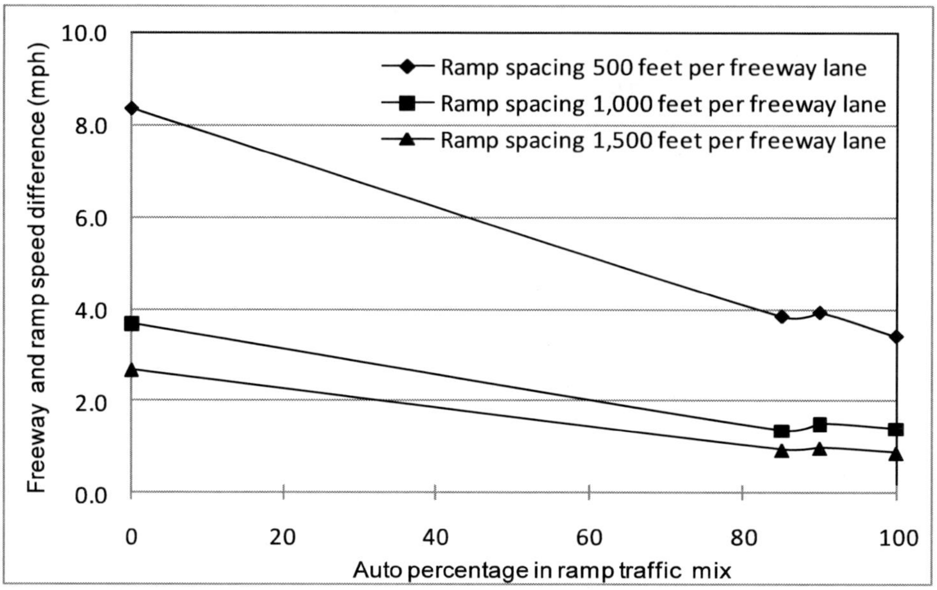

Figure 5. Freeway and ramp weaving speed difference with various auto percentages for ramp with auxiliary lane.

CONCLUSION AND RECOMMENDATIONS

From the results obtained from this analysis, the following observations were made:

- Freeway mainlane speeds and ramp weaving speeds were higher where the ramp merge featured an auxiliary lane;
- As freeway truck percentage increased, freeway through lane speeds and ramp weaving speeds decreased for both ramp merge conditions analyzed;
- Freeway mainlane speeds are observed to increase with increased spacing between the managed ramp and the managed lane access ramp;
- Freeway mainlane speeds were observed to increase with an increase in the proportion of automobiles found in the managed ramp traffic mix; and,
- The speed differential between freeway mainlane and ramp weaving speed is inversely proportional to percentage of autos in the ramp traffic mix and ramp spacing.

Generally, the results showed that for areas with high truck percentages (greater than five percent) on the freeway mainlanes and short spacing (about 500 feet (152.4 m) per lane between a managed ramp and a downstream entrance into a managed lanes facility), any intended managed ramp strategy should prevent a high percentage of trucks (greater than five percent) from using the managed ramp as this will lead to undesirable speed differentials along the weaving section of the freeway.

Further research is recommended to explore in more detail other managed lane strategies and how they might be used to improve freeway operations. These strategies might include the use of various vehicle class restrictions in support of incident/special event management, whether certain vehicle restrictions improve safety for the freeway main lanes, and if vehicle classification restrictions can be used on ramps as a partial means of ramp metering to improve freeway operations.

REFERENCES

[1] American Association of State Highway and Transportation Officials (AASHTO). 2004. *A Policy on Geometric Design of Highways and Streets, 5th Edition.* American Association of State Highway and Transportation Officials, Washington, DC.

[2] Gomes, G., May, A. and Horowitz, R. 2003. "A Microsimulation Model of a Congested Freeway using VISSIM," In *Proceedings of the 83rd Annual Meeting of the Transportation Research Board.* Transportation Research Board, Washington, DC.

[3] Kuhn, B., Balke, K., Chaudhary, N., Jasek, D., Karkee, G., Obeng-Boampong, K., Shelton, J. and Venglar, S. 2008. "*Managed Lanes Strategies Feasible for Freeway Ramp Applications*" Texas Transportation Institute, Technical Report 0-5284-2, Texas A&M University System, College Station, TX.

[4] Kuhn, B., Goodin, G., Ballard, A., Brewer, M., Brydia, R., Carson, J., Chrysler, S., Collier, T., Fitzpatrick, K., Jasek, D., Dusza, C. and Ullman G. 2005. *Managed Lanes Handbook.* Texas Transportation Institute, Research Report FHWA/TX-06/0-4160-24, Texas A&M University System, College Station, TX.

[5] Middleton, M., Venglar, S., Quiroga, C., Lord, D. and Jasek D. 2006. *Strategies for Separating Trucks from Passenger Vehicles: Final Report.* Texas Transportation Institute, Research Report FHWA/TX-07/0-4663-2, Texas A&M University System, College Station, TX.

[6] Obeng-Boampong, K., Venglar, S. and Karkee, G. 2008. "Modeling and Analysis of Ramps to Support Managed Lane Operations," In *Proceedings of the 87th Annual Meeting of the Transportation Research Board.* Transportation Research Board, Washington, DC.

[7] Stover, V. and Koepke, K. 2002. *Transportation and Land Development, 2nd Edition.* Institute of Transportation Engineers, Washington, DC.

[8] Venglar, S., Fenno, D., Goel, S. and Schrader P. 2002. *Managed Lanes – Traffic Modeling.* Texas Transportation Institute, Research Report FHWA/TX-02/4160-4, Texas A&M University System, College Station, TX.

[9] VISSIM 4.1 User Manual. PTV Planung Transport Verkehr AG (PTV AG), Karlsruhe, Germany, Copyright 1991-2006. (Distributed in North America by PTV America, Corvallis, Oregon.)

In: Knowledge Management and Communication … ISBN: 978-1-62808-250-0
Editor: Rafiqul Islam

Chapter 3

STIFFNESS EVALUATION AND DAMAGE QUANTIFICATION OF BEAMS USING KNOWLEDGE BASED FINITE ELEMENT APPROACH

***S. Shifana Fatima*[1], *C. Natarajan*[1] *and A. Rajaraman*[2*]**
[1]Department of Civil Engineering,
National Institute of Technology, Tiruchirappalli, India
[2]Department of Civil Engineering, Indian Institute of Technology, Chennai, India

ABSTRACT

A typical beam element in two dimension is chosen which incorporates the damage parameters (spread, location and intensity of the damage) in its stiffness matrix. A program in MATLAB (7.0) was generated to analyze the system using finite element technique. For that element, the intensity of damage is given as a main parameter, without quantifying or qualifying the type of damage. A general methodology is presented for identifying stiffness reduction by incorporating the results obtained from the finite element method with fuzzy logic.

Keywords: Fuzzy logic; MATLAB; Stiffness reduction; Damage

1. INTRODUCTION

Improved design methodologies have significantly increased the reliability and safety of structures. However, due to a wide variety of unforeseen conditions and circumstances, it is practically not possible to design and build a structure with zero probability of failure. There is an increasing interest in the development of smart structures with built-in fault detection systems that would provide damage and/or failure warnings. Damage can be defined as the change in structural performance, which can be identified in terms of discrete cracks or a

[*] Corresponding author: Department of Civil Engineering, Indian Institute of Technology, Chennai-600036 E-mail:lc_nataraj@yahoo.com.

weakened zone formation and a consequent stiffness reduction. Damages of structures (cracks, for instance) can be defined as the stiffness reduction of one or more elements of a structure, which provides direct indication of the extension of cracked zones in the structure.

Current designs of structures are generally based on their virgin state without analyzing possible failed structures and discard possible structural stiffness degradation. Very few literatures are available in the field of damage analysis and the available literature deal with only the formulation of stiffness degradation for a particular material under a particular type of loading. For instance,Ma Ge and Lui (2005) have presented a damage detection method for the identification and quantification of damage, using finite element to model the structure in its undamaged state as well as information on the dynamic properties such as frequencies and mode shapes of the structure in its damaged state. This technique is applicable to any structure that can be accurately modelled using the finite element method and whose frequencies and mode shapes can be reliably measured. Another example was the analytical model developed by Trisha and Chandra (2006) using cracked beam finite element to assess the residual life of damaged structures. The behaviour of this beam element is analyzed and it includes the damage by stiffness reduction and will be much easier to implement in any finite element program or analysis.

Damage can be of any physical form and its growth under a particular type of loading differs based on various factors like the type of material, location, intensity etc. Therefore it is difficult to generalize the behaviour of damage in a structural element. Zadeh (1975) first introduced the concept of fuzzy set theory and found it to be a powerful tool to handle the uncertainties involved in predicting the reduction in stiffness for a damaged beam element. It uses an imprecise but very descriptive language, to deal with input data, more like a human operator. It is very robust and forgiving of operator and data input and often works when first implemented with little or no tuning (Shifana and Natarajan, 2007). Fuzzy logic control can play an important role because knowledge based design rules can easily be implemented in systems with unknown structure, and it is going to be a conventional control method since the control design strategy is simple and practical and is based on linguistic information (Ilyas and Yunis, 2006). Damage recognition and location are the key elements in the design of structural health monitoring system. Both the parameters have a lot of uncertainty, so fuzzy logic and its application is required for the design of structural health monitoring system (Qu Wenzhong et al., 2004; Rao et al., 2006; Seung et al., 2007). Many researchers from structural engineering have successfully integrated fuzzy logic with other intelligence and optimization systems, in their domains (Gang and Lily, 2006; Pourzeynali et al., 2007). The objective of the paper is to show the effectiveness of 2D damaged beam element in the reduction of stiffness due to damage and its results are integrated with the fuzzy logic to get the results in lingusitc variables which is very useful for the common man to understand.

2. Formulation of the Problem

In this paper, the damage is simulated by reduction of modulus of elasticity (E) or moment of inertia (I) and the aim is to study the intensity of damage as the main parameter without quantifying or qualifying the type of damage. Therefore, the proposed method is a generalized one and can be applied for any material. The present paper supposes modelling of local damage on element level and mapping into structural level and assessing the changes in the structural behaviour. Damage in beam is represented by three parameters (Figure 1):

α– Intensity of damage.
β– Location of damage region.
γ– Extent of damage region.

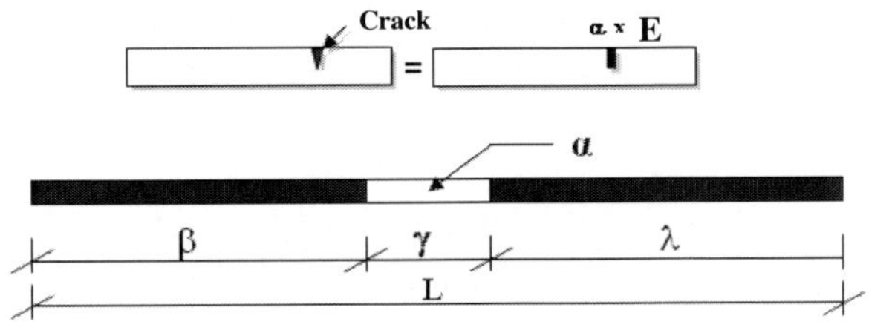

Figure 1. Beam element with damage parameters.

Intensity of damage is expressed in percentage, whereas, the location and the spread of damage are expressed as ratios of the length of the beam element. The stiffness matrix incorporating the three damage parameters were developed by subdividing the beam element into a number of elements and varying the property of the damaged portion. Consider a beam element with two nodes I and J as shown in Figure 2 with a damage of intensity α from x1 distance and spread of damage is x2. Length of the beam element is L and its flexural rigidity is EI. Each node has two degrees of freedom namely vertical displacement and rotation (V1, θ1, V4, and θ 4).This four degrees of freedom are considered as master degrees of freedom.

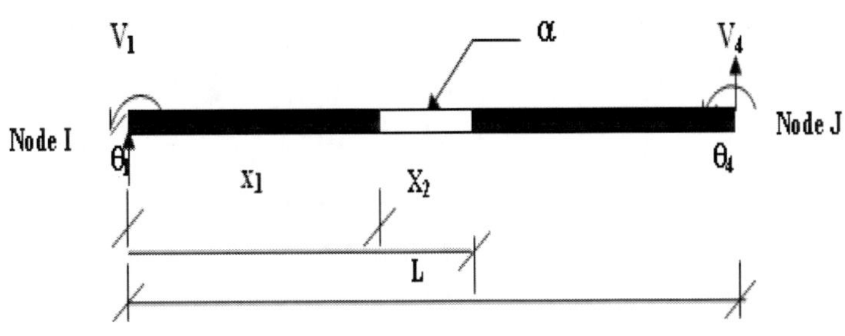

Figure 2. Beam element with damage and master degrees of freedom.

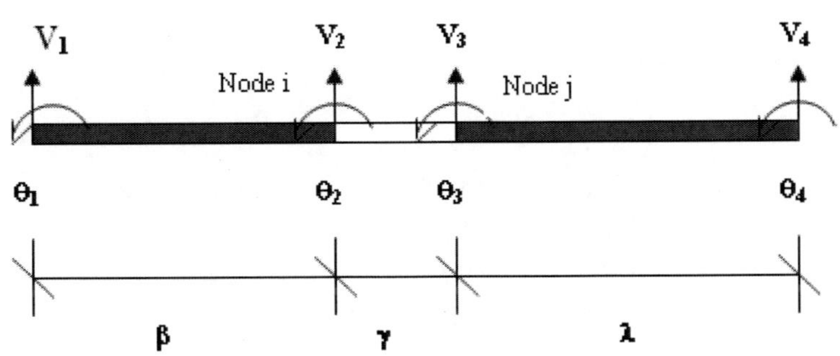

Figure 3. Dividing beam element into three sub-beam element.

The beam element is divided into three sub-beam element as shown in Figure 3 with internal nodes i and j. These internal nodes have following degrees of freedom (V2, θ2, V3, θ3) and they are known as slave degrees of freedom. The stiffness matrix of each sub-element is obtained as per the conventional procedure. The length of each sub- element can be represented in the form of ratio of the original beam element length. For example

Length of first sub- element = β x L
Length of intermediate sub- element = γ x L
Length of third sub- element = λ x L

Where,

β, known as location parameter for a damage beam element = x1/L

γ, known as spread parameter for a damage beam element = x2/L, and λ = 1-(β+γ)

The stiffness matrix of each sub matrix is shown below

$$[Ke_1]=\begin{bmatrix} \frac{12EI}{(\beta L)^3} & \frac{6EI}{(\beta L)^2} & -\frac{12EI}{(\beta L)^3} & \frac{6EI}{(\beta L)^2} \\ \frac{6EI}{(\beta L)^2} & \frac{4EI}{\beta L} & -\frac{6EI}{(\beta L)^2} & \frac{2EI}{\beta L} \\ -\frac{12EI}{(\beta L)^3} & -\frac{6EI}{(\beta L)^2} & \frac{12EI}{(\beta L)^3} & -\frac{6EI}{(\beta L)^2} \\ \frac{6EI}{(\beta L)^2} & \frac{2EI}{\beta L} & -\frac{6EI}{(\beta L)^2} & \frac{4EI}{\beta L} \end{bmatrix}$$

$$[Ke_2]=\begin{bmatrix} \frac{12EI\alpha}{(\gamma L)^3} & \frac{6EI\alpha}{(\gamma L)^2} & -\frac{12EI\alpha}{(\gamma L)^3} & \frac{6EI\alpha}{(\gamma L)^2} \\ \frac{6EI\alpha}{(\gamma L)^2} & \frac{4EI\alpha}{\gamma L} & -\frac{6EI\alpha}{(\gamma L)^2} & \frac{2EI\alpha}{\gamma L} \\ -\frac{12EI\alpha}{(\gamma L)^3} & -\frac{6EI\alpha}{(\gamma L)^2} & \frac{12EI\alpha}{(\gamma L)^3} & -\frac{6EI\alpha}{(\gamma L)^2} \\ \frac{6EI\alpha}{(\gamma L)^2} & \frac{2EI\alpha}{\gamma L} & -\frac{6EI\alpha}{(\gamma L)^2} & \frac{4EI\alpha}{\gamma L} \end{bmatrix}$$

The intensity parameter is taken into consideration in the stiffness matrix of the intermediate sub- element.

$$[Ke_3]=\begin{bmatrix} \frac{12EI}{(\lambda L)^3} & \frac{6EI}{(\lambda L)^2} & -\frac{12EI}{(\lambda L)^3} & \frac{6EI}{(\lambda L)^2} \\ \frac{6EI}{(\lambda L)^2} & \frac{4EI}{\lambda L} & -\frac{6EI}{(\lambda L)^2} & \frac{2EI}{\lambda L} \\ -\frac{12EI}{(\lambda L)^3} & -\frac{6EI}{(\lambda L)^2} & \frac{12EI}{(\lambda L)^3} & -\frac{6EI}{(\lambda L)^2} \\ \frac{6EI}{(\lambda L)^2} & \frac{2EI}{\lambda L} & -\frac{6EI}{(\lambda L)^2} & \frac{4EI}{\lambda L} \end{bmatrix}$$

Three matrixes are condensed by static condensation procedure. Such a formulated matrix gives the modified stiffness due to the presence of damage (Dinesh et al., 2006; Revathy et al., 2007).

3. MODEL DEVELOPMENT

3.1. Stiffness of the Beam with Single Damage

The stiffness matrix formulated in this paper can be used for all types of beams with different boundary condition. In this paper, two cases have been considered, a damaged beam simply supported at its ends (Figure 4) and a beam with damage whose both the ends are fixed (Figure 5), and are solved by finite element method. An M-file is developed in MATLAB (7.0) incorporating the parameters α, β, γ of damaged beam element and the pseudo code is given below. The beam is solved by dividing into two equal length elements with load at the intermediate node. If the damage is at the left element, then the formulated stiffness matrix is used for the left element alone while for the right element the original stiffness matrix is retained. Stiffness values are obtained by varying all the three parameters of damage, namely intensity of damage (α) from 0% to 99%, location of damage (β) from 0 to 0.45, with the assumption that the spread of the damage γ is constant and is equal to 0.01. Due to symmetry in boundary and loading condition, the variation in location of damage is carried only for one half of the span length of the beam. The normalized values of reduction in stiffness obtained for various combinations of damage parameters are shown in Figure 9 (a) for fixed beam and Figure 10 (a) for simply supported beam.

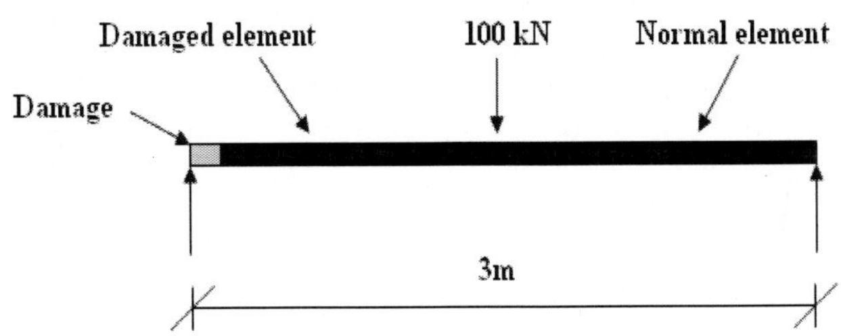

Figure 4. Simply supported beam with damage.

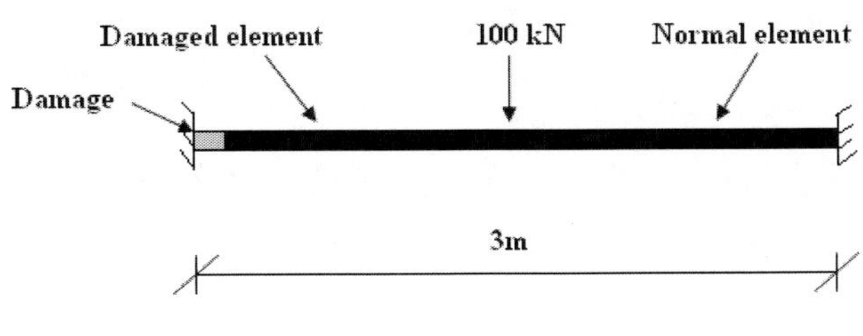

Figure 5. Fixed beam with damage.

3.2. Pseudo Code Developed for Generating Stiffness of a Beam with a Single Damage

The damage parameters (β and γ) are given in the ratio of element length and therefore it has to be twice of that of beam length ratios

1. Input parameters
   a. Young's modulus of the beam
   b. Moment of inertia of the beam
   c. Length of the beam (L1)
   d. Number of elements
   Ex: n=2; n represents the number of elements which is two in this case with a node at centre
   L=L1/n; L denotes the element length
2. Mentioning the location and Intensity of damage of damage
   disp ('Is the element damaged -- " Enter 0 for damage element else enter 1 "')
   for i=1:n
   disp ('Element no - '); disp(i);
   damage(i)=input('Enter 0 for damage element else enter 1 - ');
   end
   Enter the intensity of damage in percentage
   Enter the location of damage from left end in ratio of element length
   Enter the spread of damage in ratio of element length
3. Assigning the size of global stiffness matrix
4. Formation of global stiffness matrix
5. Specifying the boundary conditions
   disp(' The nodal degrees of freedom - ');
   disp('" 0 represents constraint and 1 represents free "');
   boundry=[];
   disp(boundary);
6. Applying boundary condition to global stiffness matrix
7. Specifying the load
8. Stiffness calculation

Using the above pseudo code, an Illustrative numerical example is presented for a simply supported beam with and without damage as shown in Figure 4.

Undamaged Case

Is the element damaged -- " Enter 0 for damage element else enter 1 "
Element no - 1
Enter 0 for damage element else enter 1 - 1
Element no - 2
Enter 0 for damage element else enter 1 – 1
Stiffness Matrix of the Damaged Beam Element:
1.0e+010 *

0.0000	0.0038	-0.0000	0.0038	0	0
0.0038	3.7793	-0.0038	1.8896	0	0
-0.0000	-0.0038	0.0000	0	-0.0000	0.0038
0.0038	1.8896	0	7.5585	-0.0038	1.8896
0	0	0.0000	-0.0038	0.0000	-0.0038
0	0	0.0038	1.8896	-0.0038	3.7793

The Nodal degrees of freedom -
" 0 represents constraint and 1 represents free "
0 1 1 1 0 1

The Nodal Loads –
0 0 100000 0 0 0

deflection –
0
0.0040
3.9690
0
0
-0.0040

Stiffness of the beam -
2.5195e+004

Damaged Case

Assuming the damage of intensity 25 % at location 1.2m with spread of damage 0.025L, the corresponding stiffness is obtained from the M File program.

(α= 25%, β= 0.4 , γ= 0.05)

Note :

The damage parameters (β and γ) are given in the ratio of element length and therefore the spread has to be twice of that of the given value. (i.e., $\gamma = 0.025$ x 3 = 0.075 . γ in ratio of element length = 0.075/1.5 = 0.05)

Is the element damaged -- " Enter 0 for damage element else enter 1 "
Element no - 1
Enter 0 for damage element else enter 1 - 0
Element no - 2
Enter 0 for damage element else enter 1 - 1
Enter the intensity of damage in percentage - 25
Enter the location of damage from left end in ratio of element length - 0.8
Enter the spread of damage in ratio of element length - 0.05
Stiffness Matrix of the Damaged Beam Element :
1.0e+010 *

0.0000	0.0037	-0.0000	0.0037	0	0
0.0037	3.7655	-0.0037	1.8470	0	0
-0.0000	-0.0037	0.0000	0.0001	-0.0000	0.0038

0.0037	1.8470	0.0001	7.4264	-0.0038	1.8896
0	0	-0.0000	-0.0038	0.0000	-0.0038
0	0	0.0038	1.8896	-0.0038	3.7793

The Nodal degrees of freedom -
" 0 represents constraint and 1 represents free "
0 1 1 1 0 1

The Nodal Loads -
0 0 100000 0 0 0

deflection -
0
0.0040
4.0366
-0.0000
0
-0.0040

Stiffness of the beam -
2.4773e+004
The increase in deflection at the center and the decrease in stiffnesss of the beam shows the influence of the three damage parameters. Similarly the stiffness values can be obtained for the various combinations of the three damage parameters.

4. Predicting the Percentage Reduction in Stiffness due to Damage in a Beam Using Fuzzy Logic

Fuzzy Logic is a powerful tool which models the uncertainties in the real world and the phenomena associated with human thinking and perceptions, hence considered to be a very fertile area of research (Klir and Folger, 1988). The perception of damage intensity on a particular beam varies from person to person. Often, it is difficult to express each of this information as crisp data. A better way to convey the uncertainty would be linguistic description of such variables. For example for a damaged structure one may suggest the intensity of damage is very high or the location of damage is at the left end. It is well known that the theory of fuzzy sets can handle such linguistic variables effectively (Zadeh, 1975).

4.1. Fuzzy Representation of Input Variables

Fuzzy set is a generalization of classical or crisp set. The crisp set is defined in such a way to dichotomize the individuals in some given universe of discourse into two groups: members (those that certainly belong in this set) and the non-members (those that certainly do not). A sharp, unambiguous distinction exists between the members and non members of the class represented by the crisp set. Fuzzy set introduces vagueness with the aim of reducing complexity by eliminating the sharp boundaries dividing members of the class from non-members. A fuzzy set can be defined mathematically by assigning to each possible individual in the universe of discourse a value representing its grade of membership (μ) in the fuzzy set (Figure 6). The grade corresponds to the degree to which that individual is similar or compatible with the concept represented by the fuzzy set. The membership grades belong to the closed interval between 0 and 1 (normalized) (Klir and Folger, 1988). For the problem under consideration, appropriate variables are represented as triangular fuzzy set for simplicity (Figure 7). Five fuzzy sets are used to define the input variable -intensity of damage - namely, very low, low, medium, high and very high, and the universe of discourse for this variable is from 0 to 99% as shown in Figure 7 (a). The second input variable-location of damage- is defined by five membership function as, at support, near to support, middle, near to center and at center. Due to symmetry, the location parameter is defined only for one half span of the beam and is shown in Figure 7 (b).

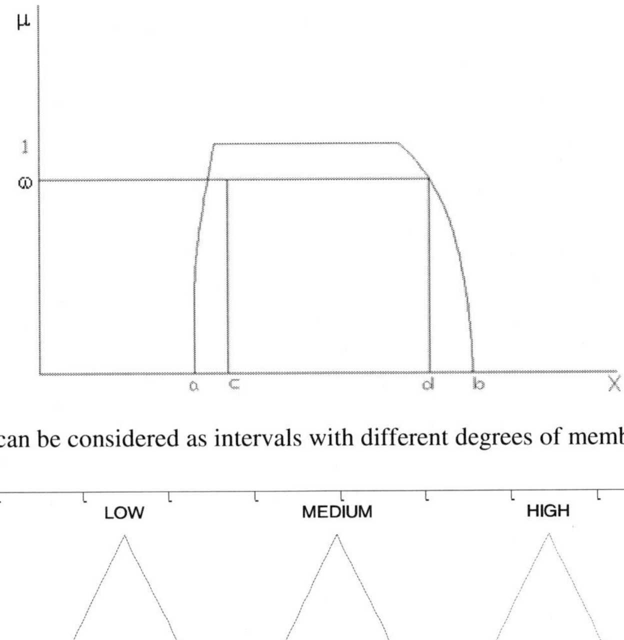

Figure 6. Fuzzy sets can be considered as intervals with different degrees of membership.

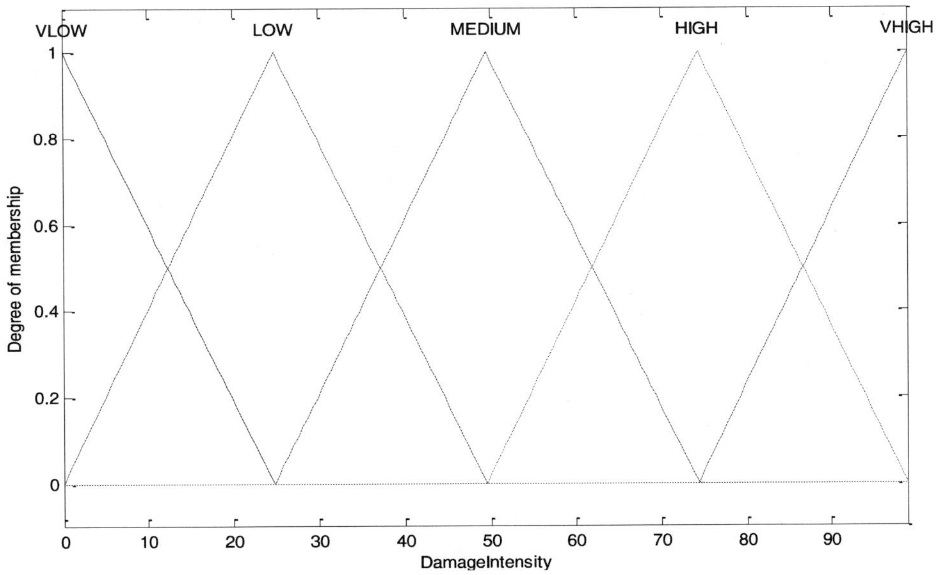

Figure 7 (a). Membership function of the input variable – Intensity of Damage.

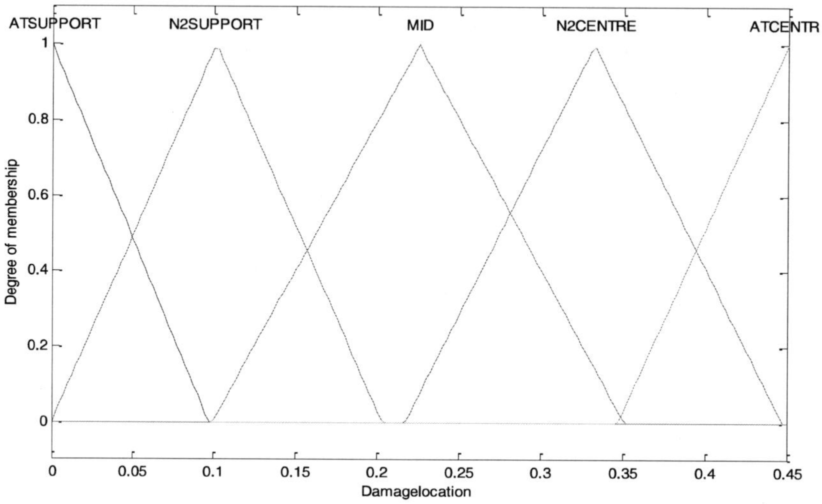

Figure 7 (b). Membership function of the input variable – Location of Damage.

4.2. Defuzzification of Output Variables

When the input is fuzzy, the output will be fuzzy and quite often it is essential to defuzzify it, since the action cannot be fuzzy. Many criterions are available for the above purpose (Sundareswaran, 2005). The choice of criteria is context dependent. The preference to each of the criterion depends on the individual. But, the most important aspect is that, once a criterion is selected and employed, it should be followed throughout.

The gravity centre criterion is a commonly used method and is adopted here. The output gets the linguistic description of the template set, whose gravity centre is closest to that of the output set. The percentage reduction in stiffness is the output variable and is defined by five membership functions, namely Not much Reduced-NMR, Much Reduced-MR, Fairly Reduced-FR, highly reduced-HR, and severely reduced-SR, and is shown in Figure 8.

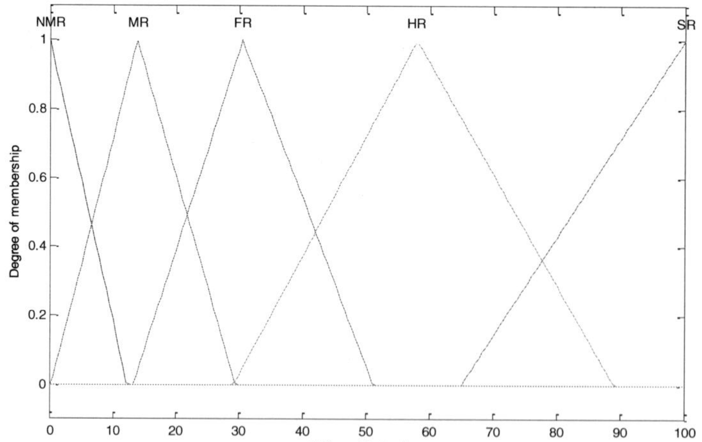

Figure 8. Membership function of the output variable – percentage reduction in stiffness.

5. CASE STUDIES

5.1. Simply Supported and Fixed Beam

Two cases have been considered to explain the robustness of the fuzzy logic. A beam, fixed at both ends and another beam simply supported at its ends were analyzed by considering the damage in the left beam element of spread 0.01, using fuzzy logic and the surface viewer of percentage reduction in stiffness was compared with the numerical results obtained from the MATLAB program developed for the 2D beam element, explained earlier (Section 2.0 and 3.0).

For both the cases, damage is defined by two input variables – intensity of damage and location of damage. Twenty different rules have been generated for the fixed and simply supported beams to get the desired output. The surface viewer obtained from fuzzy logic for fixed and simply supported beam is shown in Figure 9 (b) and Figure 10 (b).

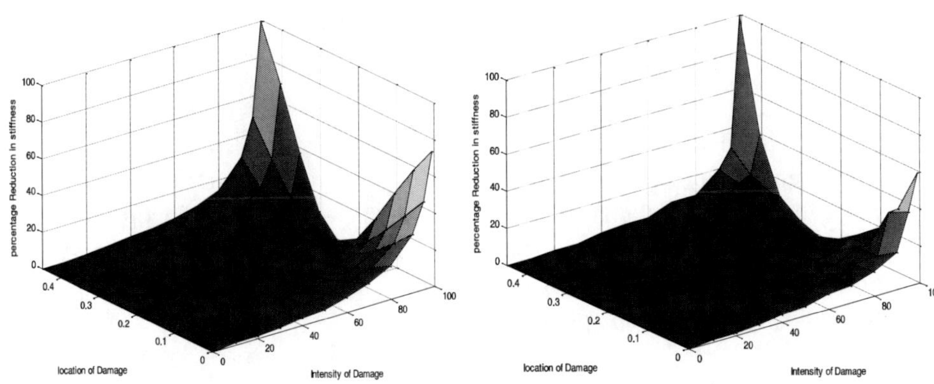

(a) (b)

Figure 9. Surface viewers obtained for a fixed beam with damage from: (a) formulated beam element and (b) fuzzy Logic.

6. DISCUSSION

The robustness of the fuzzy logic is shown by considering two cases, a beam simply supported at its ends and a beam fixed at both ends, both the beams subjected to a point load at the centre. The percentage reduction in stiffness is found out for different levels of damage and at various location of damage, for both the beams.

From the results (Figure 9 and Figure 10) it is found that the percentage reduction in stiffness is very high, when the intensity of the damage changes from 85 to 99%. This is due to the fact that the beam almost losses all its stability and at the verge of failure when the location of the damage is at the centre for simply supported beam. Similarly fixed beam loses its stability, when the damage is at the centre or at the support. The results obtained from fuzzy logic are very much closer to the model developed analytically in this paper. Triangular membership function is used for both the input and the output variable, but it can be further fine-tuned by adopting different membership functions for input and output variables.

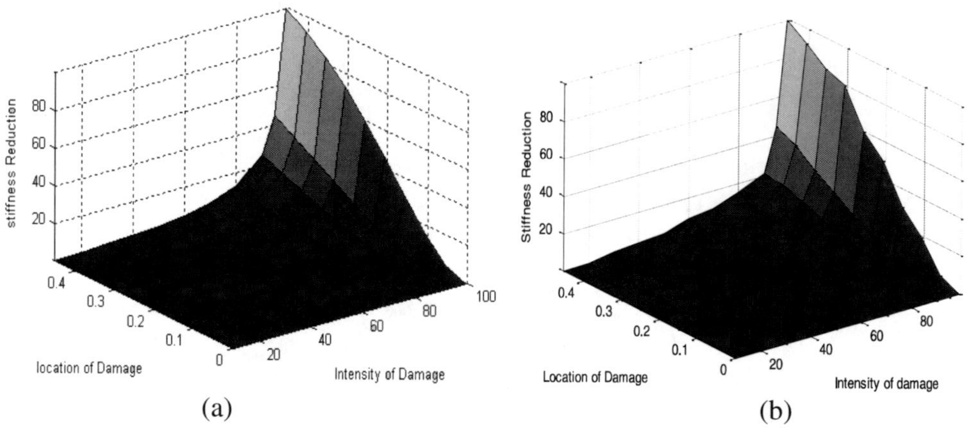

Figure 10. Surface viewers obtained for simply supported beam with damage from: (a) formulated beam element and (b) fuzzy Logic.

CONCLUSION

Fuzzy logic was applied in predicting the percentage reduction in stiffness for damaged beams with different boundary conditions and it gives a realistic or natural way of understanding the behaviour of damage. It is one of the powerful tools, incorporating all the uncertainties in the input variables. The method developed in this paper is applicable for any damaged beams, once the damage can be correlated in terms of three damaged parameters, intensity, location and spread of damage linguistically.

REFERENCES

[1] Dinesh, P., Raghavakumar, B., Natarajan, C. and Rajaraman, A. (2006). Damage finite element analysis using NDT for beams. *Proceedings of the Tenth East Asia Pacific Conference on Structural Engineering and Construction*, 114-125.

[2] Gang Yan. and Lily, L. Zhou (2006). Integrated fuzzy logic and genetic algorithms for multi-objective control of structures using MR dampers. *Journal of Sound and Vibration,* Vol. 296 , No. 1-2, 368-382.

[3] Ilyas, Eker and Yunis, Torun (2006). Fuzzy logic control to be the conventional method. *Energy Conversion and Management,* Vol. 47, 377-394.

[4] Klir, G. J. and Folger, F. L., (1988). *Fuzzy sets, uncertainity and information*, Prentice Hall, Inc., New Jersy.

[5] MATLAB, The Language of Technical computing. http://www.mathworks.com, version 7.0. *The Mathworks*, 2004.

[6] MATLAB, Fuzzy logic toolbox user's guide. http://www.mathworks.com/products/ fuzzylogic, version 7.0. *The Mathworks*, 2004.

[7] Ma Ge and Eric, M. Lui (2005). Structural damage identification using system dynamic properties. *Computers and Structures,* Vol. 83, No.27, 2185–2196.

[8] Pourzeynali, S., Lavasani, H. H. and Modarayi, A. M., (2007). Active control of high rise building structures using fuzzy logic and genetic algorithms. *Engineering Structures,* Vol. 29, No.3, 346-357.

[9] Qu Wenzhong, Sun Jincai and Qiu Yang (2004). Active control of vibration using a fuzzy control method. *Journal of Sound and Vibration,* Vol. 275, 917-930.

[10] Rao, S. S., Ravikumar, K., Raju, S. S. and Kumar, V. S. S., (2006). Decision making under conditions of uncertainty in construction Industry using fuzzy logic approach. *Civil Engineering and Construction Review*, Vol. 19, No. 4, 48-52.

[11] Revathy, M., Natarajan, C. and Rajaraman, A. (2007). Analysis of damaged beams using stiffness reduction formulation. *Proceedings of the International Conference on Recent Developments in Structural Engineering*, 186-194.

[12] Seung-Yong Ok, Dong-Seok Kim, Kwan-Soon Park and Hyun-Moo Koh (2007). Semi-active fuzzy control of cable stayed bridges using magneto-rheological dampers. *Engineering Structures,* Vol. 29, 776-788.

[13] Shifana Fatima, S. and Natarajan, C., (2007). Effectiveness of fuzzy logic in decision making. *Proceedings of the International Conference on Modeling and Simulation.*

[14] Sundareswaran, K., (2005). *A Learners guide to fuzzy logic systems*, Jaico Publishing House, Mumbai.

[15] Trisha Sain and Chandra Kishen M., (2006). Damage assessment in beams using inverse method. *Journal of Engineering Mechanics*, Vol. 132, No. 4, 337-44.

[16] Zadeh, L. A., (1975). The concept of a linguistic variable and its application to approximate reasoning. *Information Science,* Vol. 8, 199-249.

In: Knowledge Management and Communication …
Editor: Rafiqul Islam
ISBN: 978-1-62808-250-0

Chapter 4

Disturbance Data Compression Using Arithmetic Coding Approach with Fast Adaptive Wavelet Packet

N. P. Subramaniam[1*] and K. Bhoopathy Bagan[2]
[1]Department of Electrical and Electronics Engineering,
Pondicherry Engineering College, Puducherry, India
[2]Department of Electronics Engineering,
Anna University, MIT Campus, Chennai, India

Abstract

Data storage and data communication currently pose a major problem for all parties involved with power quality and power systems monitoring. The problem arises from the tremendous amount of data involved. There is a common desire in the power industry to find new techniques for high accuracy data compression and data storage. This paper introduces the details of a novel lossless data compression technique that is very suitable for application to power system disturbance waveforms. The method is developed a new fast one dimensional (1-D) convolution decimation algorithm with fast wavelet packet. The algorithm is four times faster than standard convolution decimation. The Arithmetic coding technique is then used to reduce the amount of data more effectively, thereby lowering down the cost related with data storage and transmission. Full technique implementation details are given, and test cases based on power-quality waveforms are analyzed.

Keywords: Power systems, best basis, wavelet packet, data compression, power quality

[*] Email: npsubbu@yhaoo.com.

I. INTRODUCTION

The power supply degradation is continuously increased. Hence, for utility operators, the improvement of power quality has become a crucial concern. The issue of power quality is involved in any electrical disturbance that could adversely affect customer load or power system equipment [1]. To ensure a certain level of power quality, many utilities often use dedicated monitoring devices on the stations to monitor the possible events. While all major events can be captured, this monitoring may result in an excessive amount of data as the disturbances often cover a broad frequency spectrum. For example, suppose there is a sampling- and-hold circuit that can sample 256 data points per cycle for a data length of four cycles, then it will require 73.72 kilobytes to store such a waveform. Now, by assuming 100 events in total, the storage space needs to be increased to 7372 kilobytes. As a result, high cost for storing such voluminous data become a burden and, unavoidably, degraded the quality assessment performance. In reducing such voluminous data without sacrificing any important feature, the data compression technique has emerged as a viable alternative. Concept of the data compression approach is the assignment of short codewords to more probable events and longer codewords to less probable events so that size of the information can be significantly reduced. Therefore, the storage capacity is enlarged and the bandwidth of communication system is expanded as well [2]-[3]. The arithmetic coding approach serves as the main paradigm of the proposed approach [4]-[6]. Since the arithmetic coding is able to encode arbitrary data with minimal average code length, it is also deemed nearly optimal in compression practices.

Recently, because wavelet theory provides a unified mathematical framework, it is commonly recognized as useful for signal analysis study [7]-[11]. Wavelets with many vanishing moments are very effective for coding piece wise smooth signals [12]–[14]. Wavelets, however, are ill suited to represent oscillatory patterns. Oscillatory variations of intensity can only be described by the small scale wavelet coefficients. Unfortunately, those small scale coefficients carry very little energy, and are often quantized to zero, even at high bit rates. Much larger libraries of functions, called wavelet packets, have been constructed [15] to address this problem. Because the collection of wavelet packets is over complete (there are many more basis functions than the dimension of the input space) one can construct a basis that is fitted for a target signal. In this work, a good basis should be able to describe the target signal with a very small number of basis vectors.

Our goal in this paper is to provide a fast numerical implementation of the best fast adaptive wavelet packet algorithm, in order to demonstrate that an advantage can be gained by constructing a basis adapted to a target signal. The importance of the proposed methodology is that not only is it a novel approach to data storage, but it is also capable of data compression with possibly very high compression ratios, especially with the increase in the sampling rates of monitoring equipment. The increase in the sampling rates is necessary for the capture of fast events that may be essential for a proper PQ analysis process. The compression of the data in the new proposed method is done with very high accuracy and very little or no loss of information. If needed, it is always possible to tradeoff accuracy for higher compression ratios. Emphasis in this paper has been placed on developing algorithms that are computationally efficient.

The data compression algorithm is divided into three parts. In the first part, we select that wavelet packet basis which is best adapted to encode the detailed signal. During the second part, the wavelet packet coefficients of the signal are quantized; we assume a Laplacian distribution and we use an efficient near optimal scalar quantizer [16]. Finally, the significance map is entropy coded using a higher order arithmetic coder [17] that relies on a context consisting of pixels in a causal neighborhood. This paper is organized as follows. In Section II, we review the wavelet packet theory. The factorization of the conjugate quadrature filters is described in Section III. In Section IV, we describe the quantization, and the context-based entropy coding. Results of experiments are presented in Section V.

II. Wavelet Packet Theory

Let h={h_n} and g={g_n} be biorthogonal filters. Let ψ^0 be the scaling function and ψ^1 be the wavelet associated with {h_n},{g_n}. Let us define the basic wavelet packets

$$\psi^{2n}(x) = \sum_k h_k \psi^n(2x-k) \tag{1}$$

$$\psi^{2n+1}(x) = \sum_k g_k \psi^n(2x-k) \tag{2}$$

A multiscale wavelet packet: $\psi^n_{j,l} = \psi^n(2^j x - k)$ is located at $l2^{-j}$, has a support of size 2^{-j}, and roughly n oscillations (a frequency 2^{jn}). Let us associate the dyadic frequency interval $[2^j n, 2^j(n+1)]$ to the wavelet packet $\psi^n_{j,l}$. If a collection of intervals $[2^j n, 2^j(n+1)]$ provides a cover of the time-frequency plane, then the set $\{\psi^n_{j,l}\}$ is an orthonormal basis [15]. If we consider dyadic subdivisions at each level, we get 24 bases for L levels. Clearly, we have an extremely large amount of freedom for the construction of orthogonal bases from the wavelet packet $\{\psi^n_{j,l}\}$. This greater flexibility is exploited to increase the efficiency of the representation. Coifman and Wickerhauser [15] suggested to use a fast dynamic programming algorithm with order $N\log(N)$ (N is the number of samples) to search for that best basis which is optimal.

III. Fast Convolution-Decimation

In any practical situation, the best basis algorithm is only applicable if the computation of the coefficients at each node of the wavelet packet tree does not require an absurdly large amount of time. This section presents an efficient numerical algorithm for computing multidimensional convolution-decimation that divides the number of operations by a factor 2 in 1-D.

A. Factoring the Biorthogonal Filters

Several methods [18] have been proposed in the literature for the fast implementation of convolution and decimation. FFT implementations [18] are useful for large filters (of length 64 or 128), but do not bring any improvement over a straightforward implementation for short and medium size filter [18]. Recently, several authors [19] have proposed efficient implementations of 1-D biorthogonal filters using a factorization of the filters into smaller filters. All biorthogonal filters can be factored into a sequence of ladder steps. A ladder step maps every couple of even and odd samples

$$(x_{2k}, x_{2k+1}) \text{ into } (y_{2k}, y_{2k+1}) \tag{3}$$

$$y_{2k} = x_{2k} + L(x_{2k+1}) \tag{4}$$

$$y_{2k+1} = x_{2k+1} \tag{5}$$

The proof of the factorization of 1-D biorthogonal filters into ladder steps [15] relies on the use of the Euclidean algorithm to factor polynomials. The ladder step is a lifting step, and similar results were derived using the same methods. We now introduce the notations and recall the key results. Let , h, g be any symmetric biorthogonal filters of length 2m+1. We merge h and g into an orthogonal band diagonal operator W where the only nonzero coefficients are

$$W_{2k,2k+1+j} = h_j \text{ , } W_{2k+1,2k+1+j} = g_j \text{ , j=-m,..., m.} \tag{6}$$

Let H_α be the lowpass ladder step: H_α is a band diagonal matrix, with nonzero elements given by

$$H_{2i,2i-1} = \alpha \text{ , } H_{2i,2i} = 1 \text{ , } H_{2i,2i+\alpha} = \alpha \tag{7}$$

and let G_β be the high pass ladder step: G_β is a band diagonal matrix, with nonzero elements given by

$$G_{2i,2i+2} = \alpha \text{ , } G_{2i+1,2i} = \alpha \text{ , } G_{2i+1,2i+1} = 1$$

One has the following result.

Lemma 1: W can be factored into at most m operators

$$W = H_{\alpha m} G_{\beta m} H_{\alpha m-1} G_{\beta m-1} \ldots H_{\alpha 0} G_{\beta 0} \tag{8}$$

The results extents to orthogonal filters. We note that the computational complexity of each ladder step (7) or (8) is three operations per two samples.

Fast Convolution-Decimation: A fast algorithm for convolution-decimation implements the ladder steps(7) and (8). After the calculation of the first lowpass and high pass coefficients $(y_0, y_1)^T = (h,g)(x_0,x_1)^t$, each new low pass and high pass coefficients (y_{2k}, y_{2k+1}) only require a cascade of m ladder steps (7) and (8) given in Figure 1. The complexity of the computation of (y_{2k}, y_{2k+1}) is therefore 3m. We compare this result with a standard implementation that requires 8m+2 operations. The fast algorithm divides the number of operations by a factor ((8m+2)/3m) > 2.67.

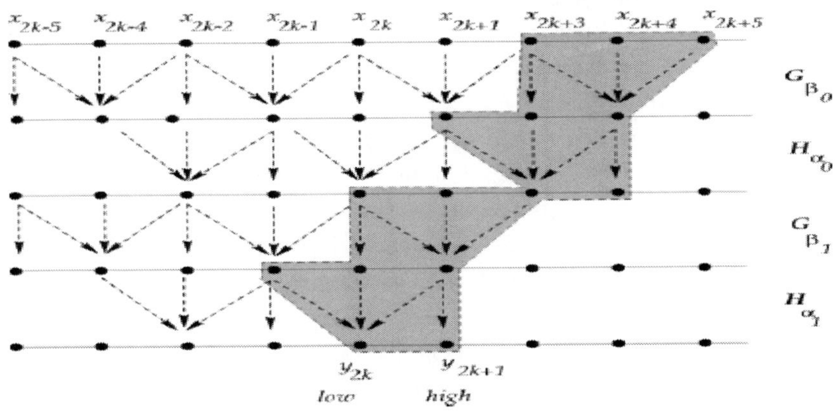

Figure 1. Four ladder steps (shown in the shaded regions) are needed to compute the low pass and high pass filtered coefficients y_{2k}, y_{2k+1}.

IV. Quantization and Entropy Coding

A. Laplacian Based Scalar Quantization

Within each subband the distribution of the wavelet packet coefficients is approximated to a Laplacian distribution [20]. We use a particularly efficient scalar quantizer, with a performance that is very close to the performance of the optimal entropy constrained scalar quantizer for Laplacian distribution [16]. It is a uniform scalar quantizer, with a symmetric dead zone, $[-\Delta+\delta, \Delta-\delta]$, and a step size Δ. The optimal (for mean square error) reconstruction offset is $\delta = 1-\Delta((e^{\Delta})/(1-e^{-\Delta}))$. Finally, we apply a dichotomic search to find the optimal value of Δ that exhausts the budget.

B. Ordering of the Coefficients and Entropy Coding

After quantization, the positions of the nonzero output levels are recorded in a significance map. The lossless compression of the significance map takes advantage of the fact that large output levels often appear in clusters. If one uses a wavelet basis, then one can also take advantage of the self similar structure of subbands, using zero-trees [12], [13]. Unfortunately, the wavelet packet coefficients fail to exhibit a self similar structure. This is due to the fact that, in principle, if the basis is well "adapted" to a given signal f, then it is unlikely that the coefficients $\langle \psi_{j,l}^{n}, f \rangle$ at a given position are significant at all the scales. In other words, the coefficients of a salient feature in the signal will only appear at a single scale that characterizes the feature. Another technical difficulty comes from the fact that we usually cannot define the "parent" subband in a general wavelet packet tree. Because of all these issues, we limit the context to be a spatial context. Subbands are scanned by increasing frequency. We then scan all the pixels inside any given subband using a Hilbert space filling curve. This self similar curve guarantees that 1) all samples in an area are visited before moving on to another area, and 2) the sequentially visited samples are always adjacent in the 1-D signal plane. The spatial context is then defined as the samples that appear before the current samples pixel in the Hilbert scan. In the experiments $n_C = 3$ have been used. This efficient context modeling permits to condition the probability of significance of a given pixel on the probability of significance of its neighbors using a n_C order arithmetic coder.

In arithmetic coding, at each half open subintervals of the half-open unit interval [0,1], a number of bits are used to specify a codeword to distinguish the subinterval corresponding to the actual data set from all other possible subintervals. By using the probabilities of each data set, this corresponding subinterval is incrementally refined and the outputs are accordingly binary encoded. The algorithm for the encoding process is described as in Figure 2.

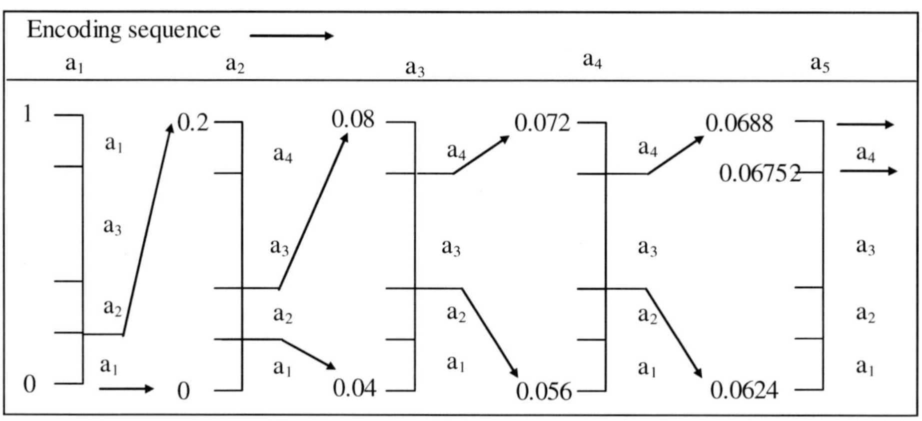

Figure 2. Arithmetic coding.

The probability table is formed from the acquired data, the probability of every event is first calculated by using histogram technique such that the distributed range of each symbol is ensured. Now to depict the process of arithmetic coding, a string of '$a_1a_2a_3a_4$' is provided here as an example. Based on the chance of each symbol appeared in the string first lists the probability of each symbol and their corresponding distributed range. Figure 2 explains the remaining procedure of arithmetic coding technique.

V. PROPOSED METHODS

Figure 3 shows the framework of the proposed method. In the Figure , the acquired signal is first sampled. Then, the fast wavelet packet (FWP) is employed for coder and decoder and an actual bit stream is generated by coder. The best basis search algorithm can provided as follows. First the signal $S_k(n)$, decomposed into a higher frequency component $D_{k+1}(n)$ and a lower frequency component $S_{k+1}(n)$. Then, the entropy of each signal is computed, where E_{sk}, E_{sk+1} and E_{Dk+1} stand for the entropy of the signal Sk, S_{k+1}, and D_{k+1} respectively. If the inequality equation $E_{Sk+1}+E_{Dk+1} < E_{Sk}$ is satisfied, this decomposition process is accepted; otherwise the decomposition process goes forward to other binary branch. This work continues until all of the branches are traced, and the resultant tree structure can be stored for the future signal needs reconstruction.

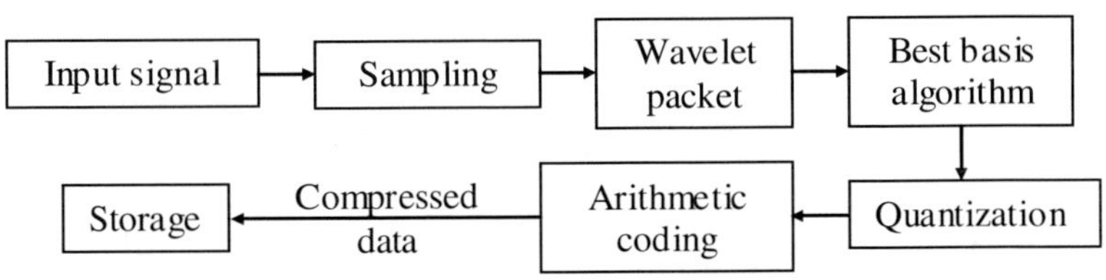

Figure 3. Framework of the proposed method.

After the above mentioned decomposition process, a series of coefficients represented in floating point will be obtained. A linear quantizer is used for transformation of decomposed coefficients within a certain range such that the decomposed coefficients can be represented by one single-byte data. Then, arithmetic coding is activated to encode every leaf in the wavelet packet tree. Note that in this paper, the Daubechies wavelet coefficients are used as high pass and low pass filters and the minimum entropy method is selected as the cost function of wavelet packet transformation with different levels of Daubechies wavelet coefficients. Next by loading the tree structure information, the compressed disturbance data can be reconstructed with the aid of inverse wavelet packet transform. A normalized mean square error (NMSE) representing the quality of the reconstructed signal in this study is also given eq.(9) and compression ratio is given in eq.(10)

$$\text{NMSE} = \frac{\|\text{original signal - reconstructed signal}\|^2}{\|\text{original signal}\|^2} \tag{9}$$

$$\text{Compression ratio} = \frac{\text{Number of input samples}}{\text{Number of arithmetic coefficients}} \tag{10}$$

VI. NUMERICAL RESULTS

The proposed method has been applied to reduce the amount of data acquired from different power system disturbances, including voltage sag, voltage swell and momentary interruption. Different approaches were used for the evaluation of compression performance, they are like:

Method 1 Huffman coding
Method 2 LZ78 algorithm
Method 3 Huffman coding with wavelet packet enhancement
Method 4 arithmetic coding with wavelet packet enhancement
Method 5 arithmetic coding with fast wavelet packet

Among these methods, it is noted that Method 5 is indeed approach proposed in this paper. The difference between Method 1 and Method 3 lies in the embedment of the wavelet packet transform with Huffman coding. Each method was tested on the following scenarios. For each test case, the original simulated signal has 502 sample points.

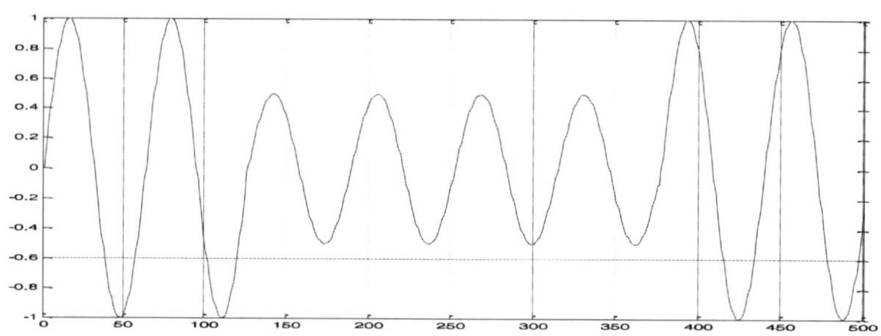

Sample point

3(a)

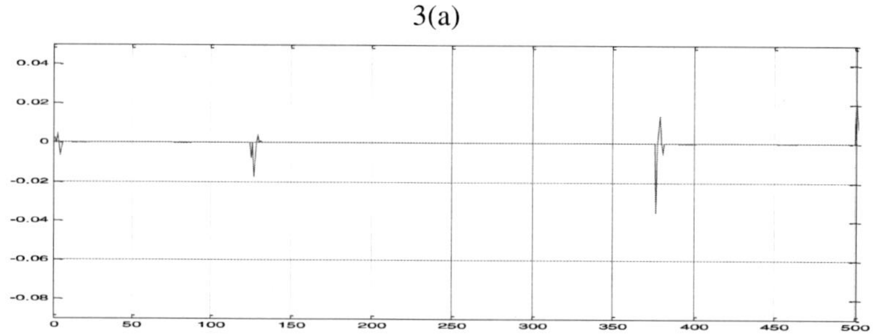

Sample point

3(b)

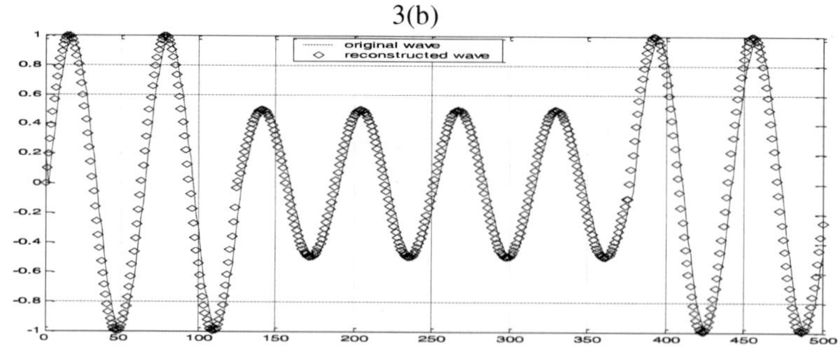

3(c)

Figure 3. (a) Voltage sag waveform. (b). Detailed coefficient. (c). Decoded wave.

A. Voltage Sag

Voltage sags are often caused by the large fault current, the switching of heavy loads or the starting of large motors. It often lasts for 0.5 cycles to 1 min. When the voltage drops 30% or more, the system status is considered severe. Figure 3(a) depicts the disturbance signal. Figure 3(b) is decomposed signals of $D_1(n)$ and Figure 3(c) shows the decode waveform of the proposed method. The reconstructed waveform match with the original waveform.

Figure 4(a) illustrates the compression ratios reached by Huffman coding with wavelet packet enhancement, arithmetic coding with wavelet packet enhancement (A-WT) and arithmetic coding with fast wavelet packet enhancement (A-FWP). In the Figure horizontal axis is given for compression ratio and vertical axis the level where decomposed wavelet coefficients are situated. This bar-chart indicates that the proposed method reached a higher compression ratio at different levels. In Figure 4(b)m the NMSE at different level is delineated, where the averaged value is computed to be $4.29X10^{-5}$. Note that as four compression methods considered in this study are all categorized as lossless ones that implies their NMSE value should be mutually similar; therefore, only the NMSE of the proposed approach is listed here. Figure 4(c) depicts the compression of different methods. As this plot reveals, the proposed method obtains a highest compression ratio when compared to other three methods. This outcome supports the feasibility of the method.

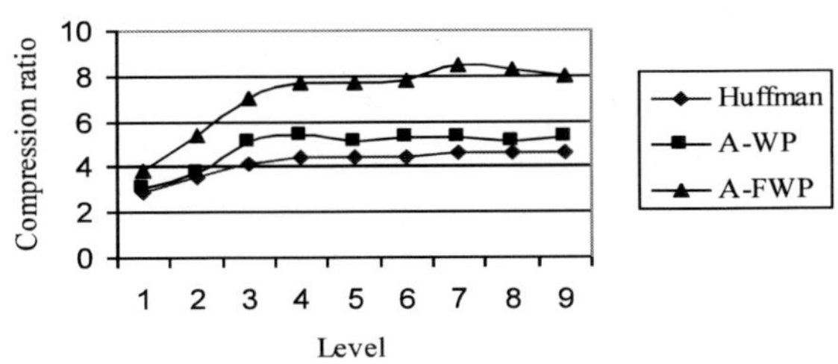

4(a)

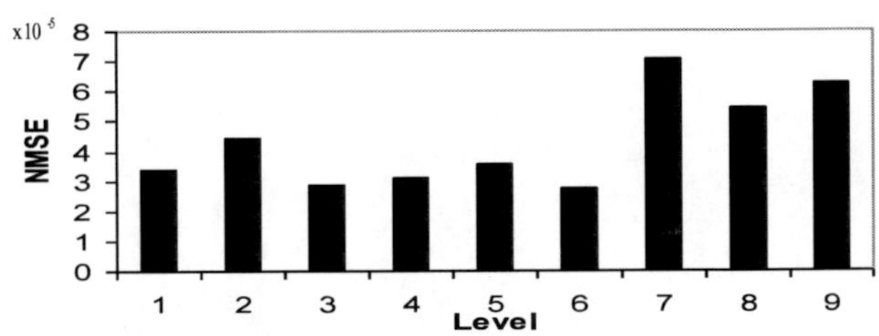

4(b)

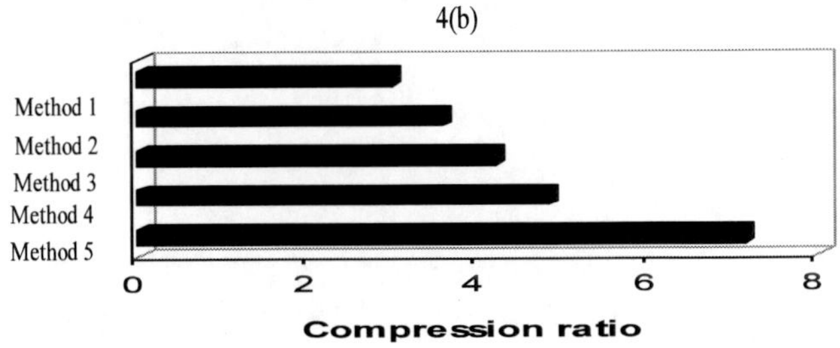

4(c)

Figure 4 (a). Compression ratio at different levels. (b). NMSE values (c). Compression of different methods.

B. Voltage Swell

During a single-line-to-ground fault, a brief increase of the rated system voltage may take place on the un-faulted phases of three phase system. This scenario is often seen as voltage swell, whose magnitude is related to the system grounding. Swell may stress the delicate equipment components to premature failure. For this study, a voltage swell waveform is plotted in Figure 5(a), where a 20% swell was seen to last for two cycles. Analysis results of using the method are seen in Figure 5(b) and 5(c) give decoded waveform. The peaks found in the Figure pinpoint the beginning and the termination of the disturbance. There was no spurious effect reported in this case.

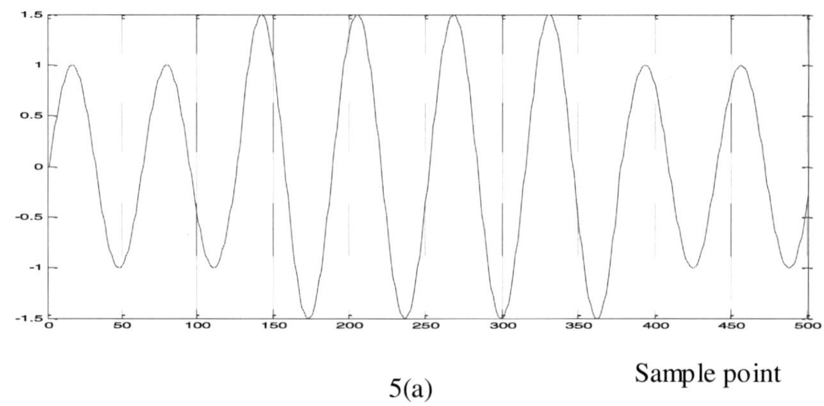

5(a)

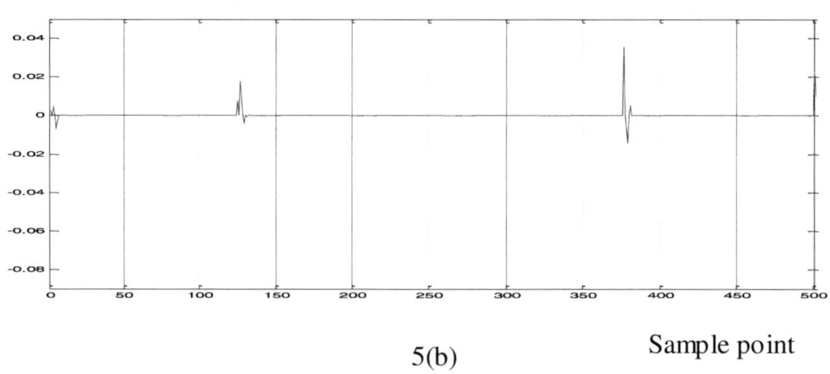

5(b)

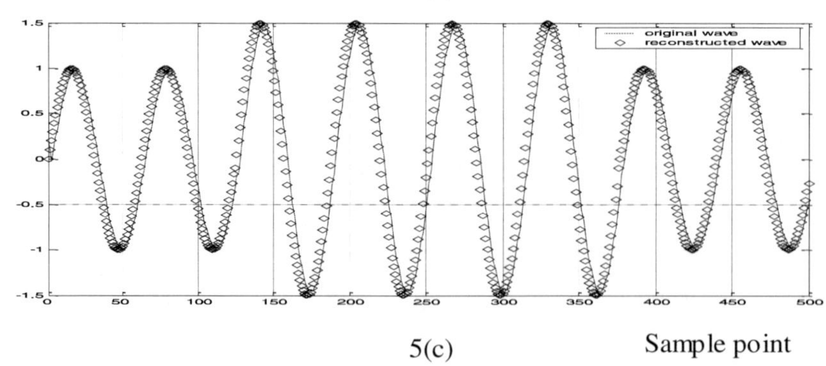

5(c)

Figure 5 (a). Voltage swell sag waveform. (b). Detailed coefficient. (c). Decoded wave.

Similar to Figure 4(a), Figure 6(a) shows the performance comparisons between the Huffman coding and arithmetic coding individually enhanced with wavelet packet transform for the this scenario. Comparison results solidify the superiority of the proposed method. Then in Figure 6(b), the value of NMSE is plotted for each decomposition level considered. From the barchart of Figure 6(b), it informs that the reconstruction error between original and reconstructed signal is considered acceptable. In order to validate the effectiveness of the proposed approach, Figure 6(c) also shows the compression capability of different coding methods. For example, by comparing Method 5 with Method 2, it is found that the file size can be almost doubly compressed. As a result, the data transfer can be thus facilitated in a higher rate, improving the disturbance monitoring capability.

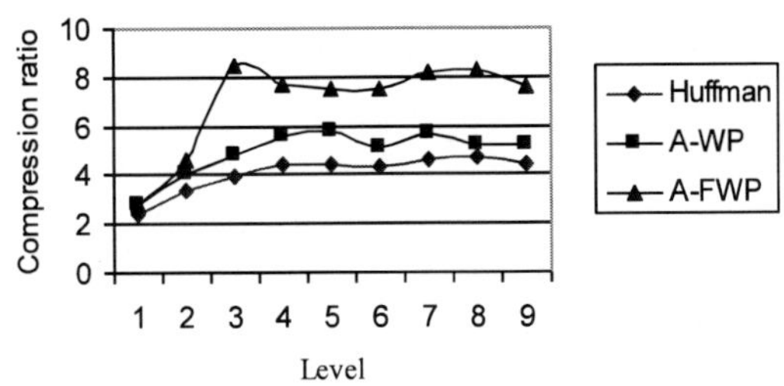

6(a)

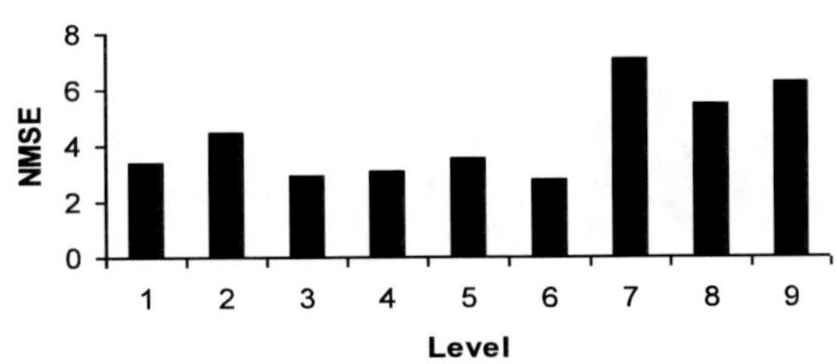

6(b)

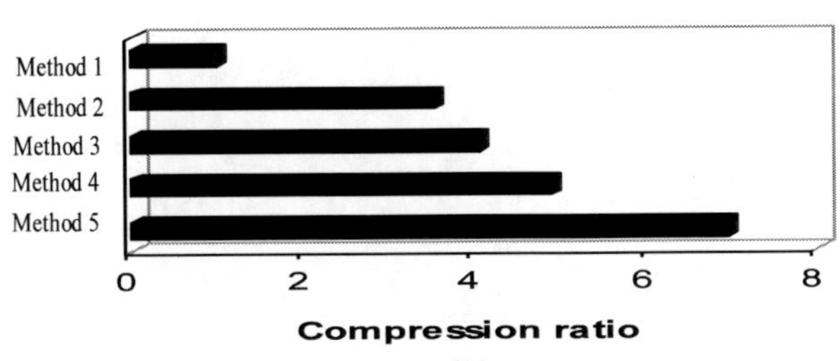

6(c)

Figure 6 (a). Compression ratio at different levels. (b). NMSE values (c). Compression of different methods.

C. Momentary Interruption

From a customer's point of view, interruptions are any loss of power. By most utility industries, this disturbance scenario is often deemed the drop of 90% - 100% of the rated voltage lasting for 0.5 cycles to 5 min. They are often measured associated with the operation of re-closing or automatic throw over devices. Figure 7(a) is the recorded momentary interruption waveform, where the interruption lasts about 256 samples. Test results using the proposed method are depicted in Figure 7(b) and 7(c) shows the decoded waveform, which are beneficial to the utility engineers in anticipation of better alleviating the effects caused by this nuisance. Then, for this case, the compression ratios of wavelet packet transform-enhanced Huffman coding and arithmetic coding under different levels are again evaluated in Figure 8(a). For all of the decomposed levels, the proposed approach exhibits a higher compression ratio. The low value of NMSE depicted in Figure 8(b) further consolidates the reconstruction capability of this method. Then in Figure 8(c), compression performance of the proposed method is also assessed and compared with that of the other four methods. For all of the methods considered, the proposed method approach is most confirmed. The test results of different cases are shown in Table I. The chosen test cases display a variety of sampling rates and one of transients at different frequencies to help establish the success of new technique.

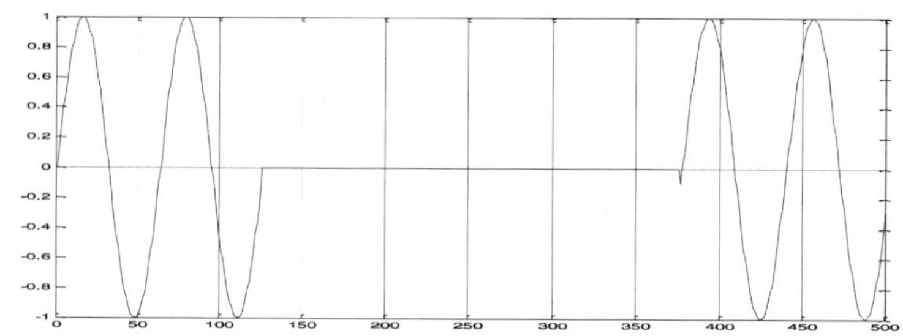

7(a)

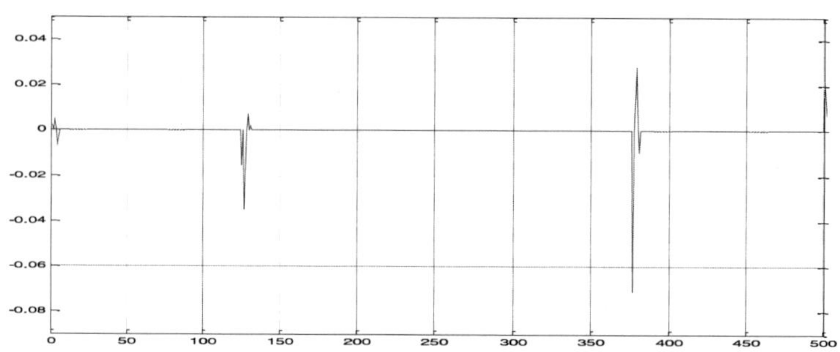

Sample point

7(b)

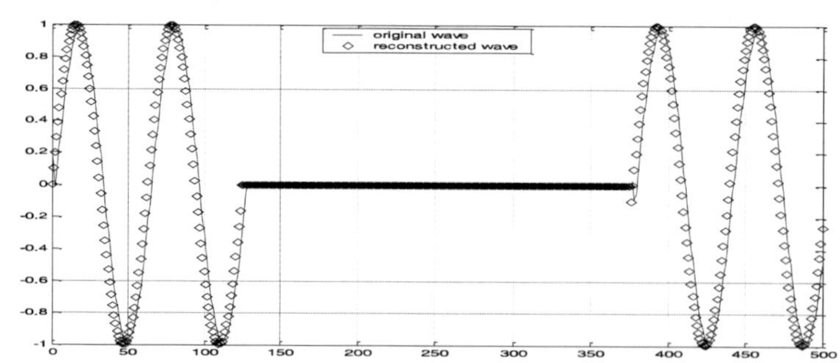

Sample point

7(c)

Figure 7 (a). Momentary interruption waveform. (b). Detailed coefficient. (c). Decoded wave.

Table I. Summary of Results

S.No	Sampling rate	NMSE	Compression ratio
1.	128	3.23×10^{-5}	4.01:1
2.	256	4.29×10^{-5}	5.33:1
3.	1024	2.54×10^{-5}	5.95:1
4.	16640	2.93×10^{-5}	19.23:1

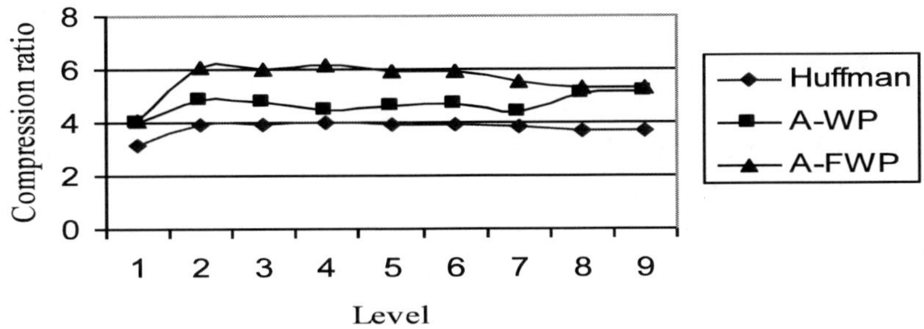

8(a)

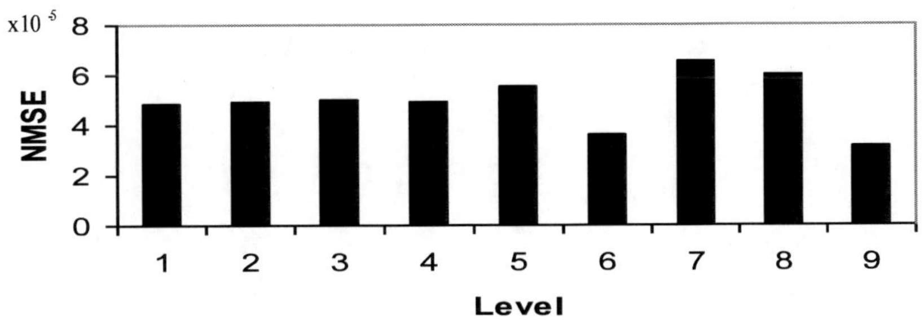

8(b)

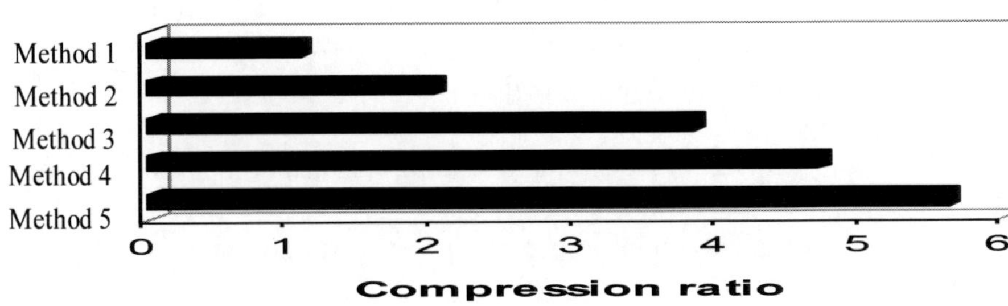

8(c)

Figure 8 (a). Compression ratio at different levels. (b). NMSE values (c). Compression of different methods.

In general, the test cases mentioned here all demonstrate the successful implementation of the new technique for a variety of waveforms carrying a variety of transients out of which some are severe. The results also show a variety of sampling rates to help establish the flexibility of the new system as well as its improved data compression capabilities at higher sampling rates. The successful implementation of the proposed technique is greatly supported by the results.

Conclusion

With a growing demand to ensure the quality of the power supply, the amount of data acquired for the monitoring and control of a modern power system may become enormous. An efficient approach that serves the role of data reduction for power quality analysis is, therefore, urgently required. In this paper, a data compression framework that includes the arithmetic coding technique enhanced with fast wavelet packet (FWP) is proposed to reduce the amount of power system disturbance data. The work establishes the high-compression ratios and excellent maintenance of data integrity offered by the proposed technique. The technique was successfully implemented and tested using a wide variety of waveforms including real-site waveform data. The test cases also establish the adapting ability of the new technique to handle a wide range of sampling rates.

The test results indicate that the proposed approach not only effectively reduces the file size; but also possesses the ability to reconstruct the compressed signal back to the original one with only a small error. Hence, the storage requirement can be significantly reduced, while the data transfer over network can be better realized.

References

[1] Domijan, A., Heydt, G. T., Meliopoulos, A. Venkata, S. S. and West, S. "Dirctions of research on power quality," *IEEE Trans. Power Delivery,* vol. 8, pp. 429-436, Jan. 1993.

[2] Davis, E. J., Emanuel, A. E. and Pileggi, D. J. "Harmonic pollution metering: Theorectical considerations," *IEEE Trans. Power Delivery,* vol5, pp. 19-23, Jan. 2000.

[3] Santoso, S., Powers, E. J. and Grady, W. M. "Power quality disturbance data compression using wavelet transform methods," *IEEE Trans. Power Delivery,* vol. 12, pp. 1250-1257, July. 1997.

[4] Howard, P. G. and Vitter, J. S. "Arithmetic coding for data compression," *Proc. IEEE.* vol. 82. no. 6. pp. 857-865. June 1994.

[5] Dtreams, S. D. "Arithmetic coding in lossless waveform compression," *IEEE. Trans. Signal Processing*. vol. 43. pp. 1874-1879. Aug. 1995.

[6] Zhu, B., Yang, E. H. and Tewfik, A. H. "Arithmetic coding with dula symbol sets and its performance analysis," *IEEE. Trans. Image Processing*. vol. 8. pp. 1667-1676. Dec. 1999.

[7] Rioul, O. and Vetterli, M. "Wavelets and signal processing," *IEEE. Signal Processing Mag*., vol. 8. pp. 14-38. Oct. 1991.

[8] Hess-Nielesen, N. and Wickerhauser, M. V. "Wavelets and time-frequency analysis," *Proc. IEEE*. vol. 84. no. 4. pp. 523-540. Apr 1996

[9] Daubechies, I. "The wavelet transform time-frequency localization and signal analysis," *IEEE. Trans. Inform. Theory*, vol. 36. pp. 961-1005. Sept. 1990.

[10] Huang, S. J. and Hsieh, C. T. "Application of morlet wavelets to supervise power system disturbance," *IEEE. Trans. Power Delivery.* vol. 14. pp. 235-243. Jan. 1999.

[11] Meyer, F. G., Averbuch, A. Z. and Stromberg, J. O. " Fast adaptive wavelet packet image compression," *IEEE. Trans. Image Processing*, vol. 9. pp. 792 -800. May. 2000.

[12] Said, A. and Pearlman, W. A. "A new fast and efficient image codec based on set partioning in hierarchical trees," IEEE Trans. Circuits Syst. *Video Technol.,* vol. 6, pp. 243–250, 1996.

[13] Shapiro, J. M. "Embedded image coding using zerotrees of wavelet coefficients," *IEEE Trans. Signal Processing*, vol. 41, pp. 3445–3462, Dec. 1993.

[14] Sriram, P. and Marcellin, M. W. "Image coding using wavelet transforms and entropy-constrained trellis quantization," *IEEE Trans. Image Processing,* vol. 4, pp. 725–733, 1995.

[15] Coifman, R. R. and Wickerhauser, M. V. "Entropy-based algorithms for best basis selection," *IEEE Trans. Inform. Theory,* vol. 38, pp. 713–718, Mar. 1992.

[16] Sullivan, G. J. "Efficient scalar quantization of exponential and Laplacian random variables," *IEEE Trans. Inform. Theory*, vol. 42, no. 5, pp. 1365–1374, Sept. 1996.

[17] Nelson, M. and Gailly, J. -L. *The Data Compression Book.* New York: M&T, 1996.

[18] Rioul, O. and Duhamel, P. "Fast algorithms for discrete and continuous wavelet transforms," *IEEE Trans. Inform. Theory,* vol. 38, pp. 569–586,Mar. 1992.

[19] Van Dyck, R. E. and Marshall, T. G. "Ladder realizations of fast subband/VQ coders with diamond support for color images," *IEEE Int. Symp. Circuits Syst.*, pp. I-670–I-677, 1993.

[20] Birney, K. A. and Fischer, T. R. "On the modeling of DCT and subband image data for compression," *IEEE Trans. Image Processing*, vol. 4, pp. 186–193, Feb. 1995

In: Knowledge Management and Communication … ISBN: 978-1-62808-250-0
Editor: Rafiqul Islam

Chapter 5

FAULTS DETECTION AND DIAGNOSIS IN A CHEMICAL PROCESS BY RBF NEURAL NETWORK

I. Manssouri,[1*] M. Manssouri[2] and B. El Kihel[3]

[1]Laboratoire de Modélisation Mathématiques et Système D'information Décisionnel, École Nationale Supérieure d'Arts et Métiers, Université Moulay Ismail, Meknès, Maroc

[2]Laboratoire de Génie Industriel et Production Mécanique, ENSA, Oujda, Maroc

[3]Laboratoire de Génie Industriel et Production Mécanique, ENSA, Oujda, Maroc

ABSTRACT

Faults Detection and Diagnosis (FDD) is a very important stake for industrial processes, since faults are often translated by unacceptable losses of productivity.

This article describes a new approach which allows detecting and diagnosing faults in a column of automated continuous distillation, which is a strongly unstable non-linear system.

It is based on the use of a RBF neuronal network (Radial Basis Function) as classifier of the various functioning modes (normal mode and abnormal mode). The determination of functions' centres of neurons' activation of the hidden layer is made by means of the K-means algorithm.

Principal Component Analysis (PCA) is employed for the graphic representation of the various functioning classes on a lower dimension space.

The studied distillation is a methylcyclohexane distillation from a toluene\ methylcyclohexane binary mixture.

Keywords: Fault Detection and Diagnosis; K-means algorithm; Bayesian neural classifier; Distillation column

* Corresponding author: Laboratoire de Modélisation Mathématiques et système d'information décisionnel, École Nationale Supérieure d'Arts et Métiers, Université Moulay Ismail, BP 4042, 50000, Meknès– Maroc. E-mail: imade_mansouri@yahoo.fr.

1. Introduction

Today, chemical industry performances are measured, among others, in terms of productivity, quality, safety and environmental protection. Improvement and control of faults detection and diagnosis tools turn out necessary to progress in these four directions.

Indeed, to intervene in a process, not only it is necessary to be capable of planning the influences provoked by the modifications, but also to chose which modifications need to be made to obtain the desired result.

In order to control a chemical process, it seems essential to determine the consequences of every damage on the installation to be able to treat it as fast as possible with minimum risks. Mastering the process leads to the use of a successful and efficient method of faults detection and diagnosis.

Several detection and diagnosis methods are proposed to analyze functioning faults in processes. These methods can be classified in two main categories [1]: the first one is based on the model of process. The second one is based on the history of the process.

The methods based on the model of process are based themselves on a fundamental understanding of the process by using qualitative methods (failures trees, signed diagrams) and quantitative methods (existing mathematical relations for describing the process) [1]. These methods based on the model of process are applicable to linear systems, but with limited availability to non-linear chemical processes [1]. This is due to the fact that these methods generate a large number of hypotheses and request a lot of time of calculation once used on-line.

The methods based on the history of the process, use big quantities of data and can also be subdivided into quantitative methods (neuronal networks and statistical techniques) and in qualitative methods (tendency analysis and method based on decision rules) [1]. When it is a question of representing heuristic knowledge, these last ones are efficient. Yet, they are not able to represent time, causality and fact. As a result, it becomes difficult to check them and determine their accuracy. They also suffer from certain inconvenient such as imperfection and inflexibility [2]. Thus, it is desirable to apply a methodology of faults detection and diagnosis, with specific properties, notably the capacity to answer real-time demands, understanding and implementing, the ease to widen and incorporate new faults, the use of a weak calculation power that demands weak memory during treatment, the capacity to supply adaptability and power to adapt itself to noised data and incorporate various types of input data.

It is true that these two detection methods are efficient, but recently neuronal techniques are widely used in problems of faults diagnosis in the chemical processes. Indeed, neuronal networks turned out fine classifiers in many faults detection and diagnosis applications such as:

Hoskins et al., [3] discuss the application of a very wide neural network of FANN type for faults detection of an installation of complex simulated distillation. They also propose the use of a state space model at the same time as the neural networks of FANN type for detection by classification.

Sorsa and Koivo [4] compare the capacities of four neural networks architectures to classify the defective behaviour in a simulated heat-exchanger engine system, and conclude that artificial neural networks (FANN) and the radial functions networks (RBF) give the same classification on the defective behaviour of the system.

Baratti and Corti [5] applied the feedforward networks for modelling a simple distillation column for the control of a distillate jet. The execution of this neural controller exceeds the conventional controllers in terms of response speed.

Brydon et al. [6] worked on the application of feedforward type neuronal networks on a database simulated by a pilot installation of distillation column. Two studies were presented: Firstly, the proposed study uses a neuronal network and Principal Component Analysis "PCA" to distinguish between two functioning modes: a first is said to be normal mode and second is said to be abnormal. The neuronal network and Principal Component Analysis yielded excellent results. Secondly, the study is carried out composed of four dysfunctionning types and the normal regime that is five classes, the use of neuronal networks yielded good results on the other hand the Principal Component Analysis was not able to correctly classify data.

Sharma et al. [7] used the neural feedforward networks for the detection of six dysfunctionning faults in a distillation column of NH3-H2O.They compare two networks structures: 6-4-6 and 3-4-6. Then, they adopted the second structure 3-4-6 which could reduce the computing time and detect the faults on-line.

Manssouri et al. [8] use the multi-layer perception neural networks to detect five defaults of functioning in a continuous distillation system of methylcyclohexane from a binary mixture toluene/methylcyclohexane. The choice of the architecture of the MLP neural networks is catered by applying different tests of robustness such as Akaike Information Criteria, Root Mean Square Error, Bayesian Information Criteria and Maximum Average Percentage Error. A back-propagation algorithm is used to fix the weights and the bias of the neural network.

In order to control the classification of a distillation column behaviour, which is a strongly unstable non-linear system, we have used a RBF neuronal network as classifier of the various functioning modes (normal mode and abnormal mode). The determination of the functions 'centres of neurons' activation of the hidden layer is made by the K-means algorithm. The Principal Component Analysis is used for graphic representation of various functioning classes on a lower dimension space.

2. Network Topology with a Radial Basis Function

The typical structure of an RBF neuronal network [9] is schematized in figure 1:

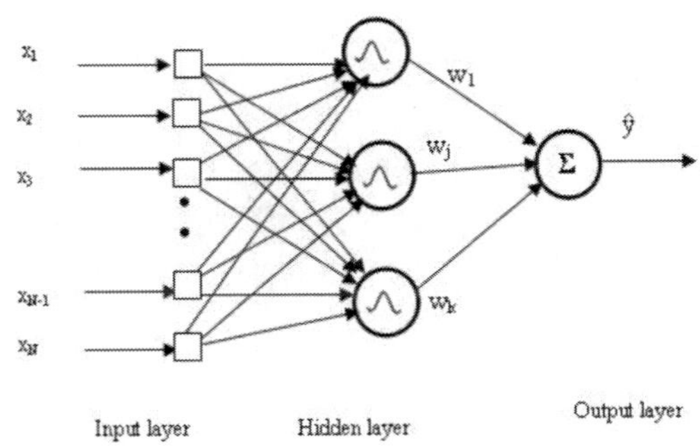

Figure 1. Typical structure of RBF neural network.

Input layer nodes are used to make cross exclusively input data to the hidden layer. No calculation is executed in the nodes of the input layer. The hidden layer contains k nodes, which apply a nonlinear transformation to the input variables. More specifically, each node j=1, 2........k has a centre c_j; where c_j is a vector whose dimension is equal to the number of inputs.

For each new vector of input X = [x_1, x_2..., x_N], the standard of the Euclidean distance between the vector of input and the centre of node is calculated as follows:

$$v_j(X)=\|c_j-X\|=\left(\sum_{i=1}^{N}(x_i-c_{j,i})^2\right)^{1/2} \tag{1}$$

Output nodes of the hidden layer are determined by a non-linear function, called the network activation function. A typical choice for the activation function is the Gaussian function:

$$f(v)=\exp\left(-v^2/\sigma^2\right) \tag{2}$$

with σ as the standard deviation of the activation function. An empirical rule [10] consists of taking:

$$\sigma_j=\left(1/N8^{1/2}\right)\sum_{i=1}^{N}\|x_i-c_j\| \tag{3}$$

Thus, the output of j node of the hidden layer is given by:

$$Z_j=f(v_j) \tag{4}$$

A set of synaptic weight w_j, j = 1.2,......., k is applied to connections between the hidden layer and the output layer. Nodes in output layer serve only as units of addition that produce the final network output. The output hidden layer will be presented as follows:

$$y=\sum_{j=1}^{j=k}(w_j Z_j+b_j) \tag{5}$$

Synaptic weights which exist between the hidden layer and the output layer are calculated according to the target vector Y= [y_1, y_2,...... y_N] and Z matrix which contains the different outputs of the hidden layer neurons. Weights are calculated in the following way:

$$w=\text{inv}(Z^T Z)\,Z^T Y^T \tag{6}$$

Through the mathematical formulas cited previously, we have presented the topology to be followed in the formation of the RBF neuronal network. Then, we shall provide a general outline for training methods non-supervised by the network.

3. Non-Supervised Classification Methods

There are numerous techniques of data classification. In this section we aim to obtain, an efficient classification of data by an automatic process. For that, we shall give more interest to the non-supervised methods of classification from which we cite some as example:

The Random method that consists in, randomly, choosing basic functions centres among the points of the whole supplied training.

The LVQ method (Learning Vector Quantization) aims to select points of the input space where basic functions are placed. These are, therefore, placed more on training points. The principle is, as it follows [11]: Centres are initialized randomly among the training database points. In each iteration, the whole training vector is presented to the selected center. A correction is applied. This one aims to improve centres 'position so that they would better reflect training points distribution in the input space.

The simple incremental method aims to obtain an RBF network containing a maximum number of basic functions fixed by the user. The principle is to place basic functions one by one on database points. In every stage, the point that gives (in the slightest squares direction) the weakest error is selected among the candidates. Basic functions are added until criterion satisfaction (for example an error value to be obtained, or a maximum number of the basic function).

The K-Nearest Neighbours method [12] determines a set of centroids on which we place basic functions. It is also about a decreasing approach because we start from an initial network where a basic function is placed on every training point. Then a K-Nearest Neighbours approach is applied where K is a parameter chosen by the user. The principle is to include every point and its K-Nearest Neighbours within the same class. Therefore, there is a distance notion to be specified: Euclidian distance, Mahalanobis distance...

K-Means [13] is one of the simplest non-supervised algorithms. This algorithm is strongly used in miscellaneous applications [6, 14, 15]. The principle consists in fixing beforehand a K number of classes centres. That is to know in priori the number of classes. We begin with the random choice of K centres. Then we calculate the distance of every element of the database with regard to each of the centres. Later, we allocate the element to the class center which is the closest to it. Once, the database is completely used, we recalculate again the new classes of centres. We repeat three distance calculation operations, allocation and determination of the new centres until satisfaction of a stop criterion. The most used stop criterion consists in stopping the process when the changes of the centre positions become insignificant.

This algorithm aims to reduce at least an objective function, in this case the square error function is:

$$J = \sum_{j=1}^{k} \sum_{i=1}^{N} \left\| x_i^{(j)} - c_j \right\|^2 \tag{7}$$

where $\left\| x_i^{(j)} - c_j \right\|^2$ is a measure chosen by distance between a point of reference $x_i^{(j)}$ and centre c_j.

The algorithm consists of the following steps:

1. Place the K points in the space represented by objects. These points represent the initial classes' centres.
2. Assign each object to the closest class centre.
3. When all objects are assigned, recalculate the new positions of centres.
4. Repeat steps 2 and 3 until centres do not move any more.

Each of these methods previously cited present advantages and drawbacks. However, K-means algorithm will satisfy later. The advantages of K-means algorithm are: a fast convergence [16], the ease of work and the possibility of treating a large data base. The algorithm is chosen for these reasons: these drawbacks include the obligation to define the number of classes beforehand and the obtained partition depends strongly on the initial choice of centres.

4. Experimental Device

To obtain a D mixture rich in methylcyclohexane from a toluene/methylcyclohexane F mixture, the mass composition of which we defined 23 % in methylcyclohexane, we use the distillation technique. This latter allows the partial separation of the most volatile compound methylcyclohexane contained in a liquid mixture (figure 2).

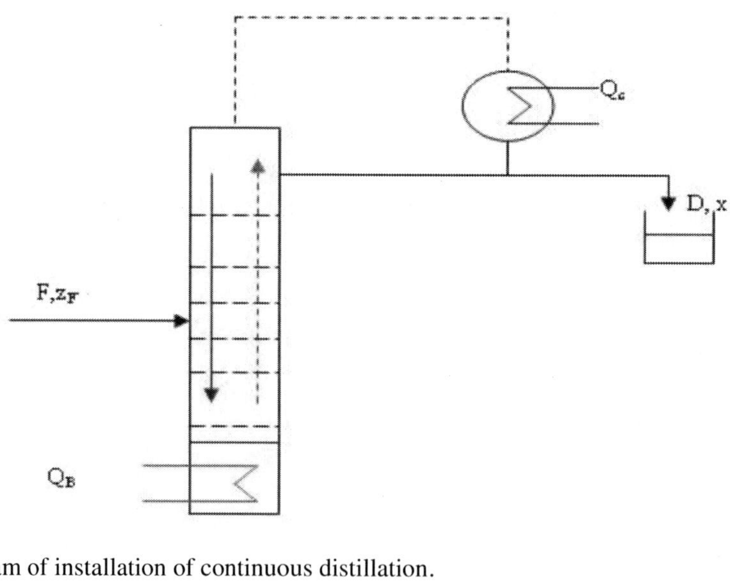

Figure 2. Diagram of installation of continuous distillation.

It consists of introducing a mixture flow of "toluene methylcyclohexane" heated in a reboiler under Q_B power where constant boiling pressure occurs. The vapour escaping from the reboiler is condensed under Q_C power, which leads to an immediate impact, which is the direct the recuperation of the condensate. The vapour stemming from the mixture is always richer in most volatile composite, so that the title X of the condensate is constantly superior to that of the title ZF of the initial mixture. Continuing the operation gradually impoverishes constituting the most volatile liquid of the reboiler. It results in a continuous decrease of the condensate title.

In order to be able to elaborate a data base serving, both as a learning base constituted of a normal mode, and as a test base constituted of normal and abnormal modes, we shall define inputs and outputs of the distillation system.

5. Definition of the Relevant I/O

Before starting the study of inputs and relevant outputs behaviour (figure 3), it is necessary to determine a working interval. For that, we assume the fact that we want to obtain a rich mixture in methylcyclohexane. Thus, starting from the isobar diagram liquid vapour of toluene/ methylcyclohexane mixture, we will determine the temperature range (of title) by using the Wilson model for mixture modelling. The selected interval is [0.48; 0.84] taking into account the fact that in an industrial environment, the purification of this binary mixture is carried out in this range. The higher limit of this interval cannot reach a final purity equal to 1, and this because of the very great difficulty of separation of these two products at low temperature. This choice determines a range for each input of the distillation system so as to have a constant temperature at the output; but the flow of cooling remains fixed a raison that's why it will not be taken into account in the database.

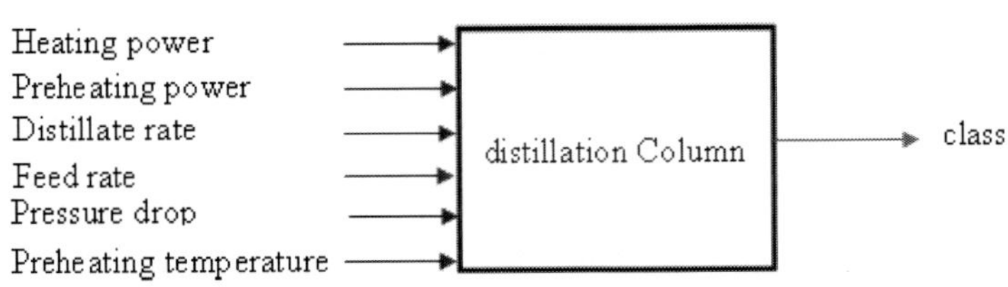

Figure 3. Determination of the inputs /outputs of the distillation column.

6. Identification of the Different Modes

6.1. Normal Mode

The distillation admits two areas of operation, a first area, considered certain, where all the parameters of operation follow the normal operational mode and the second one, considered dangerous area, where the same parameters deviate from their initial dynamics.

All input variations that keep temperature at the column head equal to 101.5°C the equivalence of 0.84 % title X of the distillate defines the normal regime of the distillation column.

Several experiments have been elaborated for data collection reflecting the behaviour of each input for a normal functioning. Figures (4, 5, 6, 7, 8, and 9) represent these variations of normalized inputs.

This data will serve to the elaboration of a learning basis rich in frequency so that of the whole spectrum of the dynamic of the process is present.

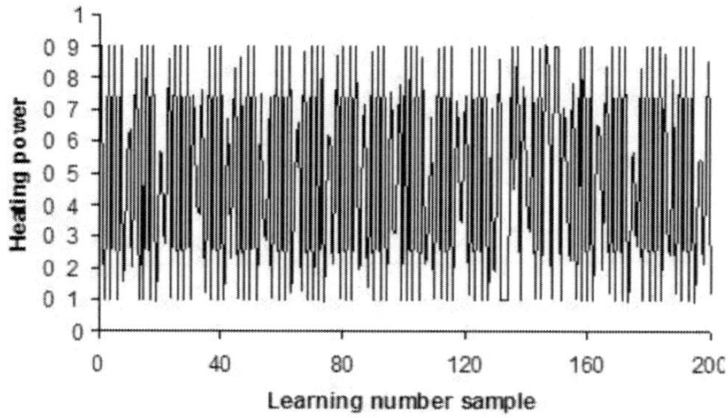

Figure 4. Heating power variation.

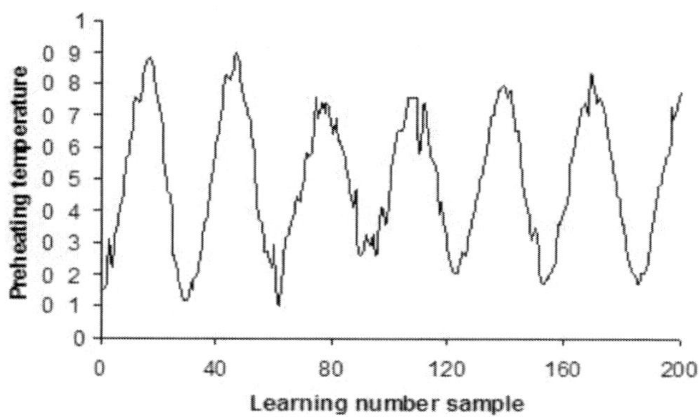

Figure 5. Preheating power variation.

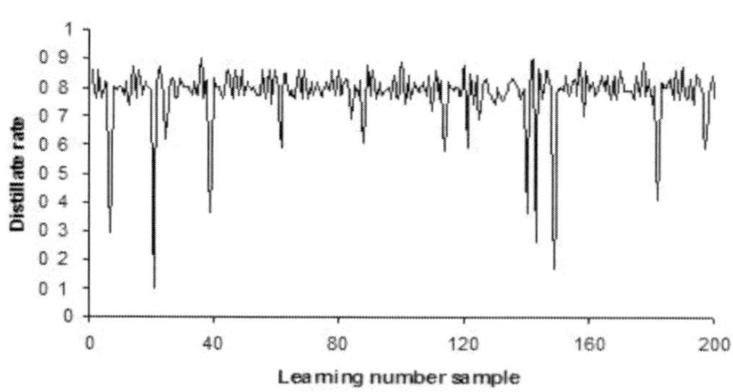

Figure 6. Distillate rate variation.

Figures (4, 5, 6, 7, 8 and 9) present the variation of the heating power, preheating power, distillate rate, feed rate and pressure difference normalized between 0.1 and 0.9 collected during a normal functioning of the distillation column with regard to a definite number of samples taken after having obtained the stable functioning regime.

The normalization of every input x_i is specified by this formula:

$$x_i = 0.8 * \frac{\max(x_i) - x_i}{\max(x_i) - \min(x_i)} + 0.1 \qquad (8)$$

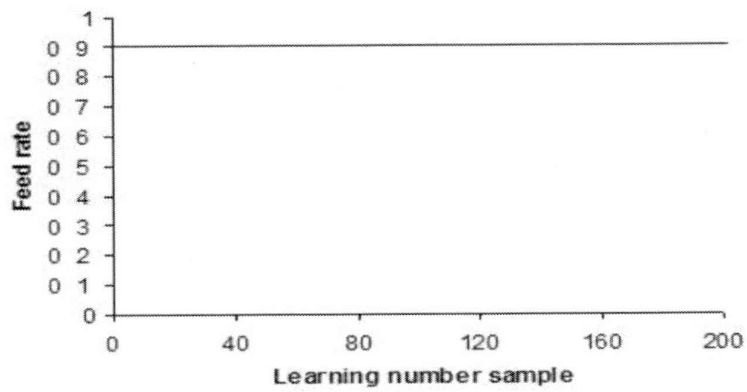

Figure 7. Feed rate variation.

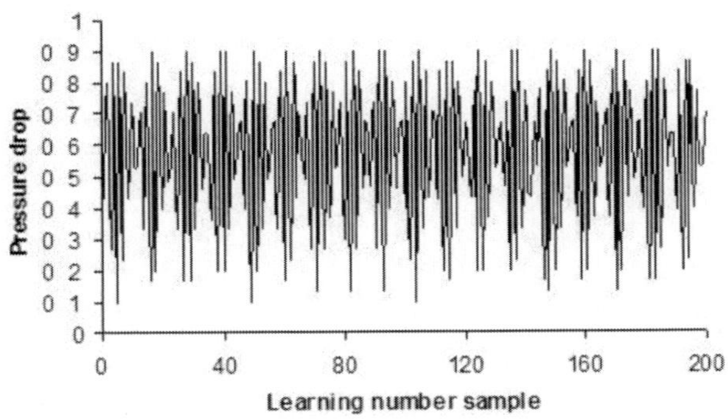

Figure 8. Pressure drop variation.

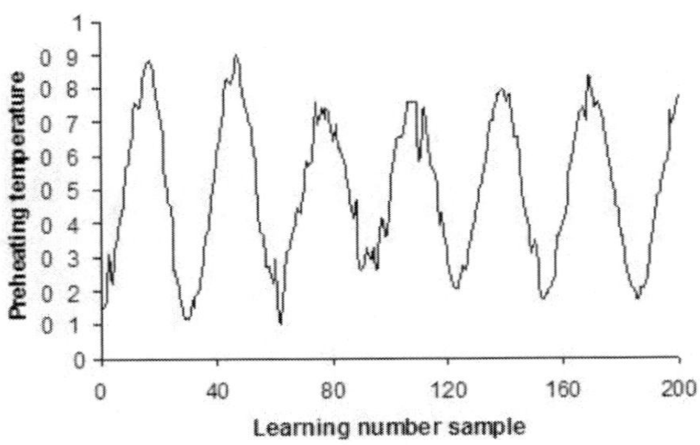

Figure 9. Preheating temperature variation.

6.2. Faults Selection

The learning base not only does not contain data reflecting the behaviour of each input for a normal function but it also contains data presenting different dysfunction that's why we have implemented the functioning faults based on the relevant inputs so as to observe the reaction of output parameters. The different dysfunctions are provoked by: the increase of heating power in 100 % of its capacity, preheating power in 90 % of its capacity, feed rate in 100 % of its capacity and distillate rate in 100 % of its capacity (table 1).

For example, working on a normal mode, is defined in a range of 35% and 60% of its functioning capacity of heating power, and between 20% and 60% of its capacity of preheating power. For this reason we have chosen 100% of heating power and 90% of preheating power in order to simulate the Dysfunction.

Table I. Selected malfunction

Malfunction	Fault number	Amplitude (of capacity)
Heating power	2	100%
Preheating power	3	90 %
Feed rate	4	100 %
Distillate rate	5	100%

These dysfunctions are obtained by varying in opened buckle a single input leaving the other constants, let us take in details the functioning faults caused by the heating power variation as an example(figure 10).

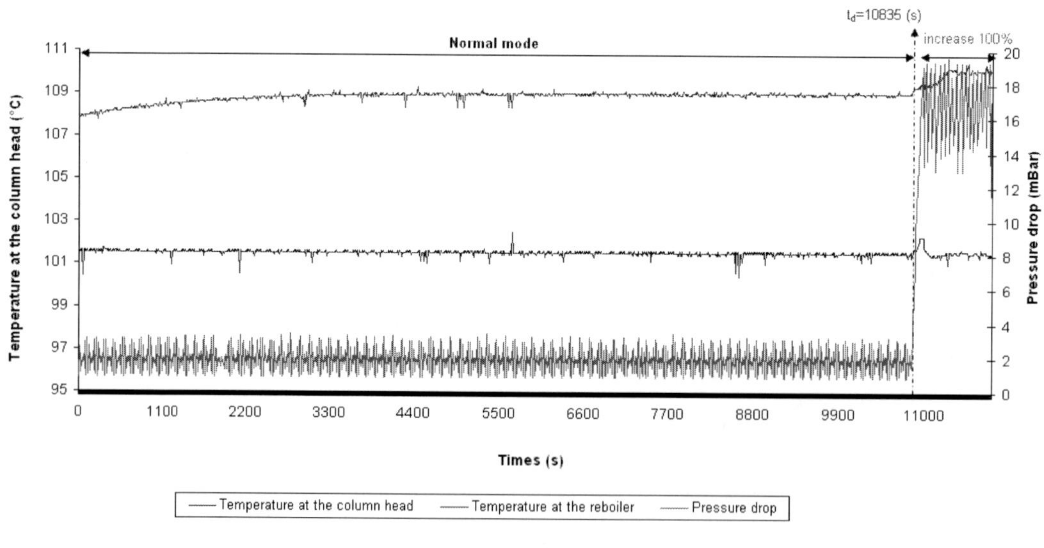

Figure 10. Dysfunction on of the heating power in accordance with time.

Let us note that during heating power variations (which come along with vapour flow variation), we obtain a variation of the distillate flow, given the fact that distillate rate bear no changes. We observe afterward that the heating power variation is echoed on the pressure difference in the column. We also notice an identical variation of the temperature at the column head (when heating power increases, temperature at the column head also increase). Indeed, the increasing vapour flow provokes a fast impoverishment of the reboiler in methylcyclohexane (the most volatile compound), and thus the vapour approaches the boiling point of the toluene (the least volatile compound). Taking into consideration the opening of the timer remaining constant (constant distillate rate) this one does not correct the temperature variation at the column head. We also find this phenomenon for a decreasing variation (the temperature at the column head decreases). The title in distillate in a temperature of 104°C is 64 %. The variation around this title is more or less than 20 %.

7. Results and Discussion of the Neural Model

The number N_h of basis functions in an RBF network is an important parameter to determine. However, there is no systematic method to do it [10].

It is a question of finding the number of adequate neurons according to the problem handled. This can be possible in two ways: we can use the same number of basis functions and input vectors; or a number of functions on an inferior basis. In the studied case, the number N_h of basis functions is the class number of different functioning modes ($N_h = 5$).

The artificial neuronal network with radial basis functions consists of an input layer containing six neurons, a hidden layer containing five neurons and an output layer containing a single neuron only.

The learning base is composed of six vectors x_1, x_2,……,x_6 independent and normalized between 0.1 and 0.9 which are: the heating power the preheating power, the distillate rate, the feed rate, the pressure drop and the preheating temperature.

The learning base of neural network is formed of 100 samples in normal mode and 40 other samples of each functioning faults. The learning of the neural model was done with the help of the learning base composed of 260 samples.

Centres positions c_1, c_2, c_3,.., c_5 of the basis functions is a parameter to be determined as well. In our case, K-means algorithm is used to select a set of centres. The various classes can be shown by using principal components analysis (PCA) [17, 18] detailed description of PCA is beyond the reach of this work.

Briefly, principal components analysis (PCA) is a mathematical technique allowing to reduce a complex system of correlation in a smaller dimension number and to visually represent the nearness between the individuals and the variables on plans. Geometrically, the process of placing in factor means placing axes in a sphere. It generally applies to a rectangular chart of measure (data chart) in which columns represent variables to numerical values and lines correspond to the individuals on whom their variables are measured. [19, 20]

The use of principal components analysis (with SPSS 11.0 tools) enables us to show the distribution of both classes: normal and fault of the heating power (Figure 11). With the first two principal components 49.692 % of information contained in the base is represented. Both classes seem well separated with the exception of some samples. K-means algorithm is applied to the determination of every class center, we find c_{1N} and c_2 normal class centre and fault class centre of the heating power respectively.

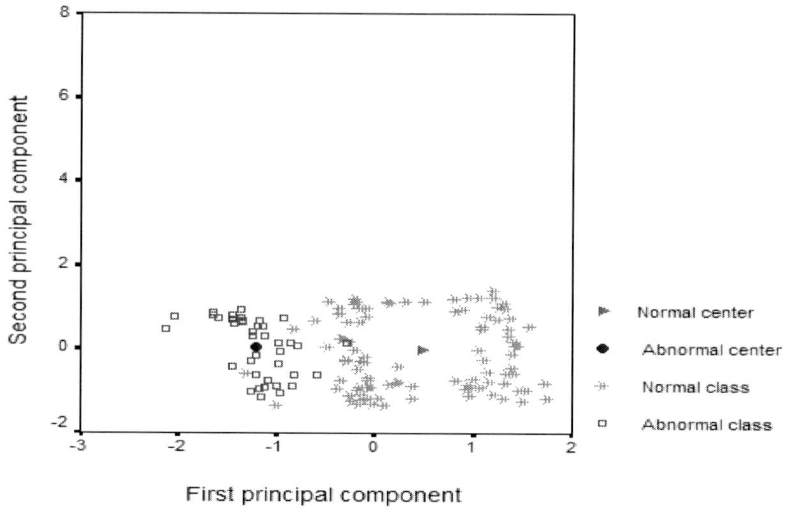

Figure 11. PCA of two-class data (normal class and abnormal class on the Heating power).

Figure (12) represents both classes distribution: normal and preheating power fault. With the first two principal components 72.288 % of information contained in the base is represented. Both classes are well separated. K-means algorithm is applied for the determination of each class center, we find c_{2N} and c_3 normal class centre and fault class centre of the preheating power respectively.

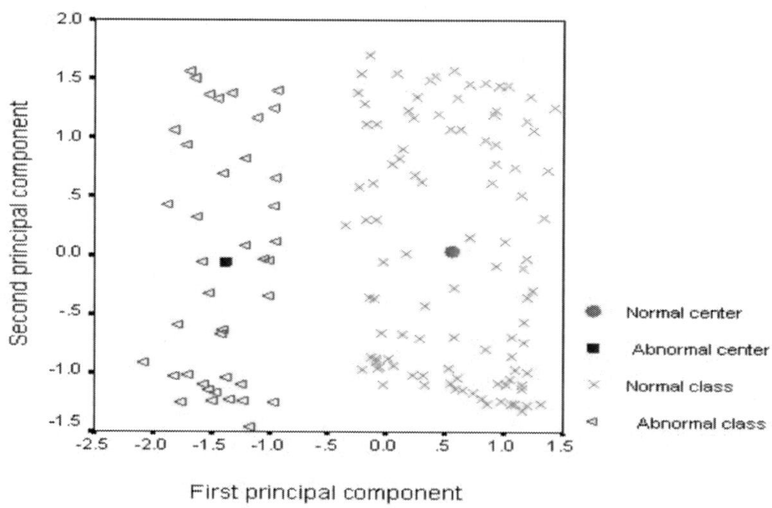

Figure 12. PCA of two-class data (normal class and abnormal class on the Preheating power).

Figure 13 represents both classes distribution: normal and feed rate fault. With the first two principal components 77.704 % of the information contained in the base is represented. Both classes are perfectly separated. K-means algorithm is applied to the determination of the centre of each class; we find c_{3N} and c_4 normal class centre and the class centre of feed rate fault respectively.

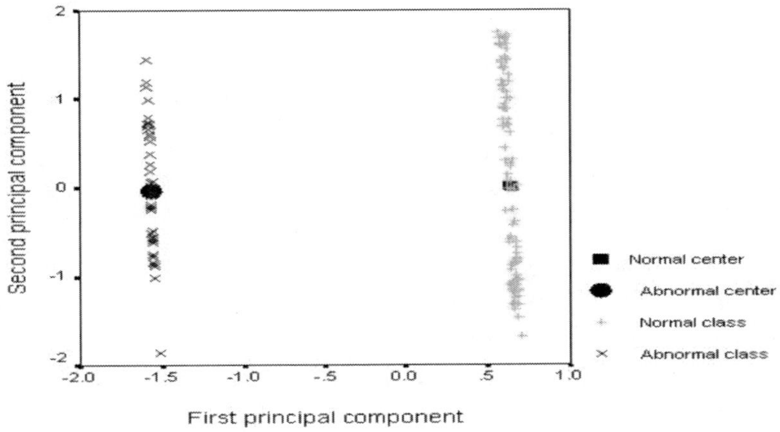

Figure 13. PCA of two-class data (normal class and abnormal class on the Feed rate).

Figure 14 represents both classes: distribution of normal and distillate rate fault. With the first two principal components 87.510 % of the information contained in the base is represented. Both classes are well separated. K-means algorithm is applied for the determination of each class centre, we find c_{4N} and c_5 the centre of the normal class and class centre of distillate rate fault respectively.

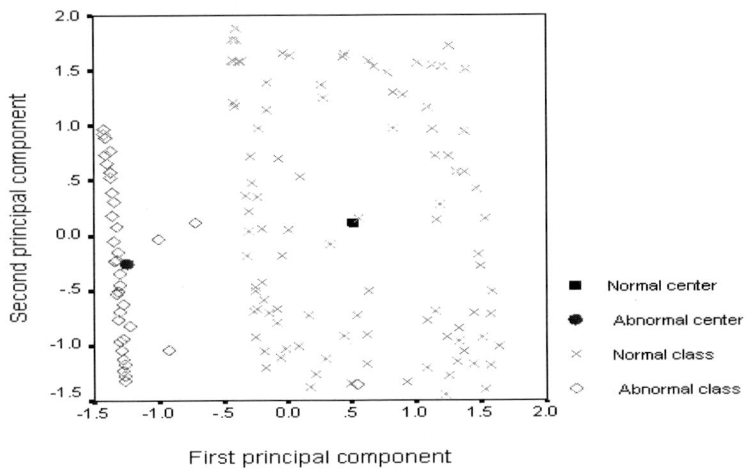

Figure 14. PCA of two-class data (normal class and abnormal class on the Distillate rate).

For the normal centre of the class c_1, we take the average of the normal centres calculated previously during every classification. That is:

$$c_1 = \frac{1}{4}\sum_{i=1}^{4} c_{iN} \tag{9}$$

Once the value of each centre c_1, c_2,c_3...c_5 is determined, we calculate the Euclidian distance between input vectors [x_1 x_2 x_3 ... x_6] and each centre. This Euclidian distance is presented to the hidden layer, to be transformed using Gaussian function.

Outputs of the hidden layer neurons will be multiplied by the weights w_i, which were calculated by means of the equation (7), then added to the bias b_i. The obtained results are passed on in the output layer. We so obtain the model of the network RBF by fixing the weights w_i and the bias b_i calculated during the learning phase.

The validation of the neuronal architecture 6-5-1 consists therefore in judging its capacity of classification by using the weights and the biases calculated during the learning phase, to apply them to another test database.

Figure 15 translates the results of this classification test. It is important to note, considering this figure, that tests data which consist of 30 samples in normal mode and 15 samples of each functioning fault are perfectly well classified except for 53 % of fourth class samples (*increasing* of the feed rate) are not well classified.

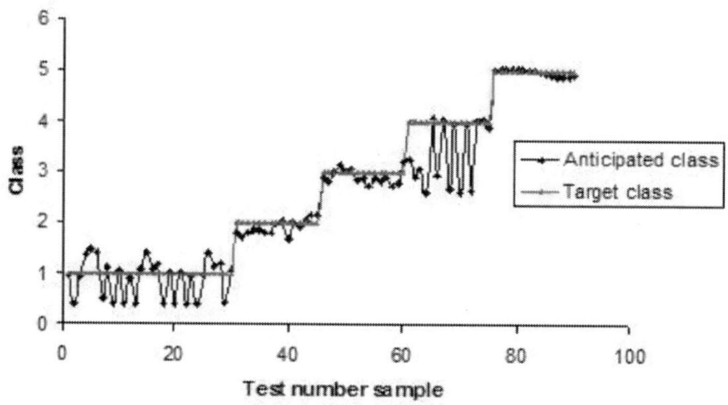

Figure 15. Result of RBF neural model classification test.

This method allows the implantation of the modules of detection and diagnosis in a column of continuous distillation to efficiently control the process by taking into account the compromise linking the safety, reliability and sensibility in the abrupt change of functioning dynamics.

However the proposed technique does not provide the limit threshold to avoid false alarms and does not allow the detection of dysfunctionning which are produced simultaneously.

CONCLUSION

This study allowed the detection and diagnosis of defects through the classification of different functioning modes (normal mode and four functioning faults) in a column of continuous distillation.

The classifier realized during this study is a neuronal classifier with radial basis functions consisted of an input layer containing six neurons, of a hidden layer containing five neurons and of an output layer containing a single neuron. The centre of the different radial functions is determined by the K-means algorithm.

In the light of the data tests' classification results, we conclude that the neuronal architecture of RBF 6-5-1 adopted for faults classification can be also adopted as neuronal model used on-line for faults detection and diagnosis in a distillation system. But this neuronal classifier does not allow the detection of dysfunctionning which are simultaneously produced and does not provide the limit threshold in order to avoid false alarms.

Nomenclature

Qc	:	Condenser Power [W]
Q_B	:	Reboiler Power [W]
F	:	Flow of binary mixture [l/h]
Z_F	:	Title of binary mixture
D	:	Distillate Flow [l/h]
x	:	Distillate Title

REFERENCES

[1] Dash, S. and Venkatasubramanian, V., Challenges in the industrial applications of fault diagnostic systems. *Computers in Chemical Engineering*, 24, 785–791, 2000.

[2] Wang, H. and Linkens, D. *Intelligent supervisory control, a qualitative bond graph reasoning approach*. Singapore: World Scientific, 1996.

[3] Hoskins, J. C., Kaliyur, K. M. and Himmelblau, D. M. Fault detection in complex chemical plant using artificial neural networks, *AIChE Journal,* 37(1), pp. 137-141. 1991.

[4] Sorsa, T., and Koivo, H. N. Application of artificial neural networks in process fault detection, *Automatica,* 29, no. 4, pp 843-849, 1993.

[5] Baratti, R. and Corti. Stefano, A., Feedforward control strategy for distillation columns. *Artificial intelligence in engineering* 11-405-412, 1997.

[6] Brydon, D. A, Cilliers, J. J. and Willis, M. J. Classifying pilot-plant distillation column faults using neural networks, Control *Eng. Practice*, Vol. 5, No. 10, pp. 1373-1384, 1997.

[7] Sharma, R., Singh, K., Singhal, D. and Ghoshd, R., Neural network applications for detecting process faults in packed towers, *Chemical Engineering and Processing* 43, pp. 841–847, 2004.

[8] Manssouri, I., Manssouri, M, Chetouani, Y. Kouddane, R and El Kihel, B., Fault Detection and Diagnosis in a Distillation Column by MLP Neural Networks. *Journal of Information, Intelligence and Knowledge* (JIIK), 2009.

[9] Ghosh, J. and Nag, A. An Overview of Radial Basis Function Networks", *Radial Basis Function Neural Network Theory and Applications,* R. J. Howlerr and L. C. Jain (Eds), Physica-Verlag, 2000.

[10] Schobbens, P. Y., *Cours d'Intelligence Artificielle: Application aux systèmes d'aide à la décision*, Facultés Universitaires Notre Dames de la Paix, Numar, Belgique, 2000.

[11] Vesin, J. M. Cours de Traitement des Signaux Biomédicaux, Ecole Polytechnique Fédérale de Lausanne, Suisse, 2005.

[12] Belur, V. Dasarathy., Nearest-Neighbor Methods in Learning and Vision, Editor *Nearest Neighbor (NN) Norms: NN Pattern Classification Techniques*, 2005.

[13] Mac Queen, J. B., *Some Methods for classification and Analysis of Multivariate Observations, Proceedings of 5-th Berkeley Symposium on Mathematical Statistics and Probability*, Berkeley, University of California Press, 1:281-297,1967.

[14] Gokhale, M., Frigo, J., McCabe, K., Theiler, J. and Lavenier, D., *Early Experience with a Hybrid Processor: K-Means Clustering*, First International Conference on Engineering of Reconfigurable Systems and Algorithms, Las Vegas, Nevada, 2001.

[15] Leeser, M., Theiler, J., M., Estlick, J. and Szymanski, J. Design tradeoffs in a hardware implementation of the k-means clustering algorithm, *First IEEE SensorArray and Multiprocessing Workshop*, 2000.

[16] Decaestecker, C. and et Saerens M., *Cours d'analyse de données et prise de décisions, Unité de Systèmes d'Information*, Université Catholique de Louvain, 2006.

[17] Johnson, R. A., and Wichem, D. W., Applied Multivariate Statistical Analysis, *Prentice Hall International*, N. J, 1988.

[18] Manssouri, I., Chetouani, Y., El Kihel, B. and Kouddane, R., La classification des différents types de *dysfonctionnement dans un système de distillation par l'utilisation de l'ACP*. 2ème Séminaire National de la Maintenance Industrielle au Centre Universitaire de Souk Ahras Institut de Science et Sciences de l'ingénieur, Algérie, 2005.

[19] Jean-Marie Bouroche et Gilbert Saporta, *Probabilité, analyse de données*, Paris, 1990.

[20] Brian Se Verritt and Graham Dunn, *Applied multivariat data analysis*, John Wiley and sons, New York, 1991.

In: Knowledge Management and Communication ... ISBN: 978-1-62808-250-0
Editor: Rafiqul Islam

Chapter 6

PROTOTYPE SYSTEM TO RELAY TRAVELER INFORMATION AND ARCHIVED BORDER-CROSSING RELATED DATA

Rajat Rajbhandari[1*], Juan Carlos Villa[2] and Rafael M. Aldrete[1]
[1]Center for International Intelligent Transportation Research
Texas Transportation Institute, Texas, US
[2]Economics, Trade, and Logistics Program, Texas Transportation Institute
The Texas A&M University System, Texas, US

ABSTRACT

Passenger and commercial vehicle crossings at the U.S.-Mexico border play an important role in the economies of the border region and both nations. The time it takes to cross the border from Mexico into the U.S. (and from the U.S. into Mexico), for both passenger and commercial vehicles, is an important element in making travel and business decisions. Unfortunately, there is no system in place to relay and archive crossing times of commercial and passenger vehicles. Public and Private stakeholders in the El Paso – Ciudad Juarez region have expressed a strong need for such a system.

This paper describes the development of a prototype Border Crossing Information System (BCIS). The objective of the prototype BCIS was to design and implement a centralized repository of archived border-crossing data and develop techniques to extract and relay traveler information. Since full-scale information system requires a considerable amount of time and resource, a prototype BCIS with limited functions was developed for this research. However, the overall goal of the prototype was to demonstrate how stakeholder needs could be fulfilled by - identifying components for a system, and integrating available technologies.

The prototype BCIS was developed and implemented using data warehouse concepts and principles, and a multi-tier architecture (client-server architecture). Using web-based tools, the prototype system provides archived border-crossing data and

* Corresponding author: Center for International Intelligent Transportation Research Texas Transportation Institute 4050 Rio Bravo, Suite 151 El Paso, Texas 79902 Email: rajat@tamu.edu.

traveler information related to travel conditions at the border to stakeholders through a single window map-based website.

INTRODUCTION

More than 80 million passenger vehicles crossed from Mexico into the United States in 2007. In addition, cross-border trade is the lifeblood of local economies on both sides of the border. Almost 14 million passenger vehicles entered into the U.S. through ports of entry (POE) in the El Paso region alone, which is second highest in the nation in terms of total annual crossings of passenger vehicles (Transborder Freight Data. U.S. Department of Transportation, Bureau of Transportation Statistics BTS). Commercial vehicle crossings at the U.S.-Mexico border play a crucial role in the border as well as regional and national economies. Truck trade between the two countries has been increasing at an annual average growth rate of 7.5 percent for the last decade.

The time it takes to cross the border from Mexico into the U.S., for both passenger and commercial vehicles, is an important element in making travel and business decisions as none of the ports of entry have similar wait times and flow volume. This is mostly attributed to the location of international bridges in relation to the bi-national urban area, toll rates, lane capacities of border crossings, etc. Unfortunately, at the present time there is no system in place along the U.S.-Mexico border to measure, relay, and archive crossing times of trucks as well as of passenger vehicles. There are similar systems in place along the U.S.-Canada border, one such system is at the Blaine-Washington border crossing [1]. The following sections describe the development of a prototype system for the U.S.-Mexico border. The researchers designed and implement a centralized repository of archived border-crossing data and developed techniques to extract and relay traveler information on the U.S.-Mexico border.

STAKEHOLDER NEEDS AND INFORMATION SYSTEM REQUIREMENTS

The researchers conducted a series of interviews with public and private stakeholders in the El Paso – Ciudad Juarez region as a part of this and other research projects. The researchers are also part of several other research projects related to border issues. For, a research project funded by the FHWA and undertaken by the researchers, a workshop was conducted, which was attended by the individuals and agencies involved with trade across the U.S.-Mexico border [2]. These individuals represented freight carriers and shippers, transport associations, and trade associations, toll collection agencies, metropolitan planning agencies from both sides of the border etc. During the workshop, the stakeholders provided information needs specially related to movement of trucks across the border.

Stakeholder needs related to enforcement of hazardous materials were obtained by the researchers from interviews with local and regional enforcement agencies. These interviews were performed as part of a research project to understand movement of hazardous materials across the U.S.-Mexico border [3]. The researchers are also part of several ad-hoc committees at the border and some of the needs were identified during conversations with the stakeholders during regular meetings.

In overall, the stakeholders expressed the need for a reliable system to relay and archive crossing times at the border. The stakeholder needs primarily focused on a reliable system to view current conditions at the border crossings and on a system to query archived data related to border-crossings operations and performance. The following stakeholder needs were identified:

- Efficient information system to relay border-crossing related traveler information and current conditions including crossing times, wait times, inspection lanes open, lane closures; and
- Centralized repository to archive border-crossing related data including performance of ports of entry, trade statistics, economic indicators of the two countries etc. The repository should provide mechanisms to compare performances of individual ports of entry for planning and policy-making purposes.

Table 1. Stakeholder Needs for Traveler Information

Stakeholder	Need for Traveler Information
Bridge Operations Agencies	Monitor incidents at border-crossings and coordinate with law enforcement agencies.
U.S. and Mexican County or Municipal Public Safety Agencies	Monitor status of incidents around the border-crossings. For example, city's fire departments respond to HAZMAT incidents inside inspection areas and outside.
U.S. and Mexican Freight Shippers, Private Commercial Carriers	Use border-crossing time, incidents, bridge closures, current roadway condition etc. for pre-trip and en-route decision-making.
U.S. and Mexican Local Media	Relay crossing times and delay at the border, bridge closures, current roadway conditions, incident locations to public.
U.S. and Mexican Private Travelers	Use crossing times and delay at the border, bridge closures, and incidents at and around border-crossings, current roadway conditions for pre-trip and en-route decision-making.
U.S. and Mexican Regional Transit Providers	Use crossing times and delay at the border, incidents at and around border-crossings, current roadway conditions for pre-trip and rn-route decision-making.
U.S. State Department of Transportation	Monitors incidents and current traffic conditions and respond to traffic incidents on state highways around border-crossings.
U.S. and Mexican Toll Operators	Monitor incidents at border-crossings and coordinate with law enforcement agencies.
U.S. Customs and Border Protection	Collects and relays current wait times at ports of entry using visual method. However, it is a well-known fact that wait times relayed by the CBP relayed are highly underestimated.
U.S. Emergency Management Agencies	Monitor and respond to major incidents in and around border-crossings.

Note: CBP = Customs and Border Protection.

Efficient Information System to Relay Traveler Information

Measuring and reporting traveler information related to border-crossings are of considerable interest to a wide-range of stakeholders at the border. For example, travel time between origins in Mexico and destination in the U.S. side of the border is of high significance to the trade industry, daily private vehicle commuters, and pedestrians. At present, only Customs and Border Protection (CBP) relays border wait times and number of inspection lanes open (for various CBP administered programs, such as FAST, NEXUS). However, CBP understands since the border wait times are measured by visual methods (looking for queue against landmarks) and motorist surveys, the wait times are inaccurate and a more scientific method of collecting wait times (and crossing times) at the border is needed.

Based on meetings with stakeholders in the region, needs for a variety of traveler information related to border-crossings were identified. Table 1 provides a list of stakeholders in the U.S. – Mexico region with needs for traveler information and current conditions of border-crossings. In addition, motorists and commercial freight carriers and shippers have to access multiple agency websites to put together a single view of the current travel conditions at and around border-crossings. Stakeholders identified this as being highly inefficient.

Archived Border-Crossing Data

Archived border-crossing data is used by private and public agencies with responsibilities to plan, operate, and manage border-crossing infrastructure. Agencies such as metropolitan planning organizations, city agencies, CBP and General Services Administration (GSA) utilize the archived data plan for future infrastructure improvements and manage resources to operate border-crossings efficiently. Table 2 provides a list of stakeholders in the U.S. – Mexico region with need for archived border-crossing data.

U.S. and Mexican federal, state and local agencies need to determine a set of performance measures and ultimately a performance management process for evaluating and improving individual border-crossings for freight as well as passenger movement. A set of travel time related performance measures has to be identified for both freight and passenger movement, which would be a basis for establishing common indices to compare performances of border-crossings throughout the U.S.-Mexico and U.S.-Canada border, irrespective of characteristics of individual crossings. Hence, developing mechanisms to compare the performance of individual border-crossings require an archive of variety of border-crossing related data. The following are the goals and objectives for implementing border performance measures:

- Provide all stakeholder agencies and users with information to assist their operations and make appropriate decisions;
- Establish a common denominator to estimate the effect of improvements and modifications in operation of border-crossings; and
- Establish performance indices, which can be easily communicated with a non-technical audience and still find the information relevant.

Table 2. Stakeholder Need for Archived Border-Crossing

Stakeholder	Archived Data Needs
U.S. and Mexican Bridge and Toll Operators	Use border-crossing volume trend data to manage, operate, and plan bridges and tolls (if present).
U.S. and Mexican Freight Shippers, Private Commercial Carriers	Need to monitor trends in border-crossing times of commercial vehicles and other economic indicators for supply chain management.
U.S. and Mexican Regional Transit Providers	Need archived data to monitor trends in border-crossing times of buses and other modes and demographics for market capture studies and service planning.
U.S. Customs and Border Protection	CBP collects and archives number of passenger vehicles, pedestrians, buses, trucks, and trains entering the U.S. and provides to other federal and state agencies. CBP uses trends in border-crossing times and volume of commercial, passenger vehicles, and pedestrians to manage and operate inspection facility at the border.
U.S. General Services Administration	GSA uses trends in border-crossing times and volume of commercial vehicles, passenger vehicles, and pedestrians to manage and plan expansion of inspection facility at the border.
U.S. and Mexican Municipal Government	Local government agencies use border-crossing trends data and other socioeconomic indicators to estimate short- and long-term socioeconomic impact of border delays, and in some regions plan and operate bridge infrastructure.
U.S. and Mexican Regional or Metropolitan Transportation Planning Organizations	Planning agencies use border-crossing trends data and other socioeconomic indicators to develop short- and long-range transportation plans for the border region.

Archived border-crossing related data also provide insights into social and economic development of surrounding communities and regions. Using data mining applications, hidden relationships between economic indicators and operational variables (economic indicators, crossing times, trade volume, etc.) can be revealed. Having a centralized repository of archived data will be highly valuable in performing such studies similar to the ones listed below:

Fullerton and Tinajero (2002) studied analysis of empirical regularities associated with monthly cargo vehicle flows from Ciudad Juarez, Chihuahua, in Mexico to El Paso, Texas, USA [4]. The study was conducted using time series data that characterized the behavior of international trucking through border crossings in the El Paso region.

Goodchild (2008) analyzed the characteristics and influence of border operations, especially delay at the Western Cascadia border, on regional supply chain. The study analyzed daily average delays for commercial vehicles and found that typical delays are not problematic for trucks enrolled in CBP's FAST program [5].

The IBI group (2008) studied the impact on commercial border-crossing volumes (both directions) due to change in U.S. gross domestic product (GDP), Canadian GDP, currency exchange rates, and value of Canadian trade [6].

Mapping Stakeholder Needs to BIFA

The Border Information Flow Architecture (BIFA) supports the planning, development, and implementation of ITS and other technology-based solutions at the U.S.-Canada border

[7]. Even though the current version of BIFA (http://www.iteris.com/itsarch/bifa/) supports ITS architecture for border regions in U.S. and Canada, information from the architecture can be easily adapted to the U.S.-Mexico border regions.

The BIFA was used to categorize services to be provided by the information system based on stakeholder needs. BIFA includes two user services - archived data management service and traveler information service that is of interest to the stakeholders. These user services include a variety of equipment packages appropriate for the implementation of border-crossing information system. Providing these user services and equipment packages are considered as high-level functional requirements. Several low-level functional requirements and data specifications described in BIFA were identified for the information system. In addition to BIFA, the National ITS Architecture [8] was also used to obtain appropriate user services and functional requirements necessary for BCIS.

Goals and Objectives of the Prototype Border-Crossing Information System

The overall goal of this research was to demonstrate how stakeholder needs could be fulfilled by identifying components for a system, data extraction and dissemination processes, and integrating available technologies. Since a full-scale information system requires a considerable amount of time and resources, a prototype border-crossing information system (referred to hereafter as prototype BCIS) with following functions was developed in this research:

- A single-window website was developed to view traffic conditions at and around border-crossings in the El Paso region. The traffic conditions include border wait times, traffic incidents, weather, highway conditions etc. for the northbound traffic, entering the U.S. However, the system has the capability to relay southbound information in the future and when such information becomes available.
- The information relayed through the website was also aggregated and archived in a centralized repository for planning and operational analysis. A web-based tool was developed to perform query of archived data and view pre-built reports and charts.

The prototype developed in this research is a highly cost-effective tool to demonstrate to the stakeholders benefits of implementing border-crossing information system at regional and state level. The prototype also sets the precedent for a full-scale information system in terms of functional requirements, resource, and technologies needed.

Design Principles and Framework

The prototype BCIS was developed and implemented using data warehouse concepts and principles. Data warehouse is suitable framework when users have to query and extract a large number of records, but on a less frequent basis. Hence, archiving voluminous border-crossing data fits perfectly into a data warehouse model.

The characterization of a database architecture as "database-centric" refers to the fact that data are stored in a relational database structure, as opposed to customized file-based data structures. In this prototype, border-crossing related data are stored in a relational database structure and business logics are built into the data warehouse as stored procedures.

Data are gathered into the data warehouse from a variety of agency sources and merged into a coherent whole. Data obtained from multiple agencies and disparate sources are filtered, aggregated, summarized, and added to the data warehouse. The most important feature of a data warehouse is the ability to integrate variety of data in scope and size, which is a crucial requirement of a BCIS.

Multi-Tier Architecture (commonly known as Three-Tier architecture) is a client-server architecture in which data extraction and storage, data fusion, and data dissemination are implemented as independent modules, most often on separate platforms. The Three-Tier architecture is intended to allow any of the tiers to be upgraded or replaced independently when required [9]. The Three-Tier architecture was used in developing the prototype and is illustrated in Figure and is described as follows:

- The DATA EXTRACTION AND STORAGE TIER consists of a series of data tables (represented by smaller cylinders in the figure below), which stores archived border-crossing data in a relational database model. It also consists of temporary tables to hold traveler information related data, which are displayed using a website;
- The DATA FUSION TIER consists of a series of stored procedures, which extract data from multiple sources, parse, filter, and aggregate the border-crossing data; and
- The DATA DISSMEMINATION TIER consists of mechanisms to disseminate archived ITS data and advanced traveler information to users through intranet, internet, and standalone applications.

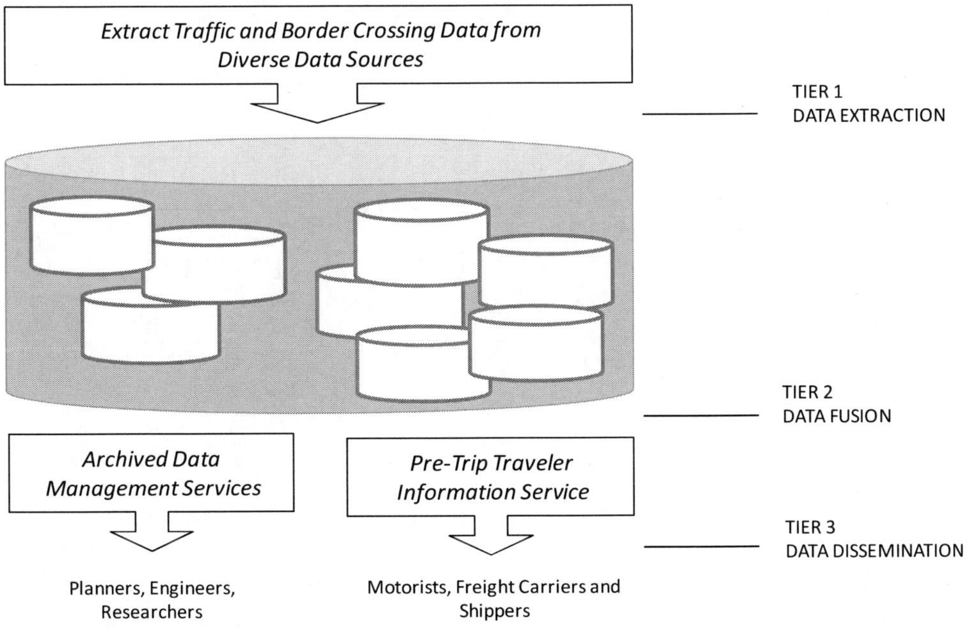

Figure 1. Multi-tier architecture consists of data storage, business logic and presentation tiers.

DATA EXTRACTION AND AGGREGATION PROCESSES

The objectives of data extraction and aggregation processes are to extract border-crossing data from multiple agencies and aggregate the data based on various temporal and spatial granularities. Table 3 provides a list of border-crossing related data that are extracted by the prototype BCIS using autonomous applications (described later in this section), purpose for extracting the data (either to relay as traveler information and/or to archive in a centralized repository), and name of the agency that produces the data.

Table 3. Sources of Border-Crossing Related Data

Type of Data	Purpose	Data Producer
Hourly Border Wait Times and Number of Inspection Lanes Open	Relay as traveler information and aggregate in the archive	Customs and Border Protection
Current and Predicted Border-Crossing Times	Relay as traveler information and aggregate in the archive	With funding from FHWA and TxDOT, TTI is undertaking several projects to install RFID based systems on at border-crossings to measure crossing times of trucks.
Hourly Homeland Security Threat Level	Relay as traveler information and store in the archive	Department of Homeland Security (DHS)
Hourly Current Bridge Closure Information	Relay as traveler information and aggregate in the archive	Customs and Border Protection
Hourly Local Weather	Relay as traveler information and store in the archive	National Oceanic and Atmospheric Administration
Instantaneous Location of Recent Roadway Incidents	Relay as traveler information and store in the archive	Local law enforcement agencies
Current and Predicted Conditions of Highway Segments	Relay as traveler information and store in the archive	State DOTs and municipal agencies provide current conditions on highways and major arterials.
Daily Exchange Rate Between Dollar and Peso	Archive only	(http://themoneyconverter.com/USD/rss.xml)
Monthly Volume of Passenger Vehicles, Loaded and Empty Trucks Under Various CBP Programs	Archive only	Bureau of Transportation Statistics (BTS)

Data extraction processes include several autonomous applications (executables) that run at predefined time intervals to extract data from agency websites and Real Simple Syndication (RSS) feeds. For example, border-crossing related data (traffic incidents, border wait times relayed by CBP, threat level relayed by DHS, hourly weather information) are extracted by "grabbing" information from agency websites, through RSS links provided by agencies, and by downloading raw data sets from agency websites or a file transfer protocol site. The autonomous applications after extracting data from agency websites transfer the "raw" data into tables in the data warehouse, where predefined stored procedures aggregate the data into different tempo-spatial granularity. The autonomous applications also convert current border-crossing information into Extended Markup Language (XML) data packets, which are read by the webpage that shows the current border-crossing condition as traveler information. This process is illustrated in Figure .

Since, there are no formalized data transfer processes between agencies and prototype BCIS, the only mechanism to extract data from these sources is to "grab" data from agency web pages, RSS feeds, and download data from the website. One of the main disadvantages in "grabbing" data from agency web pages and RSS feeds is that every time agencies change the format of data display in their web page, the application that "grabs" the data has to be modified as well. Hence, formal data exchange mechanisms between data producing agencies and the BCIS would eliminate such shortcomings.

Table 4. List of FACT Tables in the Prototype BCIS

Logical Table Name	Description
Monthly Incoming Freight Volume	Stores monthly total freight containers entering U.S. by various modes of transportation.
Monthly Import-Export Volume by Mode	Stores monthly total trade value and weight with origin as Mexico and destination states in U.S. by mode of transportation.
Monthly Import-Export Volume by Commodity	Stores monthly total trade value and weight with Mexico by commodity.
Daily DHS Threat Level Advisory	Stores daily color-coded value of threat level issued by Department of Homeland Security.
Monthly Incoming Passenger Volume	Stores monthly total passengers entering U.S. through various border regions in the state of Texas.
Monthly Incoming Vehicle Volume	Stores monthly total vehicles entering U.S. by various modes of transportation.
Daily Average Border Wait Times at Ysleta Port of Entry	Stores 24 hour average wait times at Ysleta Bridge collected from CBP website.
Daily Average Border Wait Times at BOTA Port of Entry	Stores 24 hour average wait times at Bridge of the Americas collected from CBP website.
Daily Average Border Wait Times at PDN Port of Entry	Stores 24 hour average wait times at Paso Del Norte Bridge collected from CBP website.
Hourly Border Wait Times BOTA	Stores wait times at Bridge of the Americas collected from CBP website every 20 minutes.
Hourly Border Wait Times PDN	Stores wait times at Paso Del Norte Bridge collected from CBP website every 20 minutes.
Hourly Border Wait Times Ysleta	Stores wait times at Ysleta Bridge collected from CBP website every 20 minutes.
Daily Exchange Rate	Stores daily exchange between U.S. Dollar and Mexican Peso.

Note: BOTA = Bridge of the Americas, El Paso.

PDN = Paso Del Norte, El Paso.

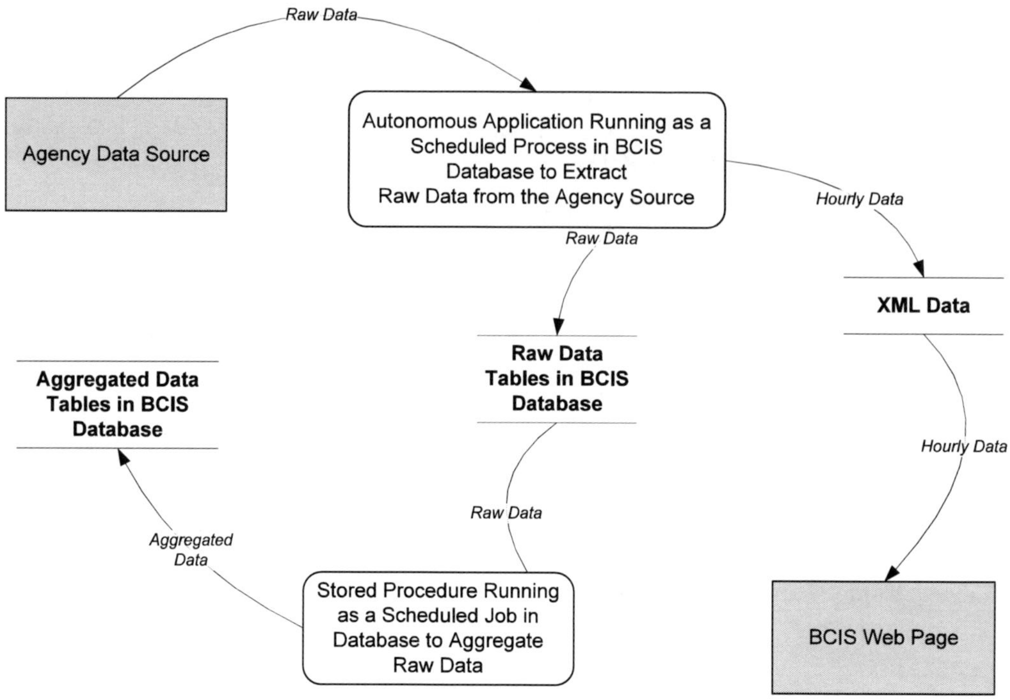

Figure 2. Data extraction and aggregation process in the prototype BCIS.

SCHEMA TO ARCHIVE BORDER-CROSSING DATA

The data warehouse in the prototype BCIS consists of numerous FACTS and DIMENSIONS tables arranged in a Star Schema. The data warehouse includes FACT tables representing monthly volume of shipments, passengers, passenger vehicles, quantity of incoming freight by mode and commodity, hourly and daily average border wait times reported by CBP, and Homeland Security daily threat advisory levels. A complete list of tables in the prototype BCIS is presented in Table 4. Tables in the data warehouse are related with DIMENSIONS tables using one or more indexes. Dimensions related to time include calendar dates and the ones related to space include name of port district, name of port of entry etc. Using common dimensions, information can be queried and retrieved from multiple FACT tables.

In Figure and Figure , DIMENSIONS tables are shown in orange (and shaded) and FACT tables that store monthly trade data, such as import-export by commodity, volume of incoming freight, import-export by mode are shown in white boxes. By storing data in relational database models, users can query data using multiple variables (provided by DIMENSION tables).

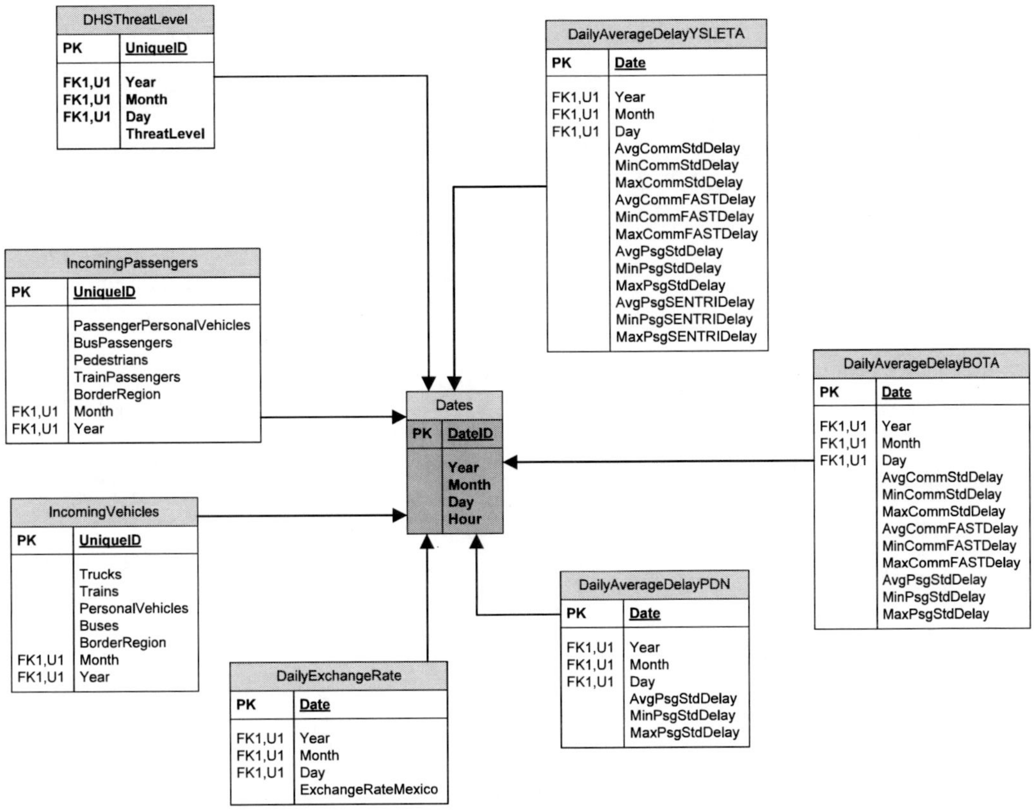

Figure 3. Sample FACTS and DIMENSIONS tables in the prototype BCIS (Several other tables exist in the final database).

INFORMATION DISSEMINATION

Traveler Information

Traveler information service enables the traveling public to make informed decisions about trips. A regional mobility information system (PDN-RMIS) developed for the El Paso region provides information to motorists and the traveling public by extracting travel related data from multiple sources and displaying travel conditions using a single window website [10]. Instead of creating a separate webpage, border-crossing related traveler information was "added" to the PDN-RMIS website that already shows current traffic conditions (e.g. snapshots from highway cameras, dynamic message signs, highway speed, incident locations etc.). The idea here was to provide a "single" window of traffic conditions around the border-crossings and beyond to travelers and freight carriers. Information read by autonomous applications convert extracted border-crossing related data from agencies (such as CBP, DHS), into XML data packets, which is read and displayed in the website. Figure shows a snapshot of the PDN-RMIS website added with border-crossing related information developed as part of the prototype BCIS.

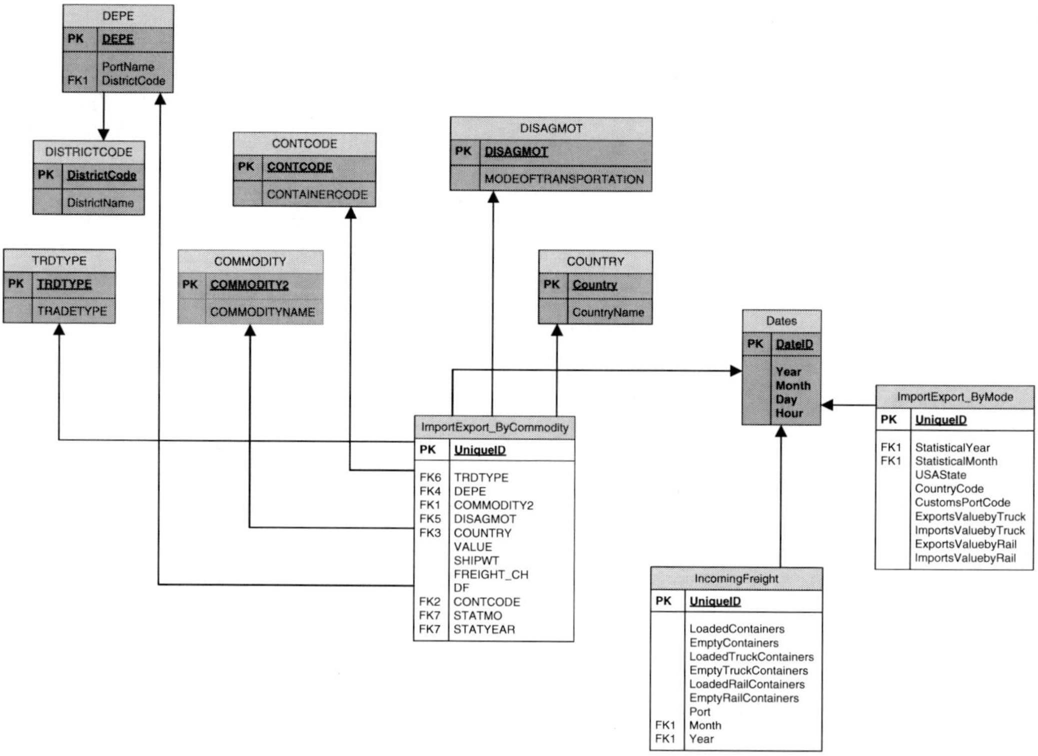

Figure 4. Monthly import export data downloaded from BTS and archived in the BCIS Prototype using FACTS and DIMENSIONS table structure.

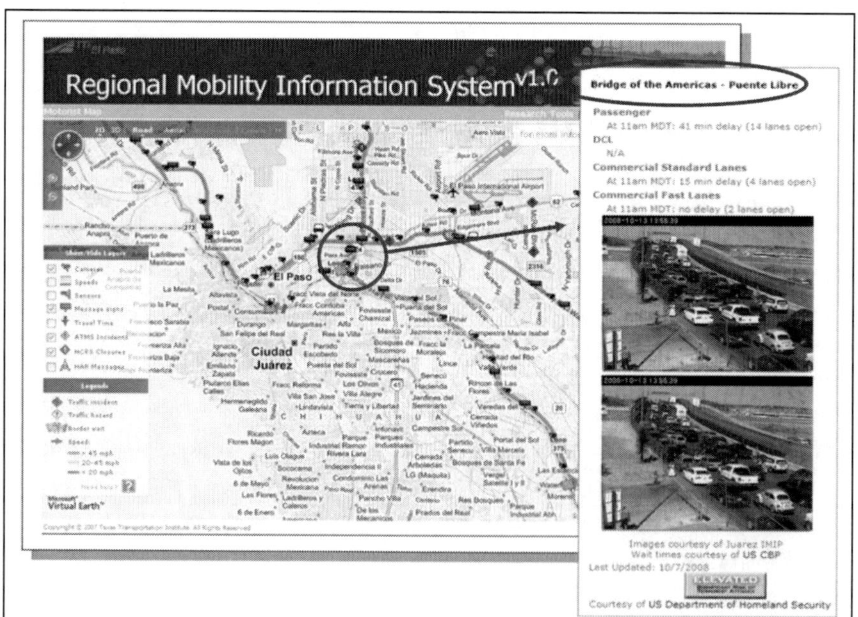

Figure 5. Snapshot of website with traveler information related to highways (developed as part of PDN-RMIS) and border-crossings in the El Paso region.

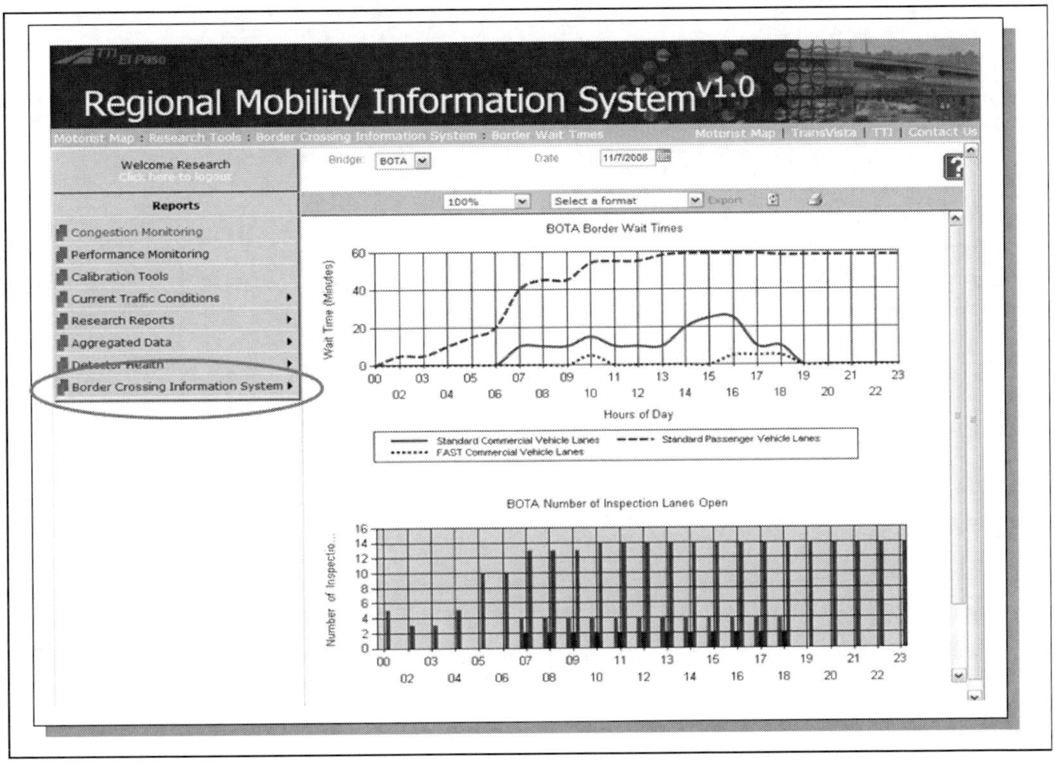

Figure 6. Website to access archived border-crossing data and pre-built reports and charts.

Archived Border-Crossing Information

Several web-based reports were developed to provide multi-granular views (temporal and spatial) of archived border-crossing data. For example, users can view hourly and daily variation of border wait times reported by CBP and compare the border wait times at different ports of entry in the El Paso region. Static and interactive reports developed as part of the prototype BCIS have been categorized as “Border-Crossing Information System,” as shown in Figure , including other mobility information related reports developed as part of the PDN-RMIS. The website also provides reports to access border trade related data (e.g. incoming freight volume, trade data, etc.), as illustrated in Figure .

INTEGRATING ITS TECHNOLOGIES WITH THE PROTOTYPE BCIS

One of the requirements of a BCIS is the ability to integrate with technologies (and field devices) to measure operational performance of border-crossings. Ideally, the BCIS could also include hardware and communication setup (field devices) deployed to measure performance of border-crossings.

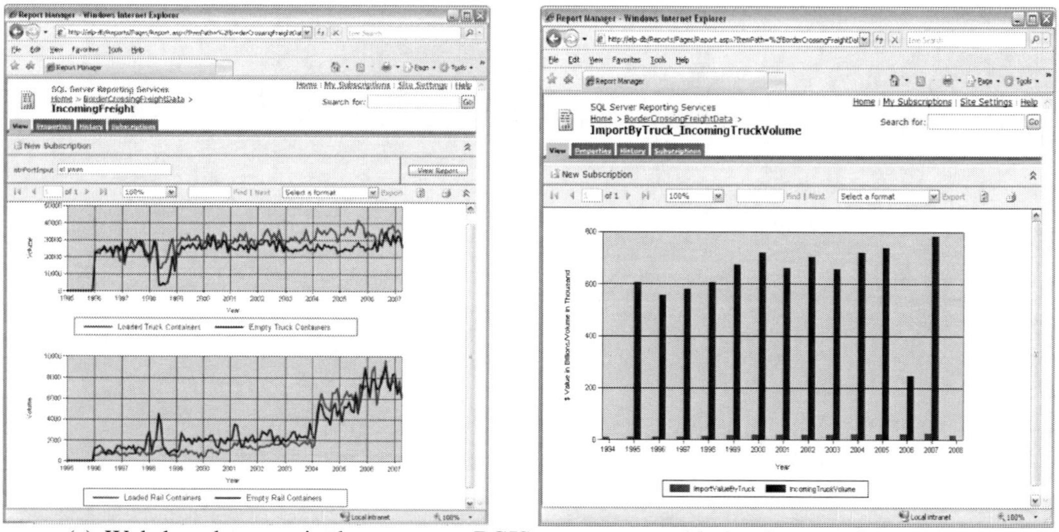

(a) Web-based reports in the prototype BCIS to query monthly border-crossing volume data.

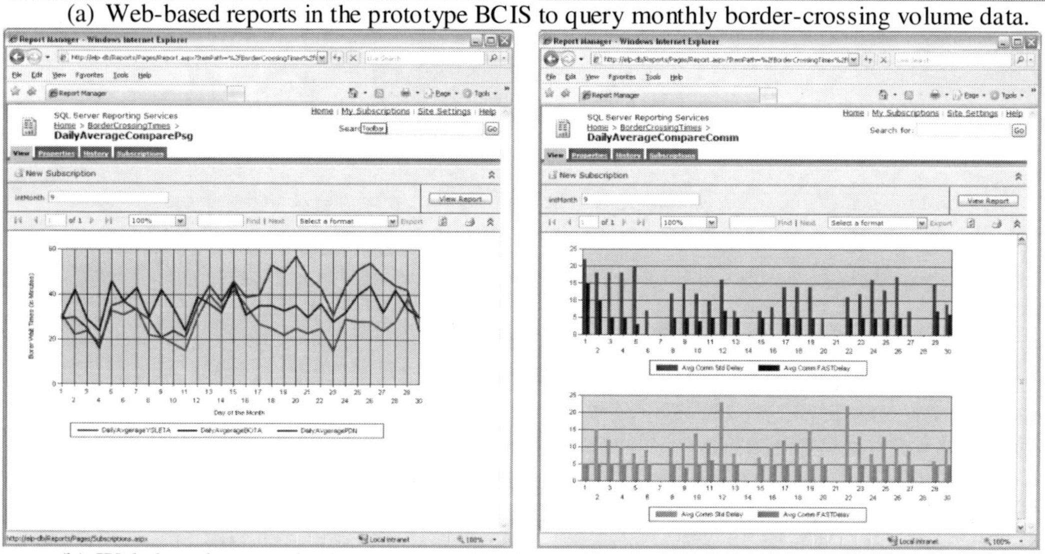

(b) Web-based reports in the prototype BCIS to query hourly and daily border wait times data.

Figure 7. Web-based reports to query archived border-crossing data from the prototype BCIS.

Several initiatives are underway within the State of Texas to deploy ITS to measure crossing times of passenger and commercial vehicles [11]. Many of these initiatives were a direct result of several tests conducted by TTI to understand the suitability of the technology to measure crossing times at ports of entry along the U.S.-Mexico border. For example, Bluetooth sensing technology is suitable to measure crossing times of passenger vehicles at border-crossing with high volume of vehicles. Similarly, radio frequency tag identification technology is suitable to measure crossing times of trucks, since a large portion of trucks entering the U.S. carry tags provided by CBP (under the Free and Secure Trade (FAST) program) and/or Mexican toll agencies.

TTI also tested if these technologies could be integrated with the database components of the prototype BCIS, which included testing the data communication between field deployments and the data warehouse. The results showed that the prototype BCIS can be directly linked with multiple and disparate ITS deployments and that the data warehouse can accept ITS data seamlessly.

Researchers believe that even though these systems could be disparate in geographical locations, an information system such as BCIS should be the one that connects to individual ITS deployments to relay current and predicted crossing times and archive border-crossing information. BCIS could be a centralized system where ITS data from individual ports of entry are gathered, processed, and archived.

CONCLUSION

Based on a thorough stakeholder analysis a prototype border-crossing information system was developed that included functions to relay border-crossing related traveler information and a centralized repository of data. The prototype developed in this research is a highly cost-effective tool to demonstrate to the stakeholders benefits of implementing border-crossing information system at regional and state levels. The prototype also set precedence for a full-scale information system in terms of functional requirements, resource, and technologies needed.

Border region stakeholders expressed a strong need for an information system that allowed convenient access to reliable border crossing wait and travel time information. The prototype BCIS has been well received by stakeholders, who have reiterated that with several ITS deployment initiatives underway to measure border-crossing performance on both sides of the U.S.-Mexico border, a tool such as BCIS would be invaluable. However, stakeholders have also pointed that data needs to be presented in much smaller temporal granularity, raising the issue of data availability and the need for commitment from data producers to provide data (in appropriate spatial and temporal granularity) to the BCIS. For example, the prototype BCIS extracts monthly northbound volume data from the BTS data, which is provided by CBP. However, stakeholders also need daily volume, which is not available either from BTS or CBP unless the bridge leading from the port of entry is tolled or some other agency has deployed field devices to count northbound volume.

Hence, the researchers believe that future implementation of BCIS should consider the fact that data in appropriate spatial and temporal granularity may not be always available from stakeholder agencies. At the same time, BCIS should be designed to receive and relay processed data from ITS deployments at border-crossings. It is more than likely that more than one agency will be responsible for deploying ITS field devices at border-crossings. Hence, buy-ins and commitments from all relevant stakeholders are crucial in implementing a successful border-crossing information system.

ACKNOWLEDGMENT

Funds for the development of the prototype border-crossing information system were provided by TTI's Center for International Intelligent Transportation Research at El Paso and the University Transportation Center for Mobility (UTCM) at the Texas A&M University.

Crystal Jones with the Federal Highway Administration (FHWA) served as Project Monitor for the UTCM funded project and provided valuable comments.

REFERENCES

[1] Cascade Gateway Data Warehouse, http://www.cascadegatewaydata.com, Accessed October 27, 2009.

[2] Measuring Border Delay and Crossing Times at the U.S.-Mexico Border, Task 1 Report Preliminary Design Document Prepared for U.S. Department of Transportation and Federal Highway Administration by Battelle Memorial Institute and Texas Transportation Institute, May 2008.

[3] Rajbhandari, R., Calzada, I., Noonchester, M. and Thatcher, B., *Pilot Information System for Cross-Border HAZMAT Transportation – Final Report*, Southwest Region University Transportation Center, Texas A&M University, 2009.

[4] Fullerton, Thomas M., and Tinajero, Roberto. Cross Border Cargo Vehicle Flows, *International Journal of Transport Economics*, 2002, Vol. 29, pp. 201-213.

[5] Goodchild, A. *Western Cascadia Border Operations: Delay and the Impact on Supply Chain*, Presented at Transportation Border Working Group Meeting, June 2008.

[6] IBI Group, *Cross-Border Traffic Demand Forecasting*, Presented at Transportation Border Working Group Meeting, June 2008.

[7] Border Information Flow Architecture, Federal Highway Administration, 2005.

[8] National Intelligent Transportation System Architecture Version 6.0, U.S. Department of Transportation, 2007.

[9] Eckerson, Wayne W. "Three Tier Client/Server Architecture: Achieving Scalability, Performance, and Efficiency in Client Server Applications." *Open Information Systems* 10, January 1995.

[10] Rajbhandari, R., Aldrete, R., *Paso Del Norte Regional Mobility Information System: Creating Archived Transportation Data User Services and Advanced Traveler Information System*, Presented to the 88th Annual Meeting of the Transportation Research Board, Washington D.C., 2009. (Accepted for Publication)

[11] Rajbhandari, R. Initiatives to Measure Crossing Times of Commercial and Passenger Vehicles Using RFID and Bluetooth Technology at U.S.-Mexico Border, *ITS Texas*, September 2008.

In: Knowledge Management and Communication … ISBN: 978-1-62808-250-0
Editor: Rafiqul Islam

Chapter 7

MAKING OF AN ARTIFICIAL INTELLIGENT DRILL MACHINE

Chironjeet Kumar Chaki*
Department of Mechanical Engineering
Rajshahi University of Engineering & Technology
(RUET), Bangladesh

ABSTRACT

Artificial Intelligence (AI) and image processing can be utilized now in a robust control system. This paper addresses a scenario where AI and Image Processing are applied to a specific machine tool to achieve control of simple drilling process. Vision based machine control provides a broad functionality when compared to existing control systems (e.g. Numerical control (NC), Computer numerical control (CNC) etc). In this study AI and Image Processing are used by computer controlled drill machine to demonstrate the potential of an AI control system. Operational reliability & accuracy, reduced production cost and rapid recovery from malfunctions are the goals this paper seek. Neural network models were implemented to find formulae to describe the motions of the various axes in the drill model. A machine bed was constructed, requiring work positioning on three axes which are controlled indirectly by a computer running imaging software which will send work bed motion commands to the job positioning controller.

Keywords: Precision drilling, AI drill Machine, Automatic control, Real time image processing, Artificial neural networks

1. INTRODUCTION

Artificial intelligence in this scenario is a system that perceives its environment and takes actions which maximize its chances of success. The need for increasing productivity and reducing production cost is pushing towards fully automatic, unmanned machining centers or

* E-mail address: titoruet@gmail.com.

intelligent machining systems [1]. In the new work environment, machine tools should be able to perform automatically the following activities: collision detection and prevention, tool condition monitoring, optimization of cutting parameters, job positioning, maintaining the safe tolerable limits, required injection of cutting fluid, proper lubrication between moving members.

AI techniques were presented first in the middle 1960's. AI is not only seen in computer science and engineering. It is studied and applied in various disciplines such as psychology, philosophy, neuroscience, linguistics, economics, control theory, probability, optimization and logic. It can be applied to model complex relationships between inputs and outputs or to find patterns in data [1]. It is clear that conventional AI, which involves methods now classified as machine learning, characterized by formalism and statistical analyses was used first in expert systems to process large amounts of data and use the results to draw viable conclusions. Recently the development of iterative techniques of learning has been very successful when applied to electro-mechanical systems.

This paper mainly discusses the development of a model automatic control drilling machine using Artificial Intelligence (AI) and Image Processing (IP) arranged in a configuration mimicking a neural network. A widely used computational model is Artificial Neural Network (ANN). Its greatest advantage depend on ability to be used as an arbitrary function approximation mechanism which 'learns' from observed data. Enhancement of the reliability of machining process, speeding the production process, reduction of required manpower, and obtaining high accuracy of the machining process are the main objectives of this system.

A vision based machine has an image acquisition device like a webcam as its eyes [2]. In this system the input from the webcam is processed by a rather powerful computer. One key point to note is that Image Processing has huge computational requirements and it is not possible to run an image processing code directly on a small microcontroller. Hence, the image processing and decision making is handled by the more powerful computer while the work bed manipulation and feedback to the computer is handled by a microcontroller system. The two are connected by a bi-directional parallel port. The code for the computer is written in a software that provides the tools for acquiring images, analyzing the content in the images and deriving conclusions. MATLAB is the software available which provides the platform for performing these tasks [2].

2. Model Development

At first a prototype model was developed for this project. A drill machine where Artificial Intelligence (AI), Image Processing (IP) and neural network was implemented to this extent has only been theorized. In this project we implement the theories and adopt them to a specific drill machine and will test its use, using the results to adjust and perfect the decision making software. For the first step a prototype model was developed and adapted to the image processing software so that any shape drawn on the work itself could be sensed and parsed. Then computations for necessary movement of the drilling mechanics could be made, and the results sent to the drilling control unit. This model consists of four individual sections which are given below.

1. Image acquisition
2. Image processing and recognition
3. Data communication through parallel port and actuation
4. Mechanical structure

2.1. Image Acquisition

In this project, Image Acquisition is accomplished using a 5 Megapixel webcam to capture high resolution frames. This camera must be mounted on a computer running MATLAB to eliminate the need for manual repositioning as process progresses [3]. MATLAB uses a software adaptor or driver to communicate with an image acquisition device and the positioning apparatus thereby enabling continuous image acquisition. Figure 1 represents the driver output result [4] of the webcam that was used in this project.

```
Command Window
>> dev_info = imaqhwinfo('winvideo',1)

dev_info =

          DefaultFormat: 'RGB24_320x240'
    DeviceFileSupported: 0
             DeviceName: 'USB PC Camera'
               DeviceID: 1
      ObjectConstructor: 'videoinput('winvideo', 1)'
       SupportedFormats: {1x10 cell}
```

Figure 1. Driver output result.

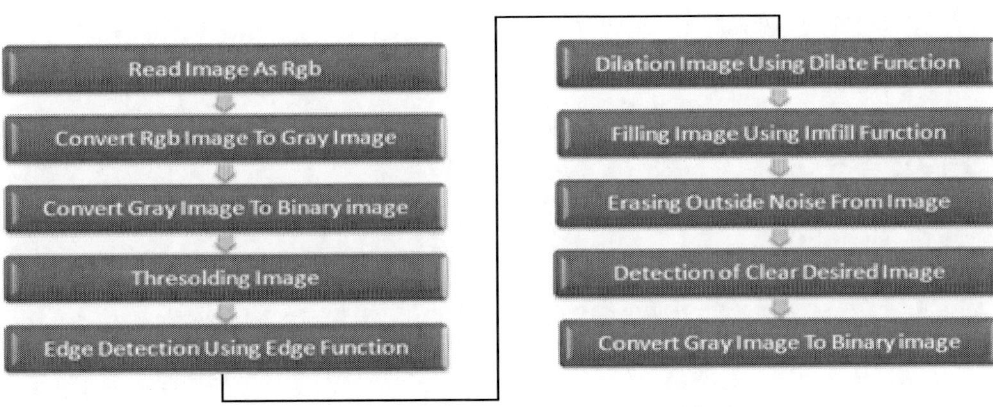

Figure 2. Image pre-processing technique.

2.2. Image Processing and Recognition

Digital image processing is used extensively to analyze the input from the camera [5]. Image processing does not increase image information content, it merely analyzes an image keeping only those parts which are pertinent to the decision processing. After the image is acquired the SHAPES will be detected using Image Pre-Processing where all information not needed, i.e. color, noise, grey scale will be discarded. This leaves only the shape of the work itself and the shapes targeted for drilling [6]. In this project five pre-processing functions are used: threshold, edge, dilate, imfill and imopen. Figure 2 represents the image pre-processing technique that was used in this project.

2.2.1. Step by Step Image Processing for Detecting Circular Mark and Taking Decision

In Figure 3 the steps to derive the final result sensing three circular shapes arranged as points in a triangle are shown below. Figure 4 shows the output of the processing.

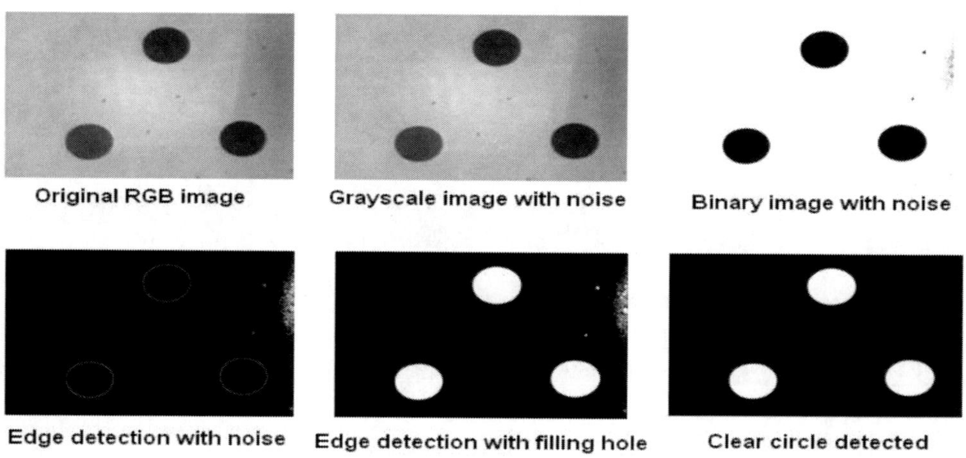

Figure 3. Image processing and detecting circular shape.

Output result for this Image processing

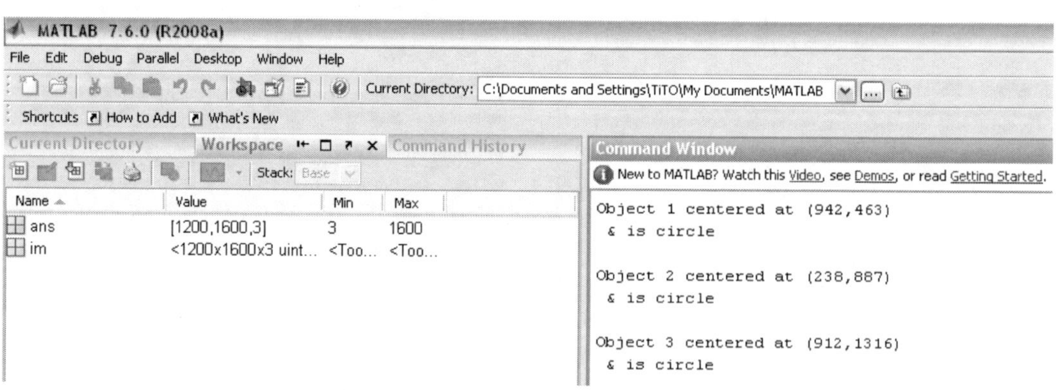

Figure 4. Recognition and center of the shapes.

2.3. Data Communication Through Parallel Port and Actuation

It consists of two different systems which are given below.

(1) Data communication through parallel port
(2) Actuation

2.3.1. Data Communication Through Parallel Port

Due to the reliable use and high rate of data transfer, parallel port was used in this project. MATLAB provides support to access parallel port (also called as printer port or LPT port) of a PC. Parallel port has 25 pins where 2-9 are bi-directional data pins (pin 9 gives the most significant bit (MSB)), 10-13 and 15 are output pins (status pins), pins 1,14,16,17 are input pins (control pins), while pins 18-25 are Ground pins [3].

There are 3 configuration modes of operating LPT port according to IEEE 1284 standard:

1. Standard LPT port configuration: Does not have any data transfer standard but used for slow communication (such as printer as earlier equipments. But now a days for faster data communication, serial@9600baud rate or USB port is used).
2. Enhanced parallel port (EPP) configuration: Transfers data @ 2 – 10 MByte/s.
3. Extended Capabilities port (ECP) configuration: Transfers data @ 2.4 MByte/s (ISA).

In this project ECP configuration was used which specification defines automatic hardware handshaking, command and data cycles and direct memory access (DMA) transfers to a FIFO location. ECP defines six registers in addition to the three standard parallel port registers. These new registers automatically place the address or data information on the parallel port data lines and then generate the parallel port handshaking signals. The ultimate goal of ECP is to improve the parallel port for plug-and-play and Windows environments by making the interface bidirectional and increasing performance [7].

IEEE 1284 defines three interface connectors: 1284-A, 1284-B and 1284-C with ECP configuration. The 1284-A connector is equivalent to the existing 25-pin DSub connector commonly used on the host side of the connection. Thus for the availability, compatibility, reliable use for this project a 1284-A connector was used with ECP configuration.

2.3.2. Actuation

Like any other software drive Mechatronic system consists of actuators controlled by the microcontroller. Here we split the chores into three groups:

1. Coupling of image processor and encoder.
2. Encoder.
3. Buffer and actuator.

2.3.3. Coupling of Image Processor and Encoder

In this project a personal computer of 1.8 GHz is used as the image processor. This processor was coupled to the image acquisition device, does computations on input and analyzes the result, then positioning requirements are sent to the microcontroller, which applies them to the actuators. The actuators have encoders attached which then send movement verification and position information, back to the microcontroller. This closed loop system closely resembles the human neural network.

2.3.4. Encoder

In this project ATMEGA32 microcontroller was used. It was programmed in such an order that it would provide some predefined signal to the buffer to drive the actuator based on the data received from image processor. This system uses input from the ENCODERS to sense tool positioning, then if movement is needed calculations are made and results fed to actuator buffers.

2.3.5. Buffer and Actuator

The microcontroller outputs cannot flow enough current to operate an actuator, so BUFFER Amplifiers must be placed between microcontroller outputs and the actuators themselves.

The characteristics of MOSFET IRF-540 have the ability of flowing 4 amp current. Here applied minimum gate to source voltage above 4 volts. In driving all actuators the required voltage is 12 volts which is above the gate to source voltage. So this device is appropriate for buffering the microcontroller signal. In this project 8 actuators were used. Their functions and rating are provided below:

1. 4 pole 8 lead 1.8 degree/step12 volt 2.5 amp for x axis movement of machine bed.
2. 4 pole 8 lead 1.8 degree/step 12 volt 2.5 amp for y axis movement of machine bed.
3. 8 pole 8 lead 1.8 degree/step 12 volt 3 amp for rotation of drill turret (Clockwise & anticlockwise).
4. DC motor 12v 500mA for revolving drill holder 1.
5. DC motor 12v 500mA for revolving drill holder 2.
6. DC motor 12v 500mA for revolving drill holder 3.
7. High torque DC motor 12v 1 amp for z axis movement of drill holder (Left side).
8. High torque DC motor 12v 1 amp for z axis movement of drill holder (Right side).

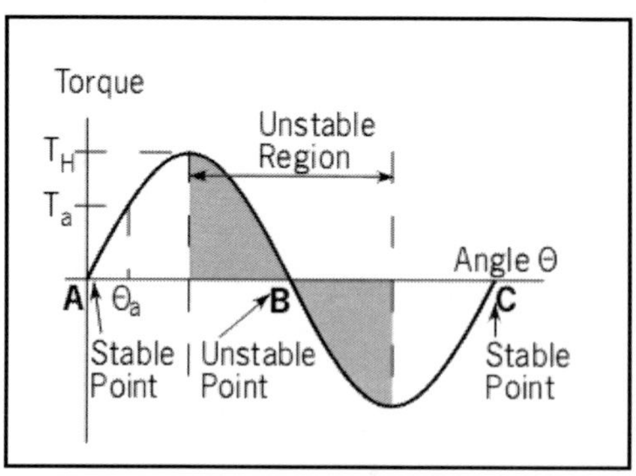

Figure 5. Stability analysis of stepper rotor.

Drilling in a very specific position the movement of the base and the turret should be very precise. In *Pulse width modulation* (PWM) the pulse top always has a magnitude of 12 volts, 2.5amp which lead the motor to operate in a very steady state which is prerequisite to maintain a high operational accuracy. The below Figure 5 shows that in steady state operation the rotor must lies between the two terminal points, A and C. So that the motor we have used must give an angular rotation at an interval of 1.8 degree and the driving order so arranged. Figure 6 represents the operational unit of the circuit diagram.

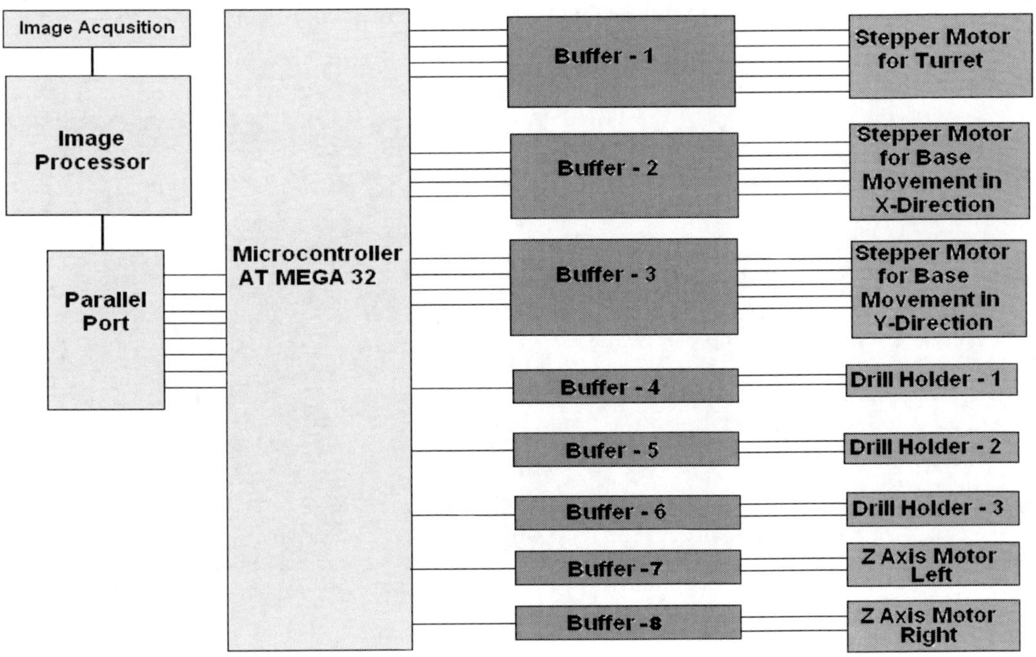

Figure 6. Operational unit of the circuit diagram.

2.4. Mechanical Structure

Mechanical system mainly consists of lead screw and internal threaded screw (collar) for the movement of the machine according to the X, Y and Z axis. The main function of lead screw-collar mechanism is to convert the rotational motion of actuators into linear motion. Inside a wooden block, a collar (internal screw) is set up by force. After that a 10 inch long right handed lead screw, with 12 Threads per Inch passes through the collar. When the lead screw rotates, according to the lead screw-collar principle, a linear motion is established to the collar which contains the wooden block. Two supporting rods near the ends of the block prevent the block from tilting. An Electronic stepper motor is connected to the one end of the lead screw and a bearing on the other end. When electric power is supplied to the stepper motor, lead screw rotates thus the wooden block moves linearly along the axis of the supporting rods. When the stepper motor rotates clockwise the wooden block moves from left to right side and whenever it rotates anti clock wise, reversely the wooden block moves from right to left [8].

A drill turret is utilized to facilitate use of different sizes of drills as needed. Here aluminum plate having one inch thickness and 4 and ½ inch diameter is used as a drill turret. At 90 degree consequent interval, 4 cavities are drawn to the plate. 3 cavities are for three various dimensioned drill bits and other one is for setting up the webcam. Center of the aluminum plate has a ½ inch diameter through drill and a 2 inch long shaft is introduced here. This shaft is connected to a stepper motor and the total attachment is directly connected to the Z axis mechanical structure. When the motor rotates with a certain angle either clockwise or anti clockwise, the turret moves radially along its axis. Figures 7, 8 & 9 respectively represent the machine bed, drill turret and total mechanical structure of this project.

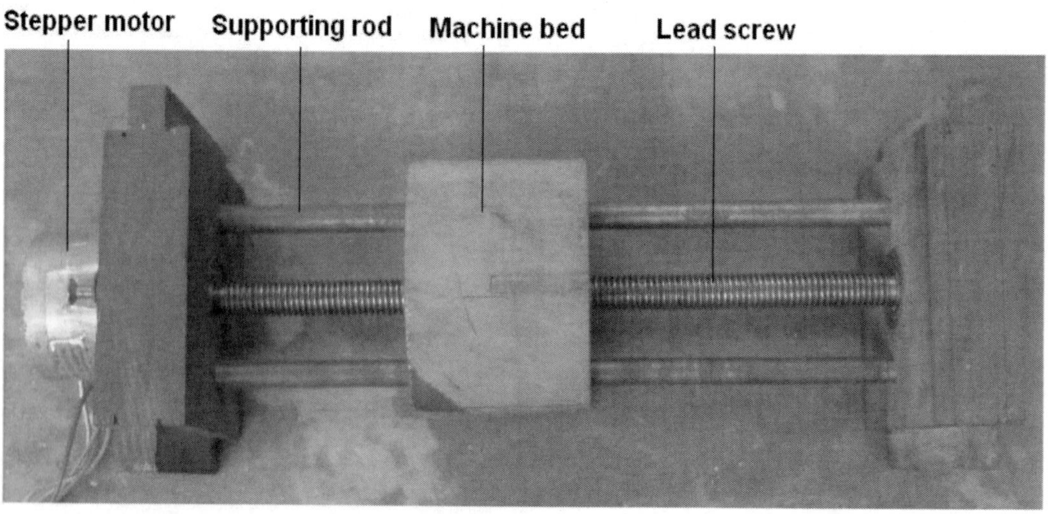

Figure 7. Machine bed.

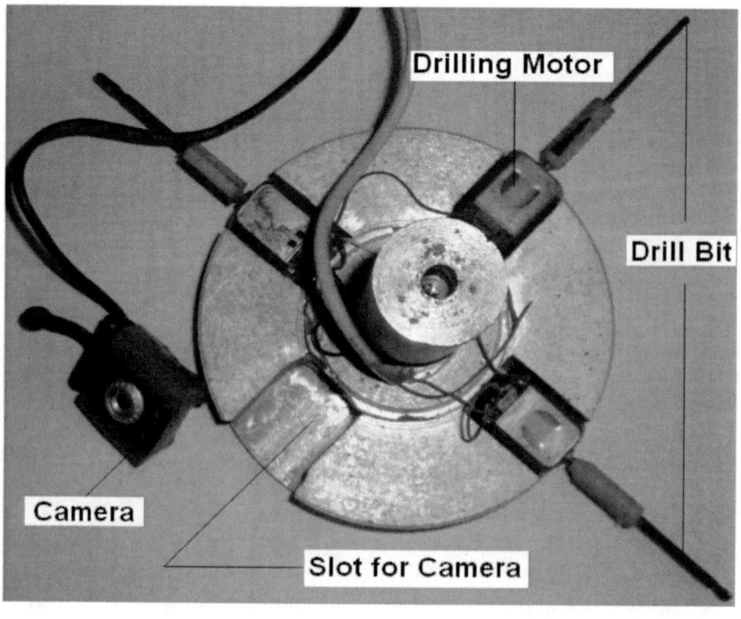

Figure 8. Drill turret.

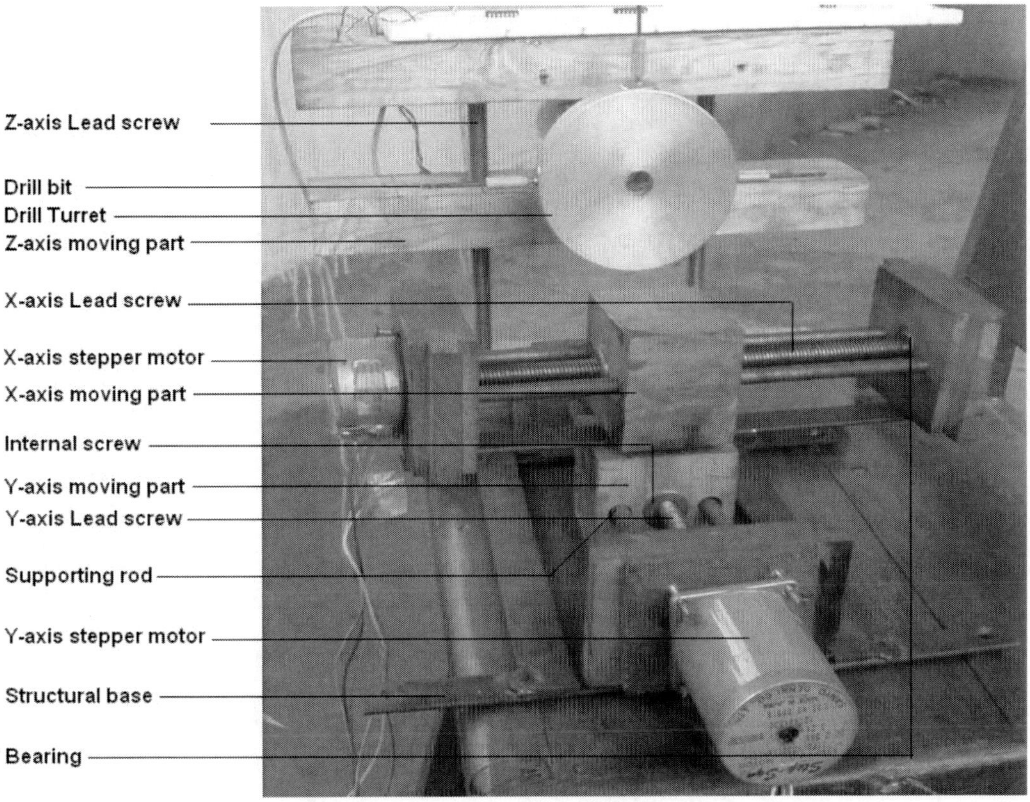

Figure 9. Total mechanical structure mentioning highlighting parts.

3. Brief Mechanism of AI Drill Machine

Artificial intelligent machine is controlled by itself without human input. Neural networking is the basic theme of its working principle. If we consider a human brain, eyes are the main sensing power of it which is connected directly to the brains neurons. In this machine a digital webcam is act as like the human eyes and it is directly connected to the image processor where image processor act as like a human brain. Here PC or personal computer is used as an image processor. Image processing is the programming which was used for sensing the object on machine bed where the drilling process is to be performed. Decisions are made concerning drill bed positioning and turret actuation and commands are sent to the microcontroller to process the job as required. All three systems are turn concurrently using decisions made by the MATLAB unit, therefore it may say that is a real time machine. Here the mechanism of AI drill machine is drawn by block diagram showing essential components which are shown in Figures 10 and 11. Also a flow diagram is drawn describing the main operation of the system which is shown in Figure 12.

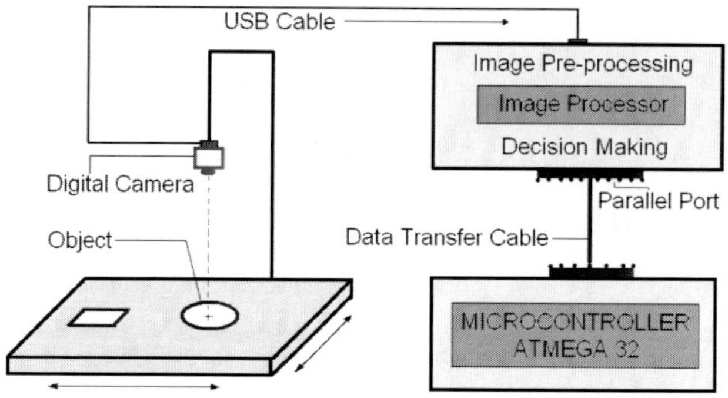

Figure 10. Mechanism up to signal transfers.

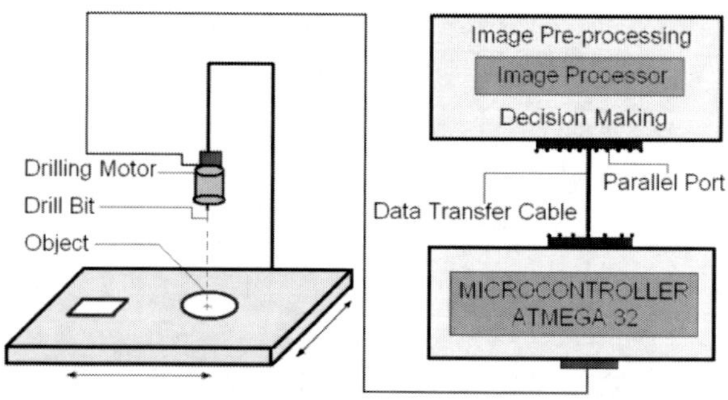

Figure 11. Mechanism during actuation.

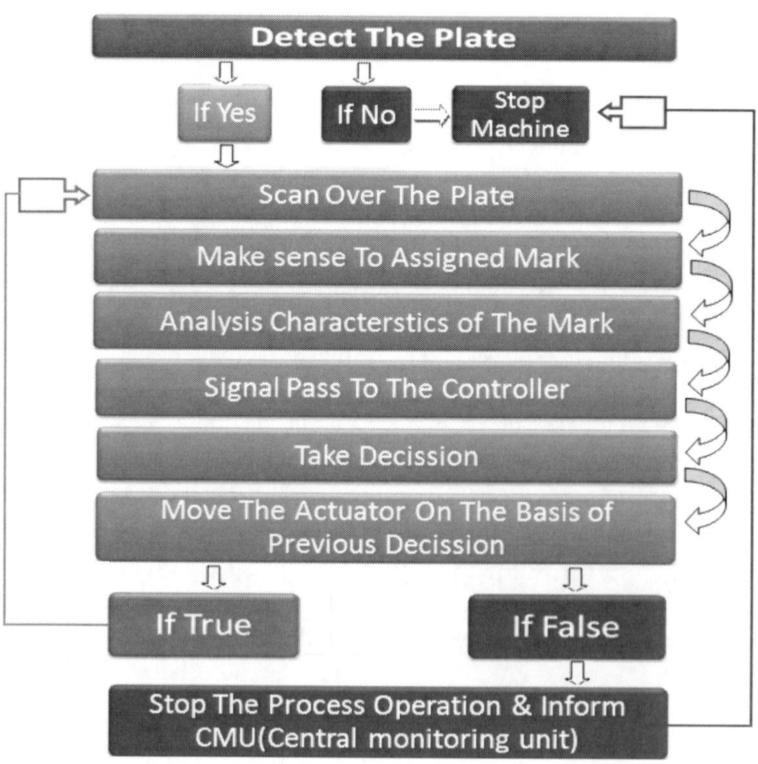

Figure 12. Mechanism of AI drill machine.

Total Experimental Set-Up of AI Drill Machine

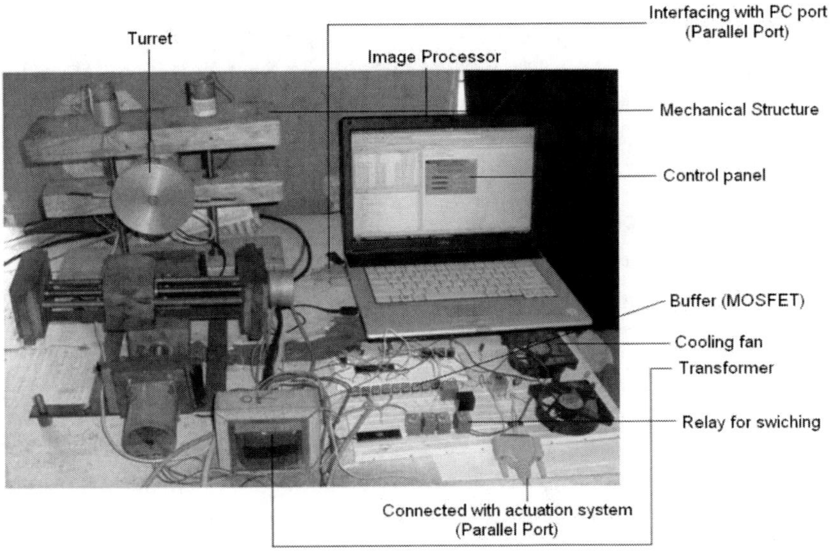

4. Control Window

A manual control window is provided by MATLAB to operate the system manually if required. Should the camera fail, the system can be run operated manually as a CNC drilling machine whereas for the automatic operation it will be acted as an Artificial Intelligent (AI) drill machine. So both facilities have been introduced here. Figure 13 represents the control window of this system.

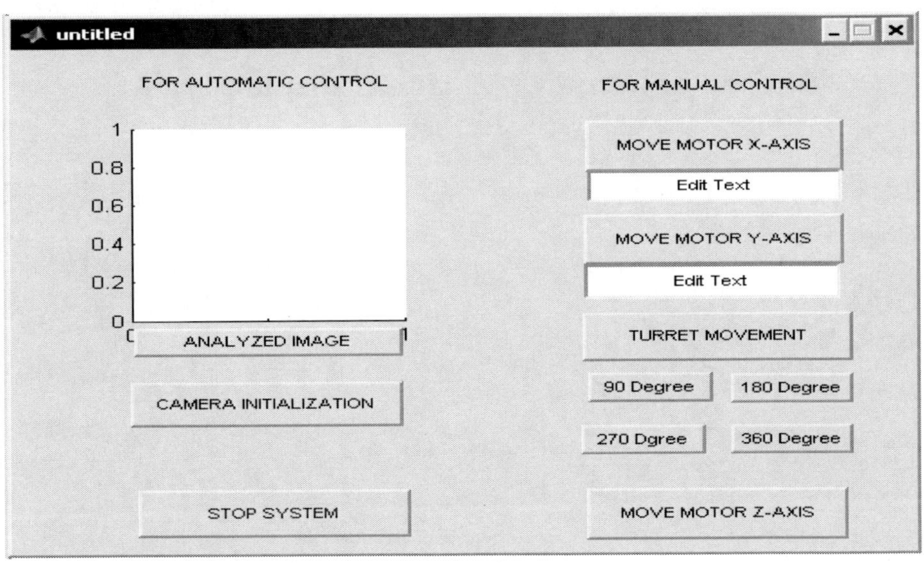

Figure 13. Control window of AI drill machine.

5. System Resolution

The concept "system resolution" is very important for this AI controlled drilling machine which is closely related to the drilling accuracy and precision. System resolution will determine how accurately or precisely a drilling machine can operate. It depends on approximately four factors which are given below:

1. Positioning accuracy of actuators.
2. Measurement accuracy of sensor.
3. Algorithms of image processing and actuators control.
4. Tool deterioration.

5.1. Positioning Accuracy of Actuators

Positioning accuracy of actuators really depends from thread of the lead screw (TPI) and the step of the stepping motor. In this project 12 TPI Acme threaded lead screw and high torque stepper motor having 1.8°/step was used. For determining the accuracy first it is necessary to measure the linear displacement of each machine bed per step of actuator or stepper motor. This calculation is given below. If the lead screw rotates 1 revolution (360°), it moves 1/12 inch linearly because it has 12 TPI.

Rotation of 360°...........linear displacement = 1/12 inch

So, Rotation of 1.8°............linear displacement = 1.8/ (12×360)

= .000416 inch

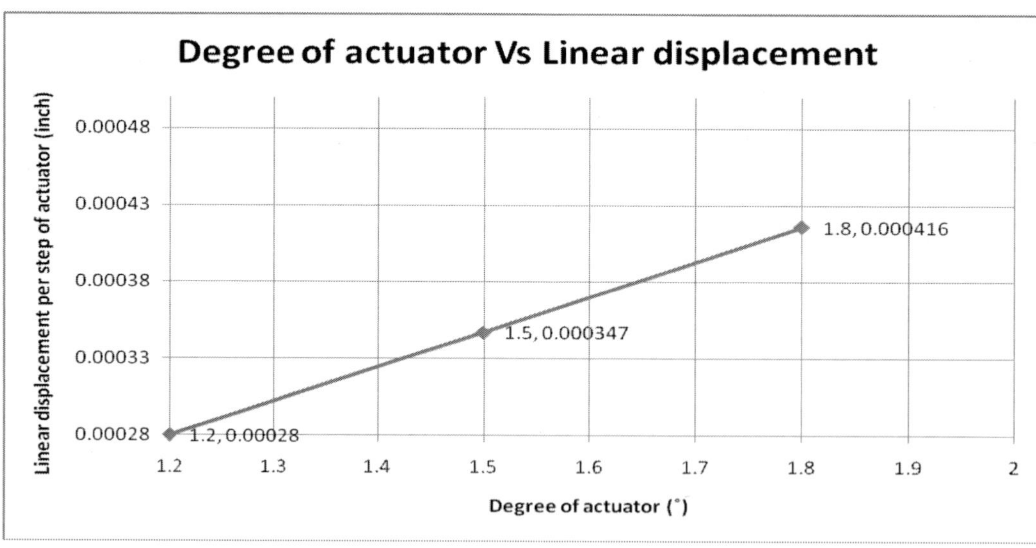

Figure 14. Displacement variation per step of actuation with degree of actuator.

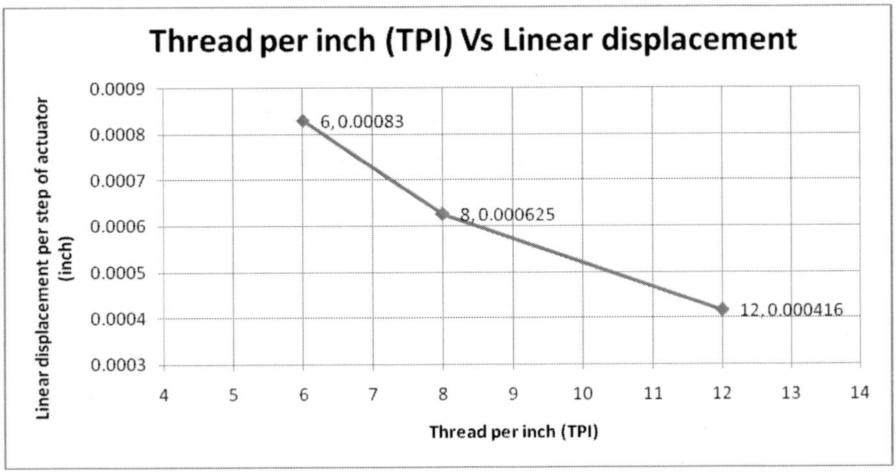

Figure 15. Displacement variation per step of actuation with TPI.

The following calculation shows that linear displacement of each machine bed according to X and Y-axis per step of stepper motor of the designed AI drill machine is .000416 inch. On the other hand, it can be said that the designed AI drill machine can scan .000416 inch of surface plate per step of stepper motor for accurate detection of desired object. So roughly it may say that accuracy of drilling depends upon how small area the system scanning per step of actuator. If the linear displacement or scanning area over the plate per step of stepper motor remains small, the detection of desired object will be more accurate. Figures 14 and 15 represent that how the linear displacement per step of actuator increase or decrease with increase or decrease of stepping degree of stepper motor and thread per inch (TPI).

Here main focus was on stepping degree of stepper motor and thread per inch (TPI) for positioning accuracy of actuators. However, no less, and often more significant factors are the backlash, Coulomb friction and elastic deformation of structural elements of the installation (drill machine).

5.2. Measurement Accuracy of Sensor

Measurement accuracy of sensor (TV camera, Webcam) really depends from the resolution of CCD matrix. In this project a 5 Megapixel (MP) webcam was used for grabbing image at 15 frames per second (FPS). Frame size is a vital factor because higher frame size provides higher resolution of images. A larger size gives greater clarity and is easier to work with, but consumes more memory and therefore is slow [2], [9]. But in a smaller image, it is difficult to differentiate between two objects. So an optimized frame size was selected for this project which given the better intelligibility and distinguish between two objects. Camera resolution and capturing image with optimized frame size, compatibility with the developed image processing (shape recognition) software were the main consideration of this project to recognize the desired object. However, focal distance (field of camera view), geometrical distortions of optical system and alignment errors are the parameters have a direct effect on the accuracy of measurements [9] were the careful consideration during assembly of the portion of image acquisition.

5.3. Algorithms of Image Processing and Actuators Control

Algorithms of image processing and control of actuators has a great influence on the drilling precision. In this project these algorithms were developed in such a way that to compensate errors of actuators and sensor (webcam). For instance, the developing image processing (shape recognition) software for AI drill machine can accurately sense the desired object under low resolution image which is captured by 1.3 MP webcam. The actuator control design was based on the algorithm for this low resolution. Technique of image processing that was used in this project, sensing desired object and Operational unit of actuators control circuit diagram has been mentioned in Figures 2, 3 and 6 in this paper.

5.4. Tool Deterioration

Tool deterioration is a factor which is also responsible for drilling inaccuracy during operation. Here simply three different dimensioned drill bit were used for demo drilling. As because the more focal point was on making an intelligent machine therefore the factor 'tool deterioration' was not in careful consideration. But manufacturing an AI drill machine for mass production this factor should be considered cautiously.

6. Experimental Results

After developing the model a number of data was collected with the purpose of how much precisely actuation system is running according to X and Y-axis. Performance table according to X and Y-axis is shown in table 1.

Table 1. Actuation performance at X and Y axis

Movement	Actuation according to X -Axis				Actuation according to Y-Axis			
	Design (inch)	Result (inch)	Error (inch)	Error (%)	Design (inch)	Result (inch)	Error (inch)	Error (%)
360°	.083	.0826	.0004	.48	.083	.08	.003	3.61
720°	.17	.168	.002	1.17	.17	.162	.008	4.7
1080°	.25	.246	.004	1.6	.25	.238	.012	4.8
.05 inch	.05	.0496	.0004	.8	.05	.0473	.0027	5.4
.1 inch	.1	.098	.002	2	.1	.097	.003	3
.5 inch	.5	.492	.008	1.6	.5	.476	.024	4.8
1 inch	1	.986	.014	1.4	1	.943	.057	5.7

Table 1 shows the performance of actuator according to X-axis and Y-axis. The performance of X-axis is better than Y-axis. From the mechanics point of view this makes sense. The reason is that the shaft of Y-axis still has little bending because of having major load on it rather than X-axis. In this project stepper motor was used having same torque for actuation according to X and Y axis. But Y-axis stepper motor need more torque than X-axis overcoming the bending load, then the performance according to Y-axis will be more accurate. Another reason is the arrangement of the Y-axis stepper motor was not positioned well.

7. DISCUSSIONS

In this project a drilling mechanism using vision based control sensing was conceptualized, developed and the performance of the developed control system was tested. Also an experimental prototype Artificial Intelligent (AI) drill machine was manufactured. For drilling operation, the most common parameters for work specification were the diameter to be drilled, depth of hole and drilling speed. Operational accuracy was considered vital but this consideration lead the other parameters to move beyond the scope of this project. Other parameters important for consideration are simplicity of design, maintenance facilitation, cost of manufacturing, operational rate, and system reliability. In order to keep pace among these parameters some sampling was made.

From sampling it became evident that system accuracy followed directly or sometime inversely proportional relationship with other operating parameters. Considering mechanical element it was found that it greatly influences system accuracy. Better resolution in work table drive may require a drive screw with greater than 12 TPI. Again it can be shown that increasing the actuator resolution results in higher accuracy to a certain level, but resolutions near zero degree per step is unrealistic. For that a considerable optimization should be drawn. A term ‘system delay’ was considered carefully while designing the machine. The system will be more accurate if we use lead screw having TPI more than 12, stepper motor having stepping degree less than 1.8, webcam having resolution more than 5 MP, but the required execution time will be increased. For executing a single and simple work more time requirement is not desirable. Therefore an optimization is also required here. During experiment three various dimensioned lead screw (12, 8 and 6 TPI), two stepper motor (1.8° and 1.2°) and two webcam (1.3 and 5 MP) were taken. After observing some result which shown in Table 2, observation 1 was selected because it gives minimum system delay with maximum accuracy.

Table 2. Calculated data using various dimensioned lead screw and stepper motor

Observation No.	Thread per inch (TPI)	Stepping degree of stepper motor	Movement of bed per step in inch
1	12	1.8	.00042
2	12	1.2	.00028
3	8	1.8	.00063
4	8	1.2	.00042
5	6	1.8	.00083
6	6	1.2	.00055

For this project the threaded element configuration was 12 TPI and camera resolution 5 MP which gives a very precise 0.000416 inch movement of base per feed per snapshot. This very small movement of job shows that this will be consistent with a highly accurate system.

Conclusion

A methodology based on Artificial Neural Network (ANN) was applied to a model in an approach to enhance machine sensing by means of image processing aiding the implementation of an artificial intelligent drill machine. Influence of various mechanical parameters on the response of the network was evaluated and is in good agreement with the physical understanding of currently available sensing elements. As AI control machining is a very new concept it provides some promising aspects. AI drill machine is totally independent on coordinate system. So coordinate system wherever does not exist properly (e.g. complicated curvature surface) this AI controlled machine can work there without any difficulty. AI control system is more accurate and reliable to use. But the main hindrances in adopting AI control system are cost and long term development project costs. As well as it is more time consuming process than the conventional drilling machine nevertheless it reduces a great extent of human input.

Acknowledgments

The author gratefully would like to thank Prof. Dr. Rafiqul Islam, *Dr. Murari Mohon Roy and Bob Ross for their co-operation to prepare this paper entirely and* also show gratitude to laboratory staffs of the department of Mechanical Engineering of Rajshahi University of Engineering & Technology (RUET) for their support to assemble this model machine.

References

[1] T. Kolodziejc zyc, R. Toslano, S. Fouvry and G. Morales-Espejel. (2009), Artificial intelligence as efficient technique for ball bearing fretting wear damage prediction, Publisher Elsevier, volume 268, Issue 1-2, pp. 309-315.

[2] Ankur Agrawal, IIT Kanpur. (2009), Elementary introduction to image processing base robot, Robotics Based article, pp. 3-34.

[3] Deepak Malani, REC Calicut, Anant Malewar, IIT Bombay, Introduction to Video analysis using MATLAB, Issued in public interest by Nex Robotics Pvt. Ltd. pp. 1-23.

[4] The documentation and demos provided in the help section of MATLAB.

[5] Rafael C. Gonzalez, Richard E. Woods, Steven L. Eddins, Digital Image Processing Using MATLAB, Book of Matlab © 2004 by Pearson Education.

[6] Milan Sonka, Vaclav Hlavac, Roger Boyle, Image Processing, Analysis and Machine Vision, third edition, book for image processing copyright © 2008 by Thomson Learning, part of the Thomson Corporation.

[7] Heidi Frock, IEEE 1284 – Updating the PC Parallel Port, IEEE 1284 standard.pdf 83 KB.

[8] H. Ferdinando, I. N. Sandjaja, G.Sanjaya, Department of Electrical Engineering Petra Christian University, Automatic Drilling Machine for Printed Circuit Board.

[9] Nasser Kehtarnavaz, Mark Gamadia, University of Texas at Dallas, USA. (2006). Real Time Image and Video Processing: From Research to Reality, Synthesis lectures on image, video & multimedia processing #5, Morgan & Claypool Publishers.

In: Knowledge Management and Communication … ISBN: 978-1-62808-250-0
Editor: Rafiqul Islam

Chapter 8

SUPERSTRATE LOADED RECTANGULAR MICROSTRIP PATCH ANTENNAS - *AN OVERVIEW*

***R. K. Yadav*[1*] *and R. L. Yadava*[2†]**

[1]Department of Electronics & Communication Engineering
ITS Engg College, Gr. Noida,
U.P. India
[2]Department. of Electronics & Communication Engineering
Galgotias College of Engg & Tech. Gr. Noida,
U.P. India

ABSTRACT

This paper provides an overview of superstrate loading effects on rectangular patch antennas with emphasis on theoretical design and practical techniques for last three decades. Several theoretical analysis methods have been summarized, including Transmission line, Method of Moments (MOM), Finite-Difference Time- Domain (FDTD) and Spectral Domain Immittance (SDI) as well as Modified Wolff Method (MWM). The characteristics of covered patch antenna have also been studied using genetic algorithm and neural network methods. In particular, effect of superstrate loading on co-polarisation and cross-polarisation as well as aperture coupled patch antenna with emphasis on radiation and directivity have also been studied. In addition to effect of dielectric loading on the patch performances, the various methods for their compensation have also been described. The paper aims those readers who desire to know about the latest developments in the field of superstrate loaded rectangular patch antennas.

Keywords: Microstrip Antenna (MSA), Dielectric Covers, Numerical Techniques, Frequency Alteration, Environmental Hazards

---

* E-mail address: ravipusad@gmail.com.
† E-mail address: rlyadava@rediffmail.com.

INTRODUCTION

Microstrip antenna has been used in air bone & space craft systems because of their low profile, conformal behaviour, planner configuration, good polarization characteristics and compatibility with integrated circuits. There are various patch geometries of microstrip antennas such as rectangular, square, pentagonal and hexagonal patches. Despite of various advantages, the narrow bandwidth (~ 3%) and poor polarization purity are the main disadvantage with these types of antennas. The resonant frequency of such antennas is function of their dimensions, permittivity of substrate and their thicknesses as well as feeding techniques. When such antennas are used for long duration, about the years or so, environmental effects such as rain, fog, moisture & snow causes variations in the performance of antennas; particularly resonance frequency shifted and polarization characteristics degraded. Due to environmental effects, when an antenna is covered with a layer of finite thickness, its characteristics such as input impedance, attenuation losses, quality factor, polarization as well as resonance frequency changes with type of layer and its thickness. As a result, since microstrip antenna is a narrow bandwidth radiator, the alteration in resonance frequency pushes antenna out of frequency and polarization range. In the view of above fact, it is essential to have prior information about the behaviour of microstrip antennas with various dielectric covers to the antenna designers. The basic geometry of a superstrate loaded rectangular Microstrip antenna is shown in Figure 1. Therefore present study is dedicated to describe the effects of dielectric loading on the performance characteristics of rectangular microstrip antennas.

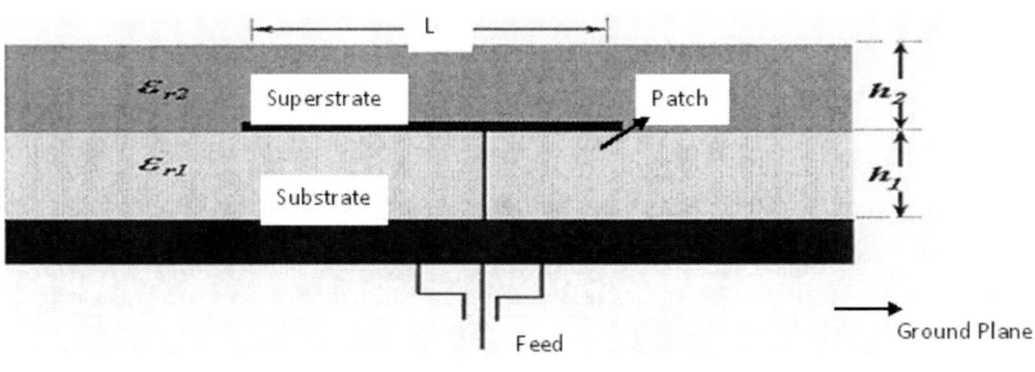

Figure 1. Side view of a superstrate loaded rectangular microstrip antenna.

PROGRESS: 1980-1990

Study of dielectric cover on microstrip lines dates back in year 1980, when I. J. Bahl & S. Stuchly proposed a variational method to describe the microstrip line covered with a lossy dielectric sheet. They showed that characteristics of microstrip covered with a thick sheet of high dielectric constant are drastically affected. The effect of dielectric cover is more pronounced for small values of *w/h* ratio, *w* is width line and *h* is height of dielectric substrate. This is because the fringing fields which interact with the covering sheet increases for smaller values of *w/h* ratio. Their studies were focused on the variation of characteristics impedance, effective dielectric constant, and conductor and dielectric losses with *w/h* ratio for different dielectric loaded materials. In addition it has been reported that above parameters also varies with *d/h* (*d*- Thickness of cover). In particular for *d/h* = 0.02, the values of effective dielectric constant decreases with increase in ratio *w/h*. However for larger values of *w/h* ratio effective dielectric constant increases for smaller values of *d/h*. The effective dielectric constant decreases with the increase in *w/h* values because in this case fringing field decreases with increasing w/h values. For large *w/h* ratio the structure behaves like a microstrip and the effective dielectric constant increases Figures 2-6, [1]. These Figures reveal that conductor and dielectric loss increases with increase in *d/h* values, whereas dielectric loss decreases with increasing w/h ratio Figure 7 [1].

Later in 1982 I. J. Bahl et al. designed a microstrip antenna covered with dielectric layer and found that covered microstrip antenna alters its performances such as *characteristics impedance, phase velocity and dielectric losses*. In particular, due to dielectric cover the resonance frequency of antenna changes, and reason behind that effective dielectric constant changes due to dielectric layer [2]. And corresponding fractional change in resonance frequency Δf_r is calculated using equation (1).

$$\frac{\Delta f_r}{f_r} = \frac{1}{2} \cdot \frac{\frac{\Delta \varepsilon_e}{\varepsilon_{eo}}}{1 + \frac{1 \Delta \varepsilon_e}{2 \varepsilon_{eo}}} \quad (1)$$

Where,

ε_e = effective dielectric constant with cover.

ε_{eo} = effective dielectric constant without cover.

$\Delta\ \varepsilon_o$ = change in dielectric constant due to cover.

Δf_r = fractional change in resonance frequency.

f_r = Resonance frequency.

And after analysis , it has been found that, the fractional change of resonance frequency for optimum design at 10 GHz are 5.8%, 7.8%, and 16% for the dielectric cover of polystyrene, ice and beryllium oxide respectively (Figure 2).

It has also been noticed that if the dielectric cover thickness increases more than 2 cm, the effect of dielectric loading on resonance frequency of patch becomes smaller (Figure 3).

In addition effect of dielectric thickness indicates that a return loss first increases with increasing the thickness of dielectric cover and then decreases. Whereas bandwidth increases with increasing cover thickness for low dielectric constant material and decreases for high dielectric material (Table 1).

In particular microstrip antenna designed on Duriod substrate for *f* = 4.1 GHz and loaded with a 0.318 cm thick dielectric cover exhibits bandwidth enhancement by 7%. Whereas, for dielectric loading of thickness less than 0.1 cm, the bandwidth is almost unaffected. N.*G.* Alexopoulous using Greens function followed by Sommerfeld method proposed that superstrates loaded microstrip antenna offers increased *gain, radiation resistance and efficiency*, provided covered material considered as desirable part of an antenna. They observed that by choosing proper thickness of cover, resonance condition may be created, whereby gain and radiation resistance are substantially improved over significant bandwidth. If thickness of the dielectric cover is about half of operating wavelength, 100% efficiency may be achieved [3]. H.Y.*Yang* and N.*G.* Alexopoulous in 1987, also analyse multiple superstrate loaded printed circuit antenna using transmission line model. They considered a Hertzian electric dipole (HED) is embedded in the substrate with multi-superstrates, and observed two resonance conditions. It was concluded that when the first resonance condition is satisfied and corresponding wave impedances are in high low-high-low sequence.

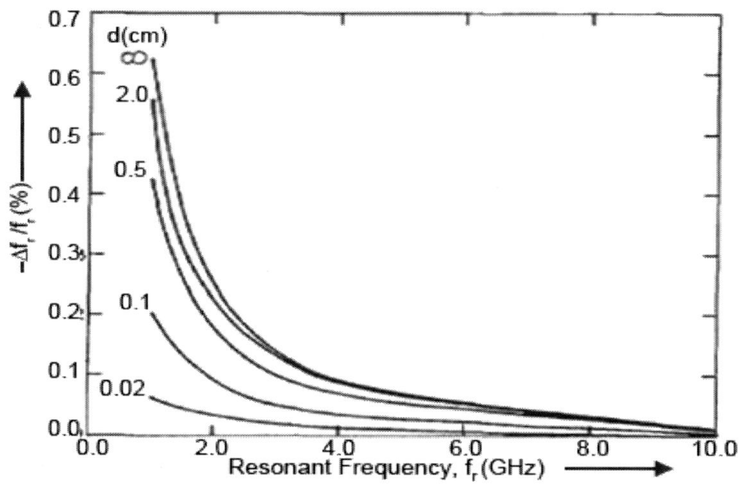

Figure 2. The fractional resonant frequency of a microstrip antenna covered with a dielectric layer as a function of resonant frequency for plasma.

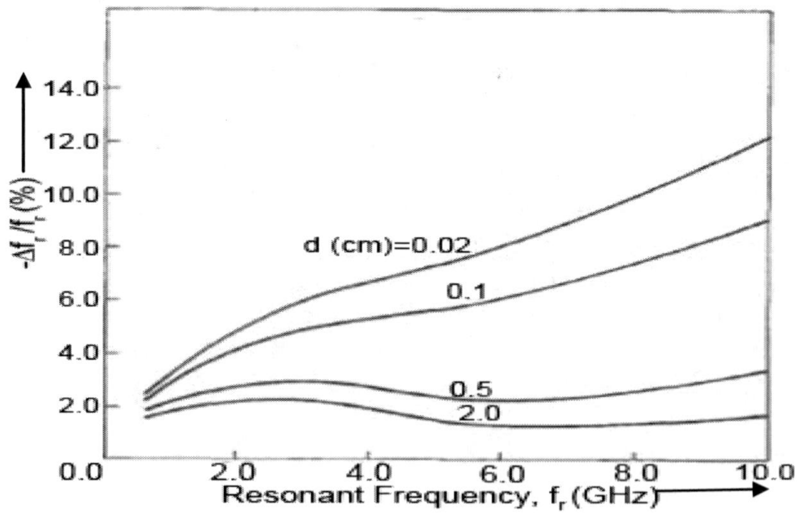

Figure 3. The fractional resonant frequency of a microstrip antenna with a dielectric covers and loaded with ice, as a function of frequency for various thicknesses of dielectric cover.

Table 1. Experimental data for the effect of dielectric loading on the characteristics of microstrip antennas

Dielectric Cover	ε_{r1}	*d (cm)*	f_r*(GHz)*	*Return Loss(-dB)*	*Bandwidth (%)*
Air	1.0	∞	4.104	32	2.17
Duroid	2.32	0.08	4.008	35	2.18
		0.159	3.934	26	2.22
		0.318	3.895	22	2.31
Plexiglass	2.6	0.112	3.952	33	2.18
		0.159	3.912	25	2.18
		0.318	3.874	22	2.20
Mylar	3.0	0.0064	4.070	37	2.18
		0.0128	4.058	39	2.18
		0.0384	4.010	40	2.20
Epsilam-10	10.2	0.0635	3.640	36	2.0
Custom High-k	10	0.154	3.482	24	1.75
		0.132	3.26	18	1.90

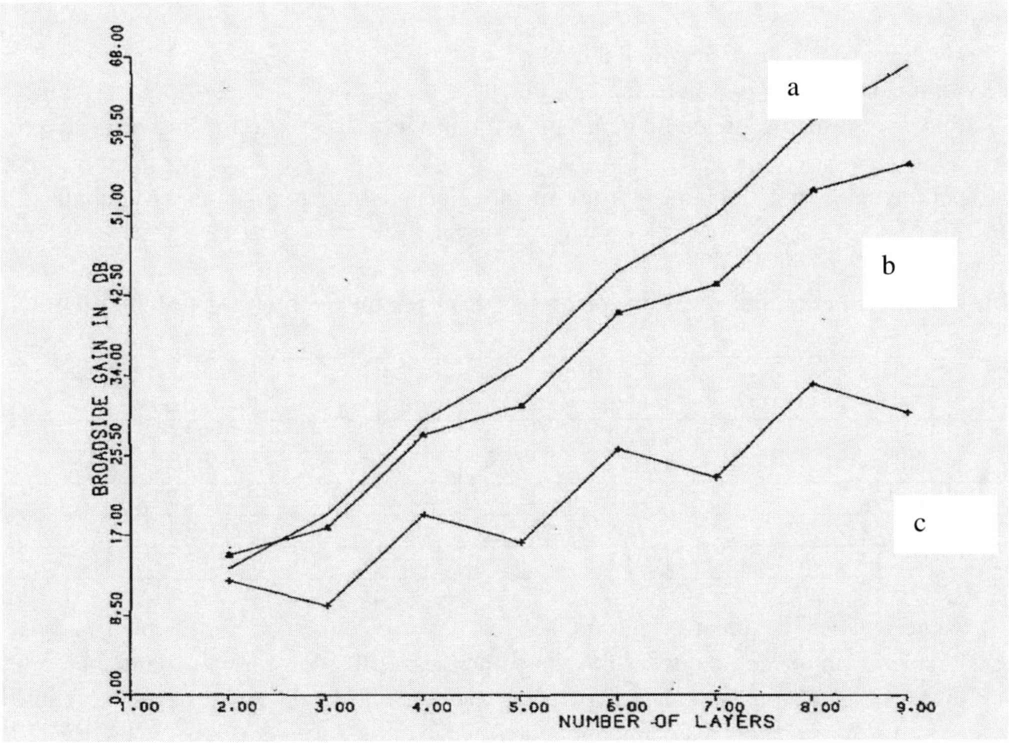

Figure 4. Broadside gain versus number of layers for type 1 resonance condition For even layers $\varepsilon = 10$ and $\mu = 1$,while for odd layers, (a) $\varepsilon = 1$ and $\mu = 4$, (b) $\varepsilon = 1$ and $\mu = 2$, (c) $\varepsilon = 2$ and $\mu = 1$.

Extremely larger resonance gain can be obtained even with moderate values of ε and μ. The same conclusion is obtained for the second case of resonance conditions except that wave impedances are in the low-high-low-high sequence is satisfied (Figure 4). The proposed method reveals that extremely high gain (= 20dB) is obtained which can be further increased by increasing the number of dielectric layers. But as the gain increases the bandwidth of antenna decreases and antenna becomes more bulky as well as sensitive to other parameters [4].

A. Bhattacharya reported that superstrate loading minimize the level of matching but it can be improved by shifting the feed location. All other parameters change usually due to loading of dielectric covers [5]. However -10 dB return loss bandwidth does not change significantly with the superstrate thickness. The impedances at resonance with and without superstrate are given as;

$$\frac{R_h}{R_o} \cong \frac{sin^2\left[\frac{2\pi c f_{rh}}{\sqrt{\mu_o \varepsilon_o \varepsilon_{eh}}}\right]}{sin^2\left[\frac{2\pi c f_{ro}}{\sqrt{\mu_o \varepsilon_o \varepsilon_{eo}}}\right]} \tag{2}$$

Where

f_{rh} = resonant frequency without superstrate.

f_0 = resonant frequency with thick superstrate A and *p* are two constants having values 9.26 and 0.89.

R_h = impedances with dielectric loading.

R_o = impedances without dielectric loading.

ε_{eh} and ε_{eo} = effective dielectric constants without/with superstrate loading respectively.

Experimental results obtained from rectangular microstrip antenna are tabulated in (Table 2).

Table 2. Experimental observation for dielectric cover on rectangular patch antenna [5]

S. No	Thickness of superstrate h (cm)	Resonance Frequency(GHz)	Return loss (db)	Frequency at -10 db loss (GHz)
1	0	2.845	-43	2.765-2.925
2	0.318	2.749	-17.5	2.668-2.824
3	0.638	2.704	-13	2.641-2.743
4	0.954	2.682	-13	2.620-2.743
5	1.272	2.670	-12.5	2.610-2.734

R.Q.Lee and A. J. Zaman et al., studied two layer electromagnetic coupled microstrip antenna loaded with dielectric cover of various thickness (0-0.16cm).It was found that in high gain region resonant input impedance increases and the 3db beam width decreases with the superstrate thickness, while resonant frequency and bandwidth decreases (17.1 GHz to 7.6 GHz). In general impedance matching becomes poor with superstrate thickness [Table 1, 6].

Multiport network model [MNM] has been extended to analyse a rectangular dielectric covered microstrip antenna **[7].** The substrate thickness is chosen to be much smaller than the wavelength (λ), so the fields underneath the patch have no variations in the Z-direction. The conductance Gr and Gs of a radiating edge are related to Voltage distribution as follows;

$$G_{r,s} = \frac{2P_{r,s}}{\frac{1}{b}\int_s V^2(s)ds} \tag{3}$$

Where the integration is along the patch edge of length *b*, *r* is radiation conductance and *s* is surface conductance and *V(s)* is the voltage distribution along the edge, $P_{r,s}$ is the power radiated for the radiation and surface waves. Particularly for the substrate thickness $d = 0.25$ mm, the percentage of the power radiated ($P / (P_r + P_s)$) is ~ 61%.Whereas edge capacitance in presence of covered rectangular microstrip antenna is found to be

$$C(b) = \frac{b}{2}\left[\frac{\varepsilon_{re}(a)}{C_o Z_o(air)} - \varepsilon_o \varepsilon_r\right] \tag{4}$$

Where all the parameters have usual meaning. The radiating-edge capacitance of a rectangular patch antenna along with resonant frequency is tabulated in Table 3 given below [7].

Table 3. Radiating- edge capacitance of rectangular patch antenna with dielectric cover ($h = 0.396$ mm, $d = 2.54$ mm, $\varepsilon_{r1} = \varepsilon_{r2} = 2.2$)

S. No.	Length (a) (cm)	Width (b) (cm)	Resonance Frequency (GHz)	Edge capacitance (pF)
1	1.3372	2.017	7.1300	0.06717
2	1.3359	4.008	7.0475	0.14141
3	1.3411	6.030	7.0450	0. 19320
4	1.3376	8.043	7.0200	0.27115

The analysis reveals that both the input reflection, the resonance frequency for one-port network, whereas input reflection, the resonance frequency as well as transmission coefficient for two-port, are in good agreement with experimental and theoretical results. In present case dielectric constant (ε_r) of substrate and superstrate are taken to equal. The measured resonance frequency is found to be 7.475 GHz, while the theoretical value is 7.485 GHz, which indicates that there is a discrepancy of about 0.134% (Figure 5(a)). However the bandwidth of the one-port rectangular patch for (VSWR < 2) is 80 MHz or 1.06%. Figure 4(b). shows a good agreement between theoretical and measured variations of input reflection coefficient $|s_{11}|$ with frequency. The measured resonance frequency is 7.485 GHz; the corresponding theoretical value is also 7.485 GHz with numerical values; -25.6 dB and -33.5 dB respectively. A good agreement between the theoretical and the measured variations of the transmission coefficient $|s_{21}|$ with frequency is also observed in Figure 5(b). At the resonance frequency $f = 7.485$ GHz, the measured $|s_{21}|$ is -2.8 dB while theoretical value is -2.64 dB. It is also observed that in spite of changing resonant frequency, reflection coefficient and other parameters, superstrate loading also changes the shape of radiation pattern. Depending on the thickness of dielectric loading at higher frequency the radiation pattern divided into two and three lobes.

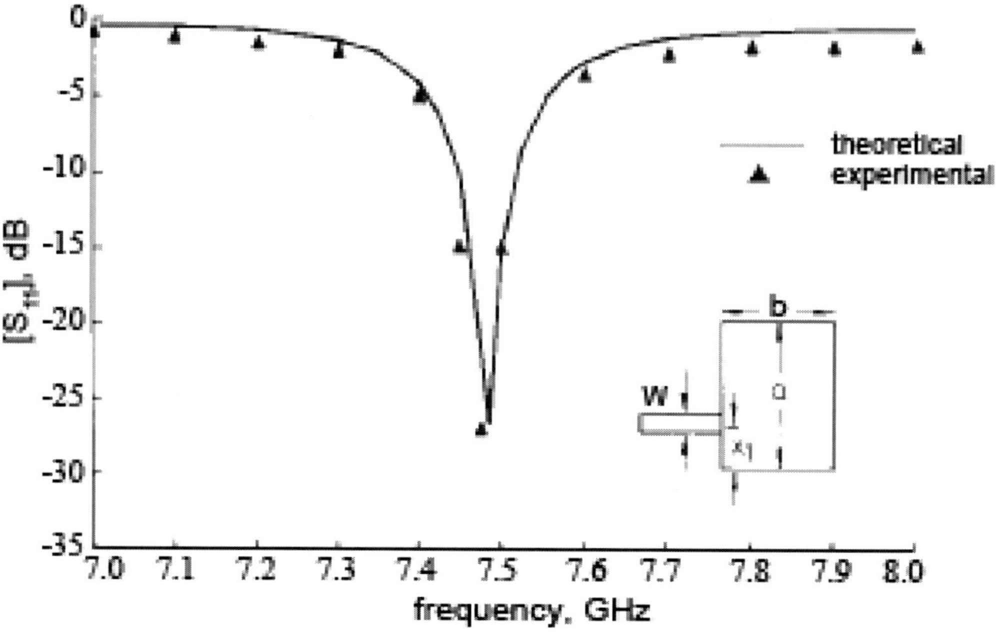

Figure 5(a). Comparison of experiment and theory for $|s_{11}|$ of a one port rectangular microstrip.

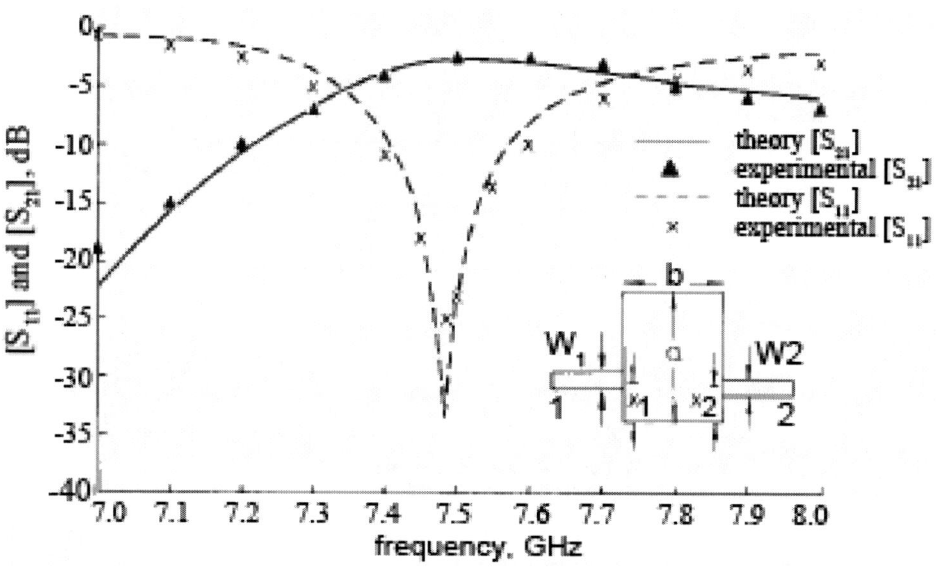

Figure 5(b). Comparison of experiment and theory for |s11| and |s12| of a two port RSMA.

Progress: 1991-2000

Experimental result obtained by R.N. Karekar, on a rectangular microstrip antenna design edon (ε_r = 9.8, h = 0.635m, tan δ = 0.0001), shows that angle of radiation pattern lobes various between ± 60^0 for superstrate relative thickness (d/λ_m) = 0.58 [Table 1, 8]. Available result on the covered patch indicates that calculation of radiation pattern has been paid less attention. In this view, Ghulam Qasim has made an attempt to calculate the radiation pattern of covered rectangular MSA using Spectral domain imittance (SDI) matrix approach after calculating the current distribution on the patch [9]. And found that for d = 0.8 mm, E-plane pattern changes significantly at $\theta = \pm 90^0$ whereas for d = 3.18 mm, a small dip occurs and the field is maximum at $\theta = \pm 10^0$ (Figure 5, [9]).The effect of superstrate loading on the performance characteristics of slot-coupled microstrip antennas have been studied in [10], and found that dielectric loading decreases the resonant frequency and input impedance, however increases bandwidth. The radiation efficiency of antenna can be made optimum, provided the superstrate is made thicker than substrate of antenna. (Figures 3, 4 & 5) [10]. A modified Wolff model (MWM) was proposed in year 1993 [11], to calculate the resonant frequency of superstrate loaded patch antenna as well as multilayer resonating structures and results obtained are found to be very close to experimental results than that of full wave analysis and almost in all cases, they are within 0.5% of published results for the fundamental mode and within 1.7% for other higher modes. It is basically a cavity model, where the dynamic permittivity is calculated using variational method. The MWM has also been applied on triangular and hexagonal patches, and found very useful tool for microwave Computer Aided Design (CAD) on patch antenna. The resonant frequency is calculated by

$$f_f = f_{mn} = \frac{V_0}{2\sqrt{\varepsilon_{dyn}}}\left[\frac{m}{W_{eff} + L_{eff}}\right] \qquad (5)$$

Where all the parameters are calculated in [11]. However for rectangular microstrip antenna, complex resonant frequency varies more significantly, when the superstrate permittivity is greater than that of substrate [12]. The variation of normalised resonant frequency with superstrate thickness is shown in Figure 6.

P. Lowes et al. [13] described the effects of laminated glass superstrate along with a layer of resistive film on the outer layer of superstrate, on the radiation pattern as well as bandwidth of a microstrip patch antenna. The study indicates that antenna offers improvement in the impedance bandwidth of the antenna. ≈ 20%, as compared to *3%* for a VSWR of 2, for an antenna with no cover. It is also found that, H-plane radiation pattern have nulls appearing at angles of ± 40°, while E-plane radiation pattern is unaffected with superstrate loading, The Effects of Superstrate Loading on the Circular polarization (CP) and Cross polarization (XP) Characteristics of a Rectangular Microstrip Patch Antenna has also been studied by W.S. Chen et al in year 1994 [14]. The effect of superstrate permittivity and thickness as well as substrate superstrate gap on Co-polarization and Cross polarization of the patch antenna have also been analysed. The optimal aspect ratio and centre operation frequency are then adjusted for producing CP radiation at the main-beam direction, In addition, the following condition must be satisfied at $\theta = 0^o$ i.e.

$$\angle E_\theta - \angle E_\emptyset = \mp 90^o$$

$$|E_\theta| = |E_\emptyset|$$

The obtained results indicate that, when the superstrate layer is spaced away from the patch, the loading effects on the antenna are considerably reduced. However the centre frequency for the CP radiation is slightly varied, on the other hand the 3dB bandwidth and optimal feed position are almost not affected. From the loading for higher air gap thicknesses, the main-beam polarization property is also better. However, ***XP*** characteristics under superstrate loading, indicate that, for the patch antenna excited at TM_{01} mode, the ratio of co polarized to cross polarized fields at $\theta = 0^o$ decreases with increasing superstrate permittivity and shows a minimum value when the superstrate thickness is a quarter wavelength or its odd multiples (Figure 7).

From Figure 7, it is clear that ratio $|E_{01}|$ to $|\Sigma E_{m0}|$ at $\theta = 0^0$ is about 20.7 dB for $t = 0$. And ratio of $|E_{01}|_{max}$ to $|\Sigma E_{m0}|_{max}$ drastically decreases from 20.7dB (no loading) to about 13.9 dB, 9.7 dB, and 8 dB, respectively, for loading the patch antenna with a $(\lambda/4)$ superstrate layer, for $\varepsilon_r = 2.32$, 4.0 and 5.6 respectively. Therefore, effect of superstrate loading on the ratio of $|E_{01}|_{max}$ to $|\Sigma E_{m0}|_{max}$ could be minimized by choosing its thickness not equal to $(\lambda/4)$ or its odd multiple.

In order to improve the gain of printed antennas L. Zhang et al. [15] have used Photonic Band Gap (PBG) like structure in cavity backed microstrip antennas. In general gain enhancement is noticed, due to strong excitation of leaky wave's modes in the PBG layer. To obtain PBG effects, the unit cell size should be comparable to the operating wavelength. In their study, they considered two types of PBG-Like substrate of (4x4) unit cells, and PBG substrate (5x5) unit cells. And they found that plane cavity backed antenna shows 5dB of gain at two resonances, however maximum gains of 7.5 dB and 9.0 dB were found for respective resonance for PBG substrate. Effects of PBG superstrate on patch antenna reveals that maximum gain over 10.0 dB is seen for both E and H Plane radiation patterns.

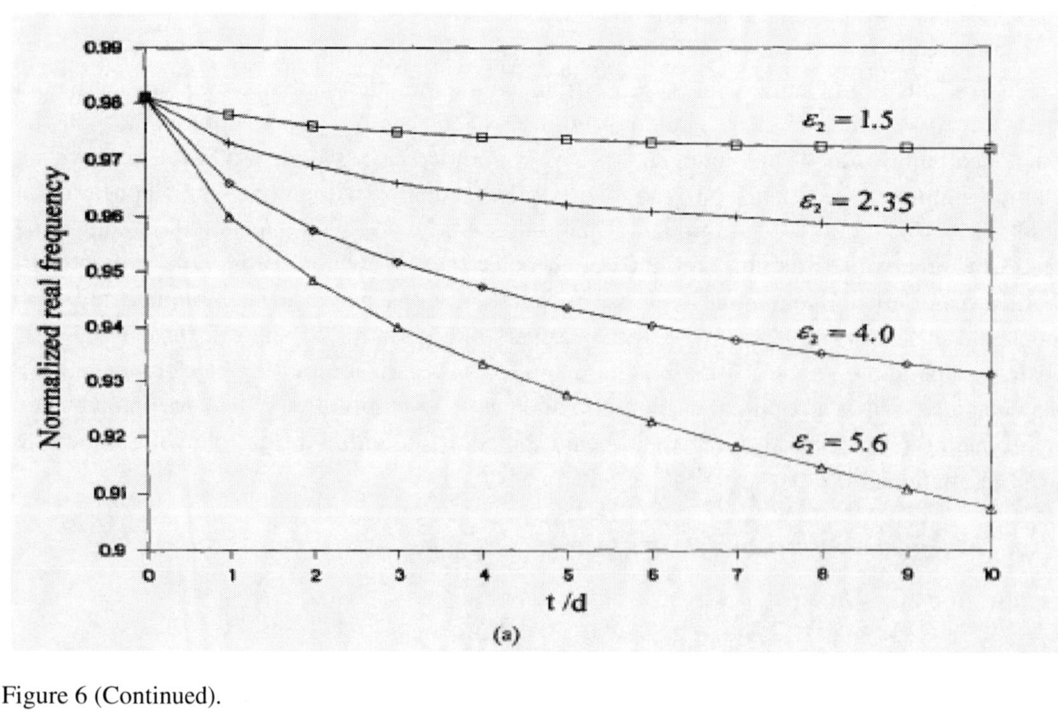

Figure 6 (Continued).

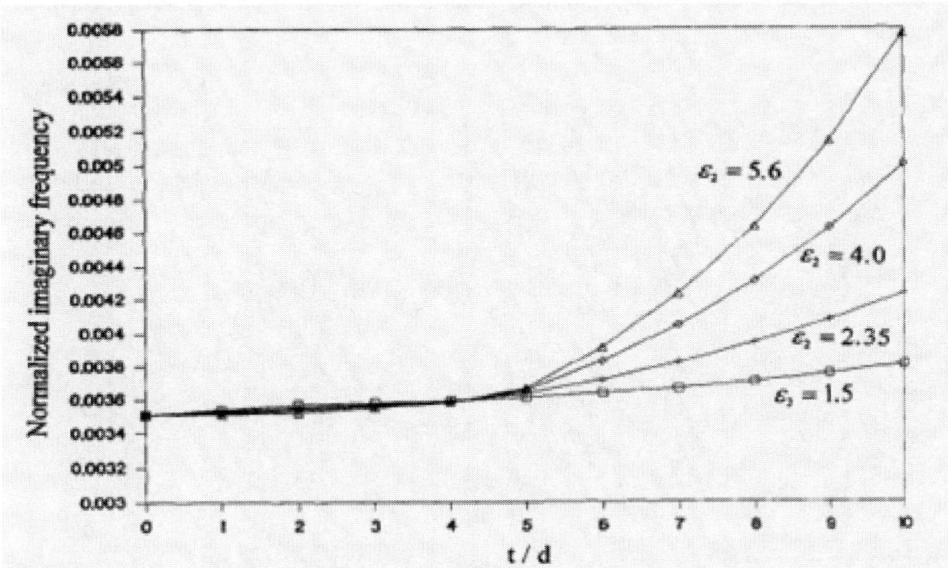

(b)

Figure 6. Normalized complex resonant frequency of a superstrate-loaded rectangular patch versus the superstrate thickness; $\varepsilon_1 = 2.35$, $\varepsilon_2 = 1.5, 2.35, 4$, and 5.6; a = 6 cm, b = 5 cm, d = 0.1 cm. (a) Real part of resonant frequency. (b) Imaginary part of resonant frequency.

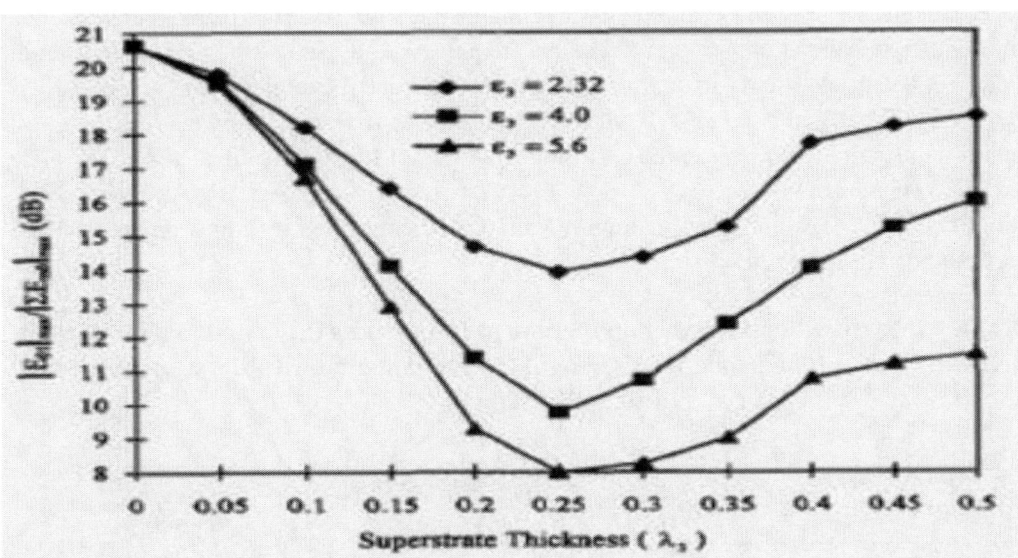

Figure 7. Variations of the ratio of co- polarized to cross polarized fields at $\theta = 0^{\circ}$ with superstrate thickness. Three cases of $\varepsilon_r = 2.32$, 4.0 and 5.6 are shown for s = 0.

J. T. Bernhard et al. developed a model that predict the resonant frequencies of rectangular microstrip antennas in terms of antenna dimensions, material parameter and antenna resonance with both flush and spaced superstrates. The desired parameter are calculated using expression (9-13) given in [16].They concluded that model is suitable for CAD and is directly applicable for the integration of microstrip antennas with protective layers in wireless equipment. The obtained results are in good agreement (less than 1% errors) for $\varepsilon_r < 3$, and slightly higher for $\varepsilon_r > 3$.).

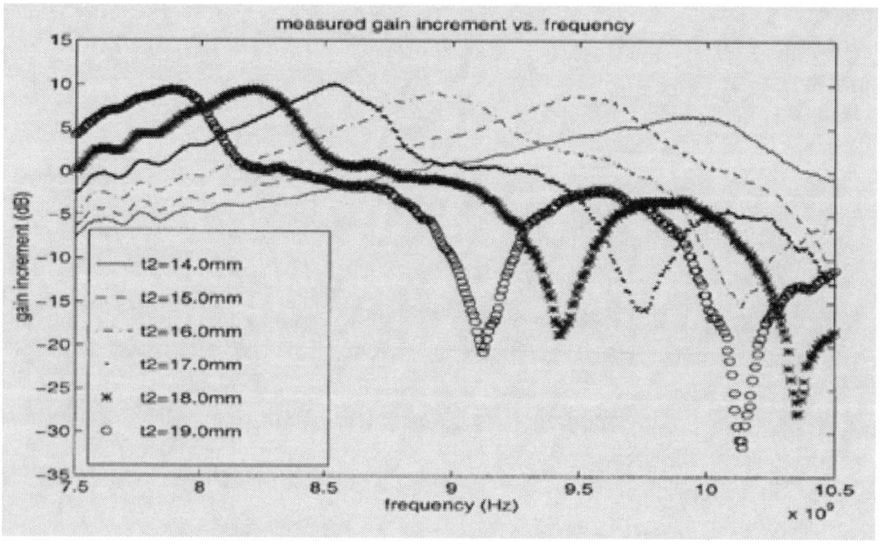

Figure 8. Measured gain increments of the probe-fed rectangular patch antenna with the superstrate versus frequency for different distances t_2.

X. H. Shen et al. have made an attempt to enhance the gain of a microstrip patch antenna with one substrate-superstrate structure between frequency 7.5 GHz to 10.5 GHz for different distances between patch and superstrates. It was found that the highest maximum gain is found as 16.3770 dB at f = 8.9170 GHz with distance 16.0 mm. However the highest maximum gain increment is observed at distance 17.0 mm of 10.060 dB at f = 8.335 GHz with ε_r =10.5 (Figure 8).

It has been also concluded that maximum gain can be controlled by adjusting the distance between patch and superstrate [17].

Table 4. Measured maximum gain increment frequencies $f_{max\ gi}$ and maximum gain increments Gi_{max} of the probe-fed rectangular patch antenna with the superstrate at different distances t_2

S. No	t_2 *(mm)*	$f_{max\ gi}$ *(GHz)*	Gi_{max} *(dB)*
1	14.0	9.9220	6.553
2	15.0	9.5170	8.832
3	16.0	8.9170	9.137
4	17.0	8.5350	10.060
5	18.0	8.2270	9.577
6	19.0	7.8820	9.486

PROGRESS: 2000-2010

Raj Kumar and P. Malathi [18] proposed a modified formula to calculate the resonant frequency of multilayered superstrate loaded rectangular microstrip antenna, and it is given by

$$f_r = \frac{c}{2(L+2\Delta L)(\varepsilon_{eff})^{\frac{1}{2}}} \tag{6}$$

Where

$$\varepsilon_e(f) = \frac{\varepsilon_r' - \varepsilon_{ef}(o)}{1 + \mathrm{P(f)}}$$

$\varepsilon_{ef}(0)$ = effective dielectric constant.

ε_r' = equivalent relative permittivity of single layer patch antenna.

P(f) = The frequency dependent factor .

It is found that, as the superstrate thickness increases from 0.5 mm to 12.72 mm, the resonance frequency decreases from 2.759 GHz to 2.670 GHz, and maximum change in f_r is around 4% when external loading thickness changes from 1 to 5 mm. The resonant frequency is shifted to the higher frequency side as the air gap increases. The % variation in resonant frequency is around 21.8% as h_1/h_2 ratio of air-gap to substrate next to the ground plane is varying from 0.065 to 1.94. Effects of dielectric cover on a rectangular MSA operating in TM_{10} modes have also been analyzed using Genetic Algorithm. The Genetic Algorithm program for this purpose is developed using C^{++} language. Two addition parameters: ε_r and h of the superstrate is also taken as input in the optimization. Obtained results reveal that there is usual variation of properties of antenna; resonance frequency and gain, with ε_r and h of the superstrate, as shown in Figures 9 & 10 [19].

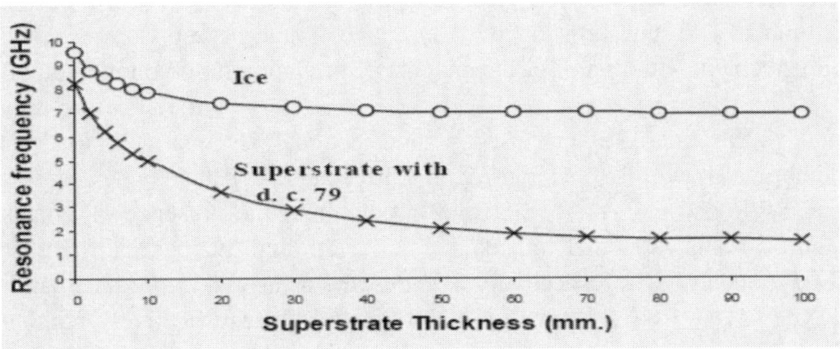

Figure 9. Variation of resonant frequency of microstrip antenna with superstrate thickness.

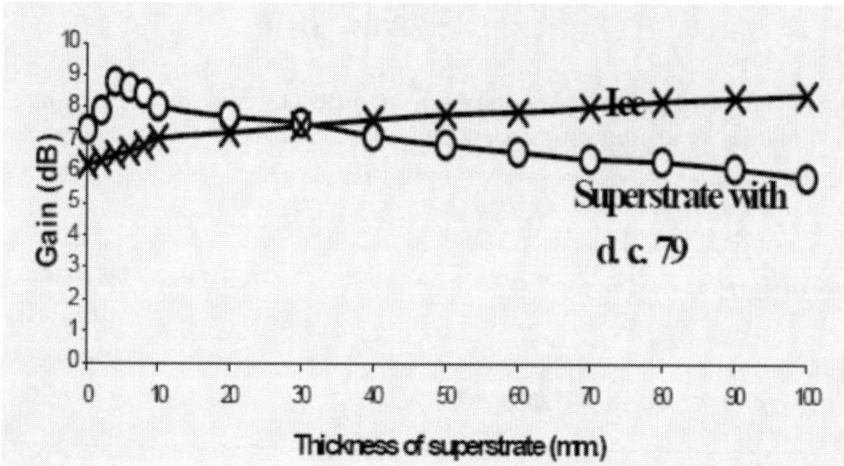

Figure 10. Variation of gain of microstrip antenna with superstrate thickness.

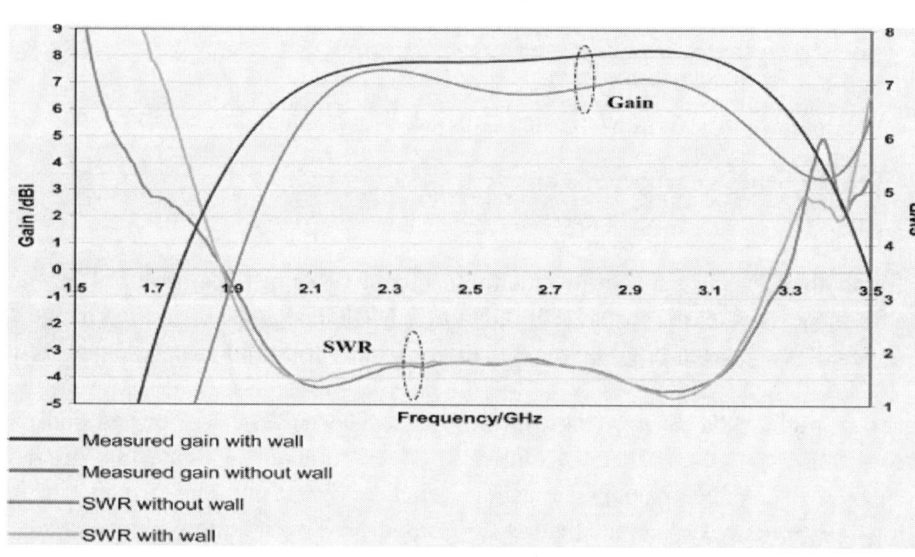

Figure 11. Measured gain and SWR of the proposed antenna with and without wall.

The shift in resonant frequency of a rectangular microstrip antenna covered with a dielectric layer has also been analysed by S.Chakraborty et al., using *a* neural network model. The proposed model is faster and accurate than conventional methods, and results are found to be in good agreement with that of electromagnetic simulation based on method of moments [20].

L.Siu and K. M. Luk [21] proposed a unidirectional antenna with loaded dielectric substrate and composed with two shorted-circuited rectangular patches operating at 2.5 GHz. The measured SWR and gain of the antenna with/without the presence of a metallic wall, shown in Figure 11, indicates that impedance bandwidth and average gain are 46.3% from 1.98 to 3.17 GHz and 7.9 dBi, respectively. Besides, the measured impedance bandwidth and average gain are 45.3% from 2.00 to 3.17 GHz and 6.7 dBi, respectively without metallic cover. However, in the presence of a metallic wall, the impedance bandwidth of the proposed is 46.3% for with an average gain of about 7.9 dBi. The advantages of the proposed antenna include: wide bandwidth, low profile, low cross-polarization level, and stable gain over the operating band (Figure 11).

H. A. Hammas [22], analyzed the effects of superstrate loading on an aperture Coupled antenna, and calculated antenna parameters such as, Radiation efficiency, Half-Power Beamwidth (HPBW) and directivity using MOM. And found that variation of resonance frequency with superstrate thickness is as usual, however superstrate thickness increases, gain of antennas decrease (Table 5).

Table 5. Variation of antenna parameters with superstrate layer thickness at resonance

	Antenna Parameters			
Superstrate Thickness(mm)	*Antenna Efficiency (%)*	*Radiation Efficiency (%)*	*HPBW (deg)*	*Gain (dBi)*
0	92.34	92.41	109.94	6.9
5	72.33	78.46	102.45	7.25
10	66.08	71.56	91.11	7.64
15	60.13	69.86	90.19	7.877

CONCLUSION

From above study it may be concluded that the superstrate loading alters the resonance frequency of rectangular patch antennas, and degree of effects vary with thickness of dielectric, types of dielectrics, shifts in feed position, impedance matching and polarizations. In general all parameters change usually due to loading of dielectric covers, however -10 dB return loss bandwidth does not change significantly with the superstrate thickness. In particular, laminated glass superstrate loaded antenna offers improvement in the impedance bandwidth ≈ *20'%,* as compared to *3% for* a VSWR of 2, for an antenna with no cover. However a unidirectional antenna with loaded dielectric substrate and composed with two shorted-circuited rectangular patches indicates impedance bandwidth and average gain of 46.3% from 1.98 to 3.17 GHz and 7.9 dBi, respectively. Therefore, in future research works all above studies are needed to be considered, particularly for modern communication technology applications, with emphasis on compensation of above mentioned effects.

REFERENCES

[1] J. Bahl and S. S. Stuchly. Analysis of a microstrip covered with a lossy dielectric, *IEEE Trans. Microwave Theory Tech.* Vol. MTT-28. No. 2, pp. 104-109, Feb. 1980.

[2] J. Bahl, P. Bhartia and S.S. Stuchly. Design of a microstrip antenna covered with a dielectric layer, *IEEE Transactions on Antennas and Propagation*, Vol. AP-30, No. 2, pp. 314-18. March 1982.

[3] N. G. Alexopoulos and D. R. Jackson, Fundamental superstrate (cover) effect on printed circuit antennas, *IEEE Transactions on Antennas and Propagation*, Vol. AP-32, pp. 807-816, 1984.

[4] H. Y. Yang and N. G. Alexopoulos. Gain Enhancement methods for printed circuit antennas through multiple superstrate, *IEEE Transactions on Antennas and Propagation,* Vol. AP-35, No. 7, pp. 860-863, July 1987.

[5] A Bhattacharyya and T. Tralman, Effects of dielectric superstrate on patch antennas, Electron. Letters., Vol. 24, No. 6, pp. 356-358, Mar. 17, 1988.

[6] R. Q. Lee and A. J. Zaman, Effects of dielectric superstrates on a two-layer electromagnetically coupled patch antenna, Ch @ IEEE, pp. 620-623, 1989.

[7] Benalla and K. C. Gupta, Multiport network model for rectangular microstrip patches covered with a dielectric layer, *Proc. Inst. Elect. Eng.*, Vol. 137, Pt. No. 6, pp. 377-383, Dec. 1990.

[8] R. Afzalzedeh and R.N, Karekar, Effect of dielectric protecting superstrate on radiation pattern of microstrip patch antenna, *Electron. Letters,* Vol. 27, No. 13, pp. 1218-1219, 1991.

[9] Ghulam Qasim and Shunshi Zhong, Radiation characteristics of microstrip patch antenna with dielectric covers, IEEE @1992, pp. 2208-2211.

[10] K. L. Wong, S. F. Sun-Fu Hsiao and H. T. Hong -Twu Chen, Resonance and radiation of a superstrate-loaded spherical-circular microstrip patch antenna, *IEEE Trans. Antennas Propagation.*, Vol. 41, No. 5, pp. 686-690, May 1993.

[11] A K. Verma and Z. Rostamy, Resonant frequency of uncovered and covered rectangular microstrip patch using Modified Wolff Model, IEEE Trans. Microwave Theory Tech., Vol. 41, No. 1, pp. 109-116, Jan. 1993.

[12] J. S. Row and K. L. Wong, Resonance in a superstrate-loaded rectangular microstrip structure, *IEEE Trans. Microwave Theory Tech.*, Vol. MTT-41, No. 5, pp. 1349-1355, Aug. 1993.

[13] P Lowes et al., Performance of microstrip patch antenna with electrically thick laminated glass superstrate, *Electronics Letters*, Vol. 30, No. 23, pp. 1903-1905, Nov. 1994.

[14] W. S. Chen, K. L. Wong, and J. S. Row, "Superstrate loading effects on the circular polarization and cross-polarization characteristics of a rectangular microstrip patch antenna," *IEEE Trans. Antennas Propagation.*, Vol. 42, No. 2, pp. 260-264, Feb. 1994.

[15] L. Zhang, H. Contopanagos, N. G. Alexopoulos, and E. Yablonovitch, Cavity backed antennas with PBG-like substrate or superstrate materials, *IEEE Antennas Propagation Soc Int Symp,* pp. 186-189, 1998.

[16] J. T. Bernhard and C. J. Tousignant, Resonance frequencies of rectangular microstrip antennas with flush and spaced dielectric superstrates, *IEEE Trans. Antennas and Propagate.,* Vol. 47, No. 2, pp. 302-308, Feb. 1999.

[17] X. H. Shen et al., Enhancing the gain of microstrip antenna at different frequencies with one substrate-superstrate structure, *Microwaves and Optical Technology Letters*, Vol. 27, No. 1, pp. 37-40, Oct. 2000.

[18] Raj Kumar and P. Malathi, Effect of superstrates on the resonant frequency of rectangular microstrip antenna, *Microwaves and Optical Technology Letters*, Vol. 49, No. 12, pp. 2946-2950, Dec. 2007.

[19] Neela Chattoraj and J.S. Roy, Applications of Genetic Algorithm to the optimization of microstrip antennas covered with a dielectric layer, *Institution of Engineers (India), Journal-ET*, Vol. 88, pp. 01-06, 2008.

[20] S. Chakraborty et al, Neural Network Modelling for the fast estimation of superstrate Loading effect on rectangular microstrip patch antennas, *Jadavpur University, Kolkatta, IEEE @2007.*

[21] L. Siu and K. M. Luk, Unidirectional antenna with loaded dielectric substrate, *IEEE, Antennas and Wireless Propagation Letters,* Vol. 7, pp. 50-53, 2008.

[22] H.A. Hammas, Radiation performance evaluation of microstrip antenna covered with dielectric layer, *Eng and Tech. Journal,* Vol. 27, No. 7, pp. 1280-1286, 2009.

In: Knowledge Management and Communication … ISBN: 978-1-62808-250-0
Editor: Rafiqul Islam

Chapter 9

IMPLEMENTATION OF DIGITAL MODULATION IN EXISTING ANALOG POWER LINE CARRIER

Bander Mahfouz Saman[1*], Mosleh M. Al-Harthi[1†], Nadjim Merabtine[1‡] and A. Said[2§]

[1]Electrical Engineering Department, Faculty of Engineering, Taif University, Taif, Saudi Arabia

[2]UAE University, United Arab Emirates

ABSTRACT

One of the alternative technologies for communications that is widely used across the globe is Power Line Communications or Carrier "PLC" technology. Power Line carrier is a term used for the power cables as communication media. The broadband PLC is used for home automation and broadband data transfer i.e. internet in a low voltage distribution network in home electrical wring or between home and transformer station. The narrowband PLC run on overhead lines network between power plant and transformer station. This survey has been concentrated on over head network between substations.

The advantage of using power line carrier (PLC) for data transfer is that it does not require the installation of new cables. But the main disadvantage of PLC is a low transmission capacity.

In this work, the complex components of existing power-line carrier are explained and modeled with respect to frequency band up to 500 kHz. Also the exact teleprotection system (TPS) unit is designed as dynamic systems model, to enable the connection of four protection command on one unit.

In this work, the data rate is also developed for the data transfer from 1200 bps using BFSK to 2400 bps using QDPSK.

In order to achieve our research goals a number of PLC and TPS systems studies were performed. Additionally, a set of tests were also performed to investigate the

[*] E-mail address: ban

[†] E-mail address: mosleh2k @hotmail.com

[‡] E-mail address: na_merabtine@hotmail.com

[§] E-mail address: Abdellatifsd@yahoo.com

consistency of the PLC and TPS models. The results showed that the models responded satisfactorily according to the expected results of the tests.

1. Introduction

Operators of long transmission lines require reliable communications for control of the power grid and, often, associated generation and distribution facilities. Fault-sensing protection relays at each end of the line must communicate to monitor the flow of power into and out of the protected line section. Protection of the transmission line from short circuits and other faults is usually so critical that common carrier telecommunications is insufficiently reliable. In remote areas a common carrier may not be available at all. Communication systems associated with a transmission project may use:

- Microwaves.
- Control Power line carrier (PLC).
- Optical fibers.
- Pilot cable.

The main disadvantage of control PLC is a low transmission capacity. One single channel reserves 4 kHz band. This channel consist (1.7-2.1) kHz band for voice and 600 baud per second (bps) for data via FSK modem (1.2 KHz band) and pilot and teleprotection signal, also the channel may be occupied by other services like embedded communication channel (ECC).

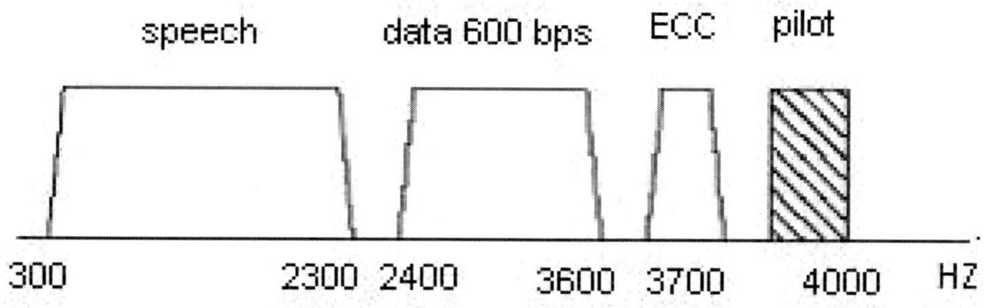

Figure 1.1. Service Structure.

The PLC bandwidth problem comes up when we are trying to put 1200 bps with voice or 2400 bps in one single channel with sufficient reliability. The available Solutions for utilizing previous applications are:

1. Double PLC channels.
2. Add new PLC system.
3. Replacing the analog PLC with digital PLC.

All above solutions are very expensive and bandwidth inefficiently. The Applying of PSK modulation in PLC will be a cheap solution because the FSK modulation resaved more bandwidth comparing with PSK.

Our target is replacing the FSK modem with PSK modem in the existing PLC. We need to reconfigure the AF bandwidth and connecting the new modem.

This paper is divided into four sections. Section one: consists of the paper introduction .Section two: explains the system models. Section three: shows the whole schematic of this research that helps us to find the effecting and improving fields. Section four: presents the main conclusions and recommendations.

2. System Model

With MATLAB SIMULINK version 6.0 (R14) dated 05-May-2004, we could be able to design BFSK modem and QDPSK modem for general and PLC use.

2.1. BFSK Modem Equipment

The FSK modem is divided in two parts, Transmitter part (FSK TX) and receiver part (FSK RX), in the real system the two parts compact in on one card or device.

2.1.1. BFSK-TX

We used BFSK, as shown below on Figure 2.1, for modulating the incoming data. The BFSK TX contains the hold circuit with hold time equal to reciprocal of incoming data rate bps. The incoming signal is uni polar binary signal. The FSK signal is sum of multiplication, the input data with FI carrier and multiplication the inverter of data with F0 carrier. The last stage is including power amplifier.

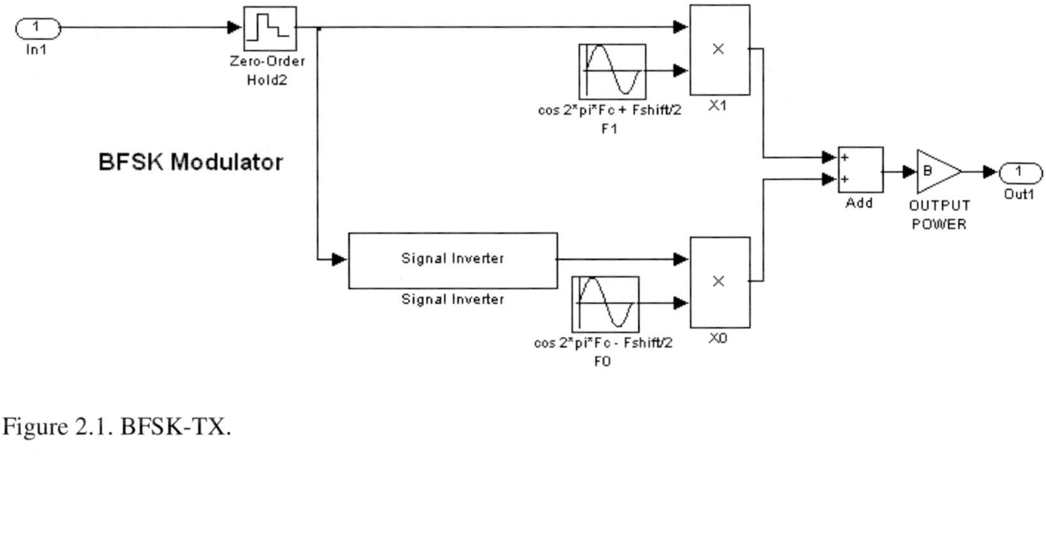

Figure 2.1. BFSK-TX.

2.1.2. BFSK-RX

We used non-coherent BFSK as show in Figure 2.2 for demodulating the incoming AF signal.

The FSK RX contains two stages, in the first stage the level of frequencies is checked by applying band pass filter (BPF F1, F0), then calculate the voltage peak for each received frequencies. In second stage the highest level is selected by subtracting the level of frequencies. The receiving bit is recognized by switch relay with 0 threshold. The holder returns the data as uni polar binary with same sending data rate.

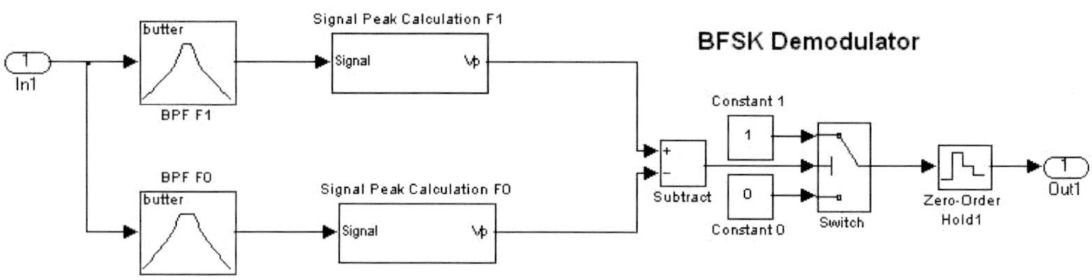

Figure 2.2. BFSK-RX.

2.2. QDPSK Modem Equipment

The QDPSK modem is divided in two parts. The Transmitter part (QDPSK TX) and the receiver part (QDPSK RX), in the real system the two parts will be compact in on one card or device.

2.2.1. QDPSK-TX

The transmitter QDPSK is shown on figure **2.3**.The modulator of QPSK is based on I & Q channels. The channel with cosine reference is called in phase (I) channel and the channel with sine reference is called quadrature (Q) channel. The data sequence is separated by the serial-to-parallel converter (S/P) to the odd numbered bit sequence for Q-channel and the even numbered bit sequence for I-channel. It is clear that both the I-channel and Q-channel signals are BPSK signals with a bit rate equal to half of incoming massage rate. The information on I & Q channels are represented by the phase differences encoder. An example is established for this choice.

$$U_K = (\overline{I_K \text{ XOR } Q_K})X(I_K \text{ XOR } U_{K-1}) + (I_K \text{ XOR } Q_K)X(Q_K \text{ XOR } V_{K-1}), \qquad (3.1)$$

$$V_K = (\overline{I_K \text{ XOR } Q_K})X(Q_K \text{ XOR } V_{K-1}) + (I_K \text{ XOR } Q_K)X(I_K \text{ XOR } U_{K-1}). \qquad (3.2)$$

Where XOR denotes exclusive OR operation, U_k and V_K are respectively coded I-channel and Q-channel bits. Then logic 1 is converted to a positive pulse and logic 0 is converted to a negative pulse by using I &Q switch relays .Pairs (U, V) are used to control the absolute phase of the carrier. U_k pulse train is multiplied to cos $2\pi fct$ and V_k pulse train is multiplied to

CONCLUSION

4.1. Summary of Key Findings

This research is based on modifying existing analog PLC. The main advantage of this study is the use of the existing equipments, which simplifies the implementation. We have been successful in implementing a working PLC model. During the design of this system we were faced with a number of difficulties due to the combination of analog and digital modulations for PLC and programming interface. Due to this reason we had to redesign and test a number of blocks. However at the end of rigorous testing we were able to come up with a working PLC model.

We have succeeded in establishing the fowling applications:

- Voice service and 600 bps data service.
- Data services for 600 bps and 1200 bps without voice service.
- Tele protection single and duel tone with voice and data services.

The required tests were performed. By changing the modulation from FSK to QDPSK the model is modified to transport the all above applications and additional 1200 bps with voice service in one single channel. This paper also provides guidance to establish a 2400 bps without voice. Our present experiments show that PLC systems are not designed for broadband applications.

4.2. Suggestions for Further Research

It should be noted that even at this stage the model still only represents a system in single line diagram. Realistic modeling of power cable and channel noise should be considered in the future to improve the models and to make it more feasible for real life situations.

Also we propose QAM and high MDPSK *modulation* to achieve higher data rates, with error correction code

REFERENCES

[1] "SWT 2000-i Siemens Power Line Carrier" By Siemens /Edition 2002- Equipment Manual.

[2] "Power Automation & Communication Systems" By ABB /Edition 2000- Equipment Manual.

[3] "Digital Communication Fundamentals And Applications" By Bernard Sklar /Edition 2001-Book.

[4] "ETL500 Instruction Manual" By ABB /Edition 2003- Equipment Manual.

[5] "Principles Of Communication Engineering" By Anokh Singh And A.K. Chhabra /Edition 2004-Book.

sin 2πfct. Finally a summer adds these two waveforms together with the output power amplifier.

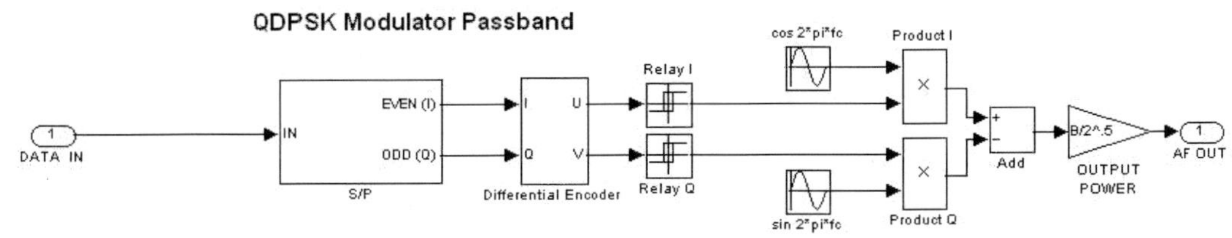

Figure 2.3. QDPSK-TX.

2.2.2. QDPSK RX

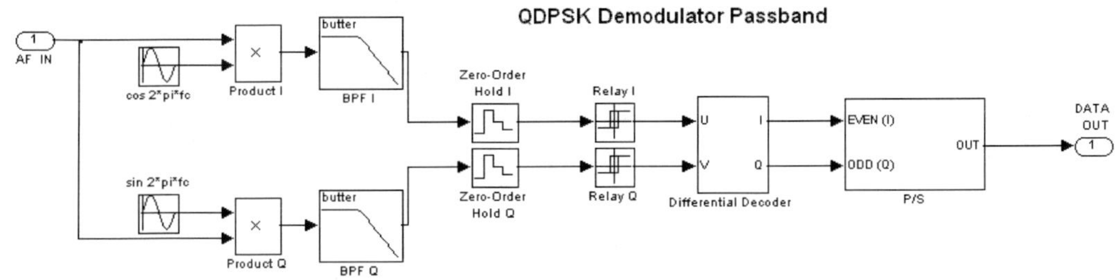

Figure 2.4. QDPSK-RX.

As shown on the above figure, the AF QDPSK signal is applied to two sinusoidal modulators that are 90° apart. The outputs from the multipliers are then applied to two low pass filters that both perform filtering high frequency parts. After that the signal is applied to hold circuit. The circuits are passed the cos and sin values of the differential phase. The relays are converted as such: a positive value to logic 1 and negative value to logic 0. Pairs (U, V) are demodulated form the relay. The differential decoder must be inserted after pairs (U, V) demodulation. The differential decoding is carried out on digital signal levels. The decoding rules are:

$$I_K = \overline{(U_K \text{ XOR } V_K)} \text{ X } (U_K \text{ XOR } U_{K-1}) + (U_K \text{ XOR } V_K) \text{ X } (V_K \text{ XOR } V_{K-1}), \quad (3.3)$$

$$Q_K = \overline{(U_K \text{ XOR } V_K)} \text{ X } (V_K \text{ XOR } V_{K-1}) + (U_K \text{ XOR } V_K) \text{ X } (U_K \text{ XOR } U_{K-1}). \quad (3.4)$$

I- and Q-channel signals are demodulated separately as two individual BPSK signals. A parallel-to-serial converter (P/S) is used to combine two sequences into a single sequence that determines the bit which will be received.

3. Model Testing

This part shows the whole schematic of this research that helps us to find the effecting and improving fields.

3.1. Existing PLC Model

The data channel configured in existing PLC for different baud depends on the application and bandwidth availability, so the module support all available applications and is designed to work with different signal to noise ratio (S/N) depend on channel parameters. In this study we configured the Matlab model for most applications use as, Voice(300-2300 Hz) with 600 bps data(2400-3600 Hz) in one channel then dynamics simulations are performed to study the BER .With following considerations 23 db line loss,50 kHz transmit and receive carrier frequency,80 watt transmitter level,75 output impedance, guard band frequency from (3.7-4) kHz.

The result are shown on Figure 3.1, the best BER is occurred when we are sending 600 bps data alone, because the voice signal is add noise at data port signal.

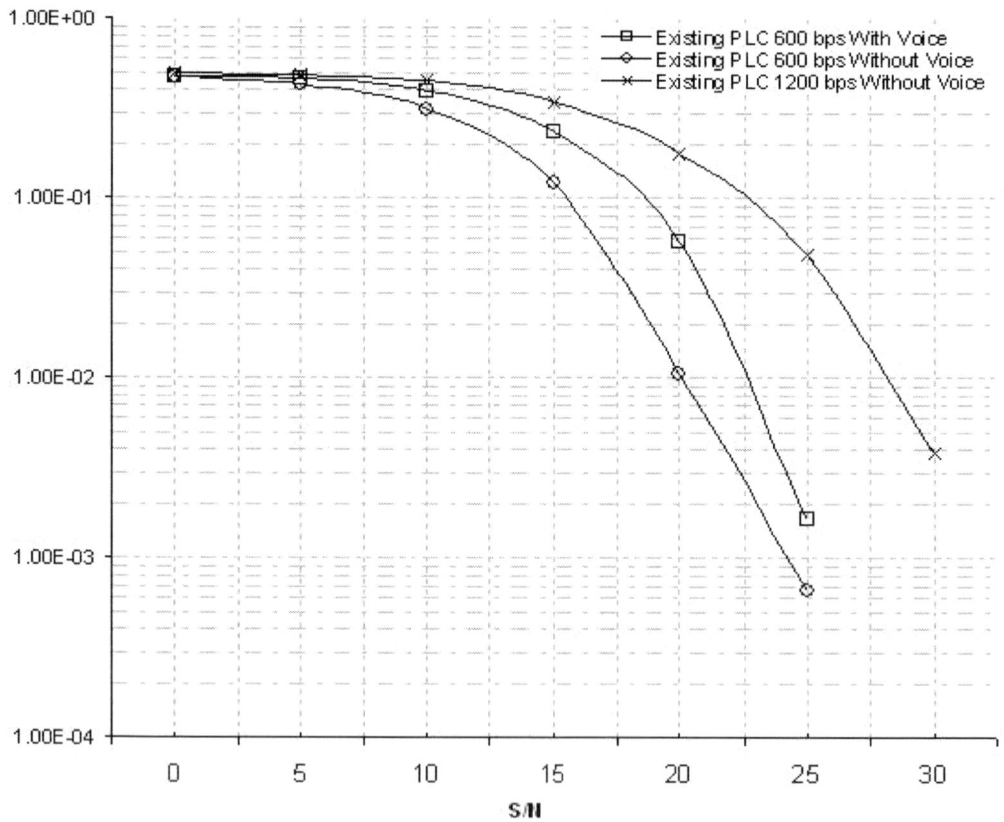

Figure 3.1. BER for Existing PLC.

3.2. Modify PLC Model

Base on the achievement required, the new model is configured and tested for following cases:

- Voice (300-2300 Hz) with 600 bps data in one channel.
- Voice (300-2300 Hz) with 1200 bps data in one channel.
- 2400 bps data in one channel.

The effect of data rate and service structure are performed, the result is shown on Figure 3.2.

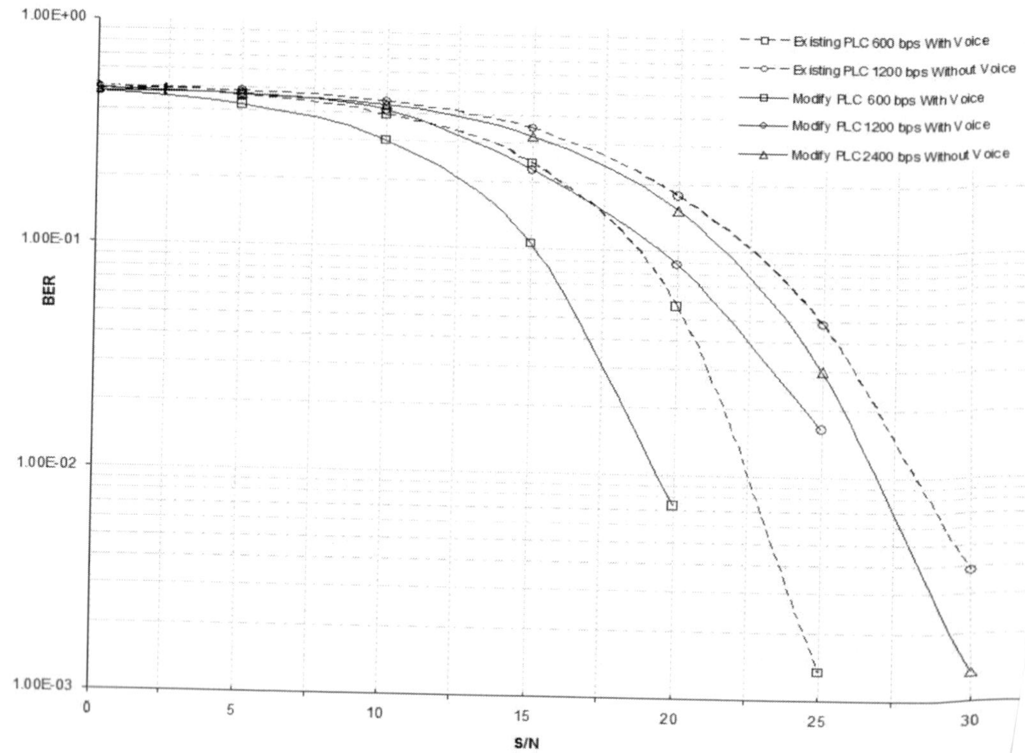

Figure 3.2. BER for Existing & Modify PLC.

In the Figure 3.2 both existing and modify PLC modes are mentioned. The best occurs when we send 600 bps data through QDPSK modem.

Bit error rate in PLC system is following the theoretical phenomenon for qua differential phase shift keying (QDPSK) and non-coherent binary frequency shift (NBFSK). However, the BER of QDPSK is better than NBFSK for the same power a rate. Finally the test has proved that QDPSK modulation can be use in PLC system.

[6] “Alstom Alspa 80-35/80-75 Power Line Carrier” By Alstom /Edition 2001- Equipment Manual.

[7] “Digital Signal Processing” By John G Proakis And Dimitris G Manolakis Edition 2000-Book.

[8] “Microelectronics” By Jacob Millman/Second Edition-Book.

[9] Web(Http://WWW).

[10] “Matlab7 Help” Release 14.

[11] Web(Http://WWW).

[12] “Introduction To Simulink With Engineering Applications” By Steven T. Karris.

In: Knowledge Management and Communication ... ISBN: 978-1-62808-250-0
Editor: Rafiqul Islam

Chapter 10

A 2.45 GHz Triangular Fractal Antenna for RFID and Wireless Applications

***R. L. Yadava*[1*], *Mukh Ram*[1] *and Ravindra Kumar Yadav*[2]**
[1]Department of Electronics & Communication Engineering
Galgotias College of Engineering and Technology
Uttar Pradesh, India
[2]Department of Electronics & Communication Engineering
ITS Engineering College
Uttar Pradesh, India

Abstract

A triangular fractal antenna with good radiation performances for multiband applications has been proposed in this paper, and found that increase in number iterations led to improvement is VSWR & return loss. It has also been observed that with increase in the number of iterations the bandwidth of the antenna increases and on second and third iterations, the antenna starts showing the multiband characteristics. The operating frequencies are found to be 8.9 and 9.6 GHz for II-iteration and 16.7, 18.3 19.6 & 22.2 GHz for III-iteration with acceptable bandwidth and S_{11}< -10 dB (VSWR < 2). The radiation characteristics, VSWR, directivity, gain and S_{11} of the proposed antenna are described and simulated using Ansoft HFSS and the obtained results have also been validated using HP810B network analyzer and found in good agreement.

1. Introduction

In high-performance aircraft, spacecraft, satellite, and missile applications, where size, weight, cost, performance, ease of installation, and aerodynamic profile are constraints, and low-profile antennas may be required. Presently there are many other government and commercial applications, such as mobile radio and wireless communications that have similar

* E-mail address: rlyadava@yahoo.com.

specifications. All these system requires antennas with wider bandwidth and smaller dimensions than conventionally possible. Now days, the size of electronic systems has decreased drastically, whereas their functionality has increased considerably, however patch antennas have not experienced the same evolution. The antenna size with respect to the wavelength is the parameter that will have influence on the radiation characteristics. But as antenna size reduces, the bandwidth, gain and efficiency of antenna deteriorated [1]. In addition the explosive growth in the wireless industry has renewed interest in multiband antennas.

The most recent multiband antenna development is the incorporation of fractal geometry into radiators and the Koch and Sierpinski gasket antenna are the prime example. Since the Sierpinski gasket has proven itself to be an excellent multiband antenna, other multiband antennas can similarly be constructed using fractal geometry. The multi band properties of antenna are due to their self-similarity of fractal geometry [2],[3] while the space filling properties [4], [5] of antenna leads to the miniaturization of antenna. The input characteristics of one such fractal design based on the hexagon has also been evaluated and found suitable for multiband usage [6], [7]. Circular microstrip fractal antenna has also been designed for ultra wide bandwidth and size reduction using self-similarity and space filling properties. This antenna exhibits the impedance bandwidth from 0.8 GHz to 10.68 GHz which is more than the FCC bandwidth 3.1 GHz to 10.6 GHz for wireless communication system. The radiation pattern of this antenna is nearly Omni – directional and exhibits the low backscattering [8]-[10]. The microstrip triangular patch finds extensive applications in the design of many useful MIC components such as resonators, circulators and filters. The triangular patches have been studied, both theoretically and experimentally [11]-[15]. They are found to provide radiation characteristics similar to those of rectangular patches, but with a smaller size. Therefore, in present paper, an attempt is made to design the triangular fractal antenna of compact size and, good radiation and multiband characteristics.

Historical Views

Dr. Nathan Cohen assembled the first true FEA in 1988 to work the 2-meter amateur band. Later he built a 10-meter dipole and established dozens of stations in Europe with a power of 1 watt. Dr. Cohen initially reported his findings at an ARRL convention in 1994, and published the first article on fractal antennas in 1995. In 1998 Carless Puente Baliarda of Universitat Politècnica de Catalunya (UPC) wins $230,000 prize for fractal antennas (GSM + DCS single base station antenna). Since then several research works have been carried out in this field.

II. Generation Fractals on the Antenna

The construction of the proposed fractal shape is carried out by applying a finite number of times an iterative process performed on a simple starting topology. According to the properties of self similarity, the fractal dimension D_s of a set A is defined as:

D_s= log (N)/log (r). (1)

Where N is the total number of distinct copies similar to A, and A is scaled down by a ratio of 1/r.

Figure 1. Basic concept of Iterations (0-3).

Ideal fractal geometries cannot be used in antenna design because they cannot be simulated and fabricated. Therefore, in this paper, only 3rd order of proposed fractal antenna is investigated.

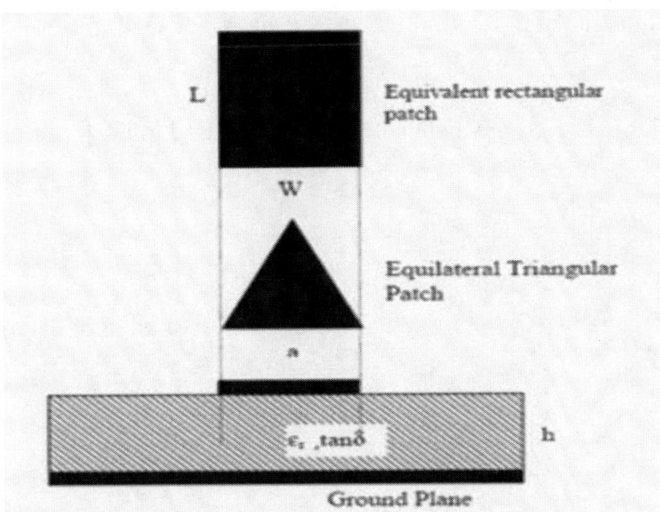

Figure 2. Equilateral triangular patch & equivalent rectangular patch.

The resonance frequency of the equilateral triangular patch can be determined by [10]

$$f_r = \frac{2\,c}{3\,L_{eff}\sqrt{\varepsilon^*_{r,dyn}}}. \quad (2)$$

Where, *c* is the velocity electromagnetic wave in free space. The *Leff* is the effective length of the equivalent rectangular patch [L = a, W = ($\sqrt{3}$/2) a] and $\varepsilon^*_{r,dyn}$ is the dynamic relative permittivity of the triangular shown Figure 2. The dynamic relative permittivity of the equivalent patch antenna is given by [14];

$$\varepsilon_{r,dyn}^{*} = \frac{C_{dyn}^{*}(\varepsilon_r^{*},h,W)}{C_{dyn}^{*}(\varepsilon_r^{*}=1,h,W)} \tag{3}$$

Where C_{dyn}^{*} is the complex dynamic capacitance of equivalent rectangular patch and;

$$C_{dyn}^{*} = \frac{\varepsilon_0 \varepsilon_r^{*} A}{h \gamma_n \gamma_m} + \frac{1}{2\gamma_n}\left[\frac{\varepsilon_{reff}^{*}(\varepsilon_r^{*}, h, W)}{v_o Z(\varepsilon_r^{*} = 1, h, W)} - \frac{\varepsilon_0 \varepsilon_r^{*} A}{h}\right]$$

$$\gamma_j = \begin{cases} 1, & j = 0 \\ 2, & j \neq 0 \end{cases} \qquad j = n, m$$

Where A is the area of equilateral triangular antenna, $Z(\varepsilon_r^{*}, h, W)$ is complex characteristic impedance and $\varepsilon_{reff}^{*}(\varepsilon_r^{*}, h, W)$ is complex relative permittivity of the eqivalent rectangular patch computed using closed form of Hammerstad [15]. But at this stage, the resonance frequency of the patch has been approximated as it does not take into account effect of the various losses occurred. If we incorporate effect of these losses on the resonance frequency by involving the effective loss tangent of the patch on single substrate. The relative permittivity of the single layer equivalent substrate is modified to [12];

$$\varepsilon_r^{*} = \varepsilon_r' - j\varepsilon_r'' \text{ where } \varepsilon_r'' = \varepsilon_r' \tan\delta_{eff}\,. \tag{4}$$

The effective loss tangent, $\tan\delta_{eff}$ of the total structure due to all kind of losses, is computed from the total Q-factor,

$$\tan\delta_{eff} = \frac{1}{Q_T} = \frac{1}{Q_d} + \frac{1}{Q_c} + \frac{1}{Q_{sw}} + \frac{1}{Q_r}$$

The Q_s,Q_c,Q_{sw} and Q_r are the Q-factors, due to dielectric loss, conductor loss (in patch, ground plane and in the co-axial feed), surface-wave loss and radiation loss respectively.

The average of effective relative permittivity of a equivalent rectangular patch on the single layer substrate $\varepsilon_{r.eff.av}(\varepsilon_r, h, W, L)$ is obtained as the average effective permittivity from both the with (W) and length (L) sides of the equivalent rectangular patch.

$$\varepsilon_{r,eff,av}(\varepsilon_r, h, W, L) = \frac{\varepsilon_{reff}(\varepsilon_r, h, L) + \varepsilon_{reff}(\varepsilon_r, h, W)}{2}, \tag{5}$$

$$\varepsilon_{reff}(\varepsilon_r, (W\ or\ L), h) = \frac{(\varepsilon_r+1)}{2} + \left\{\frac{(\varepsilon_r+1)}{2}\left(1+\frac{12h}{WorL}\right)^{-\frac{1}{2}}\right\}. \tag{6}$$

Where

ε_r = Dielectric constant of material,

h = Thickness of substrate material.

To summarize the process of accurate computation of the resonant frequency, at first we compute the approximate resonance frequency of the antenna on the loss-less isotropic substrate, i.e. $\tan\delta_{eff} = 0$, and hence $\varepsilon_r^* = \varepsilon_r$. Then we calculate correct resonance frequency by taking into account the ε_r^*. Therefore the final resonance frequency depends on various kinds of losses in the substrate and then computes all Q-factors of the structure.

III. Design Specifications

In order to calculate the side of the triangular fractal antenna the following parameters are required:

$$c = 3 \times 10^8 \text{ m/sec},$$
$$\varepsilon_{r,dyn}^* = 2.1\ (RT\ duriod),$$
$$L_{eff} = \frac{\sqrt{3}.a}{2}, \tag{7}$$
$$f_r = 2.45\ GHz.$$

Using the above formula for resonant frequency, the side of the triangle calculated is: a = 65.046 mm. The minimum distance of the radiation box from the antenna is given by:

$$\lambda_g = \frac{\lambda_0}{4\sqrt{\varepsilon_{eff}}}. \tag{8}$$

Therefore, λg = 21.124 mm.

IV. Simulation and Experimentation of the Antenna

The Figure 1 shows the 3rd order iteration triangular patch antenna. The patch antenna was printed on RT Duroid substrate of relative permittivity 2.1 and thickness 1.6 mm. All the structures have been simulated with the Ansoft HFSS 11 simulator. The procedure of simulation is as follows. The copper conductor of height 0.035mm is taken. The RT Duroid dielectric with relative permittivity 2.2, thickness 1.6mm is placed over it. For feed, a microstrip line is made to give a port for feed. Finally the antenna placed inside a radiation box about $\frac{\lambda_0}{4}$ far from the side & thereafter the set up is simulated and measured using HP810B network analyzer in Anechoic chamber (Figure 3).

Figure 3. Experimental set-up using HP810B network analyzer in Anechoic chamber.

IV. Results

The results for the three iterations performed on the triangular patch to get the desired fractal antenna are as follows:

Results for Iteration 0

The structure of triangular fractal antenna having zero iteration has been shown in Figure 4. On simulating this structure with the help of Ansoft HFSS, the following results were obtained. The values of Directivity, Gain and Radiation efficiency are found to be; 4.4048dB, 4.0205dB and 92.188% respectively.

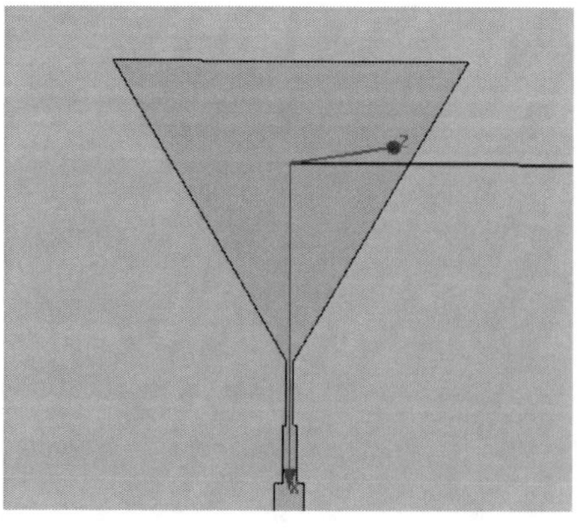

Figure 4. Iteration 0 of the antenna.

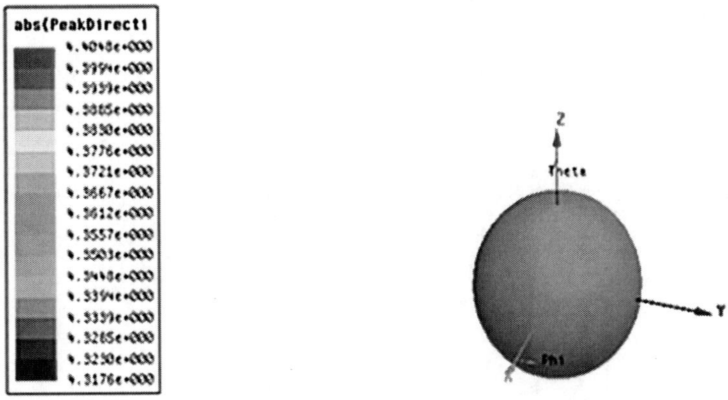

Figure 5. Directivity for Iteration 0 antenna.

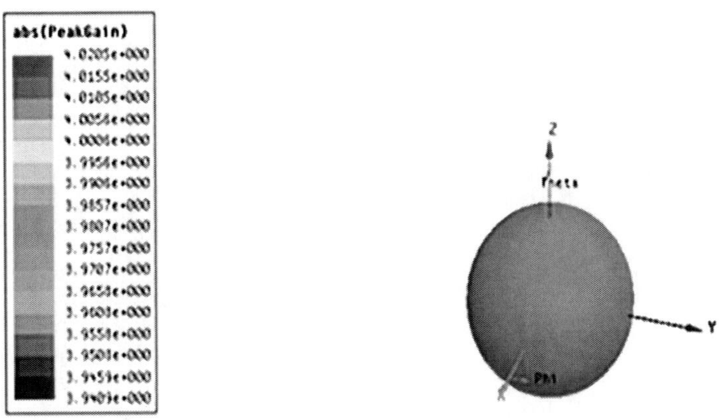

Figure 6. Gain for Iteration 0 antenna.

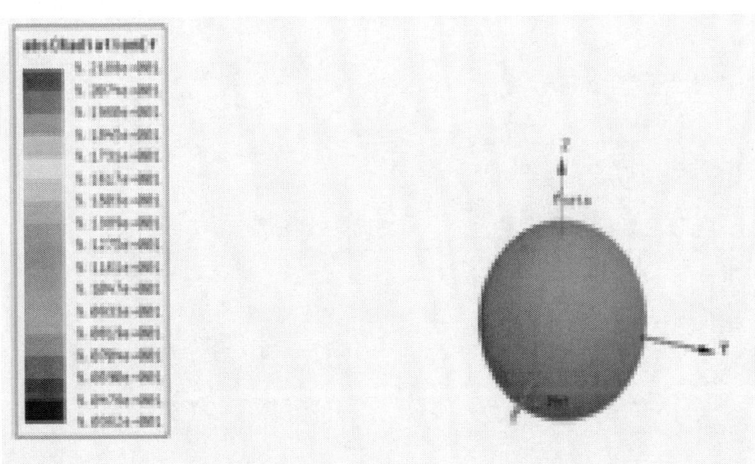

Figure 7. Radiation efficiency for Iteration 0 antenna.

Results for Iteration 1

The structure of triangular fractal antenna with first iteration is shown in Figure 8. On simulating the structure with the help of Ansoft HFSS, the following results were obtained:

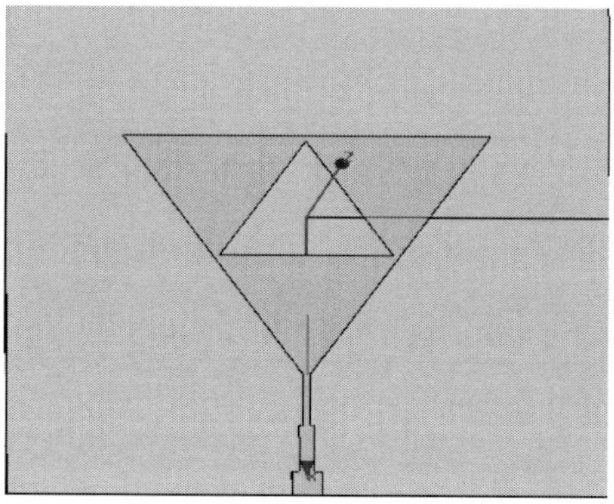

Figure 8. Iteration 1 of the antenna.

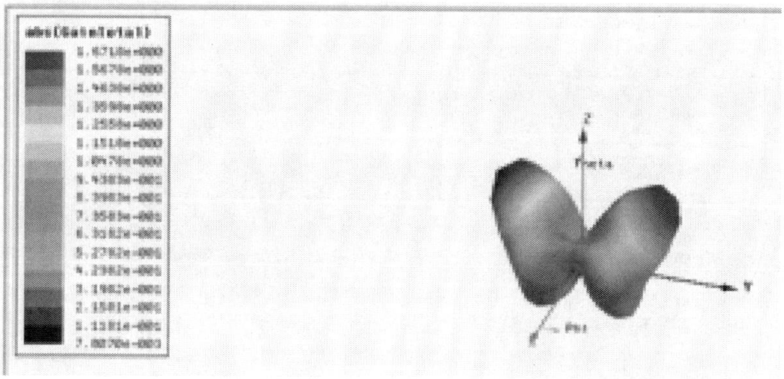

Figure 9. Gain of I-iterated antenna.

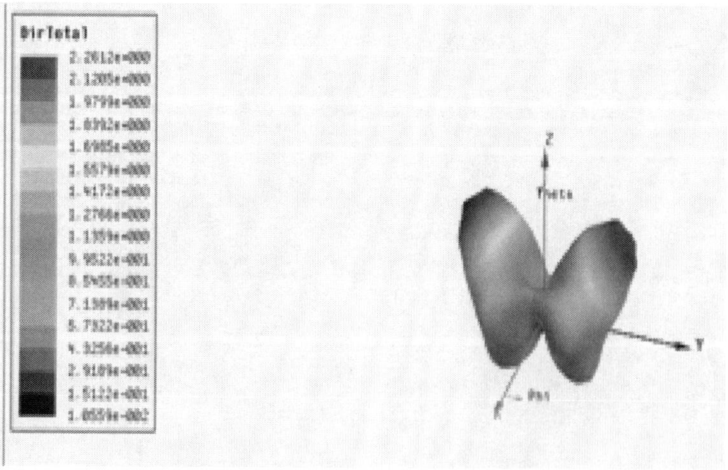

Figure 10. Directivity of I-iterated antenna.

Figure 11. Gain of I-iterated antenna.

Results for Iteration 2

The structure for triangular fractal antenna with 2nd iteration is shown in Figure 12. The results obtained by simulating the triangular fractal antenna came out to be as follows:

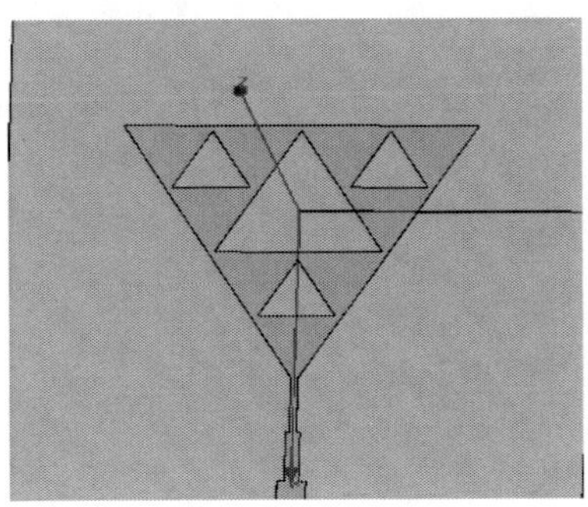

Figure 12. Iteration 2 of the antenna.

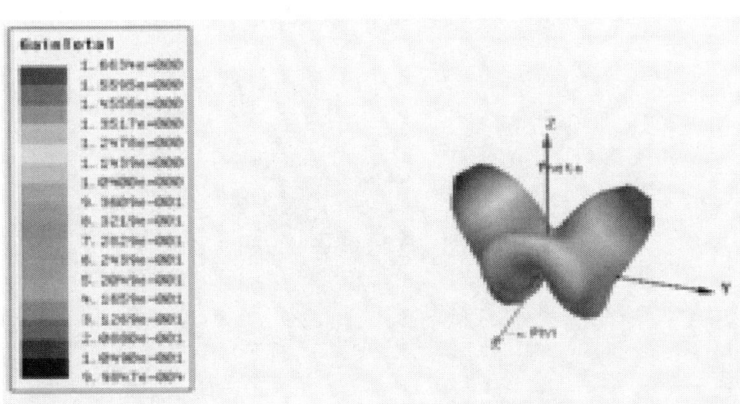

Figure 13. Gain of the II-iterated antenna.

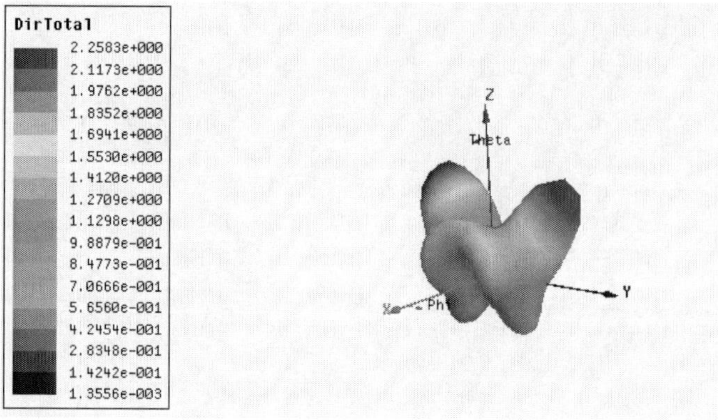

Figure 14. Directivity of II-iterated antenna.

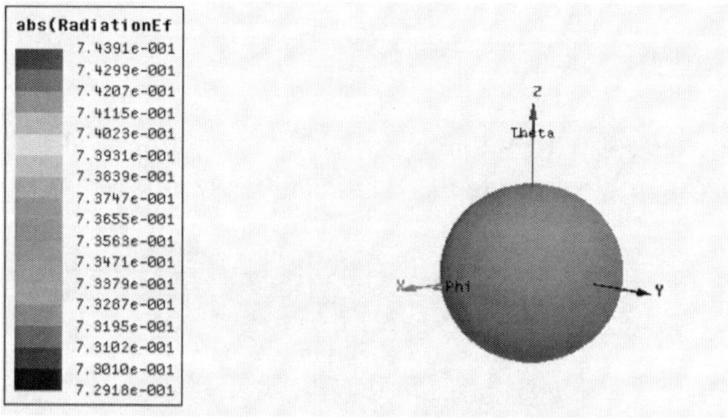

Figure 15. Radiation efficiency of II-iterated antenna.

Results for Iteration 3

The structure for triangular fractal antenna with 3rd iteration is shown in Figure 15, and the results obtained by simulating the above structure are given as follows:

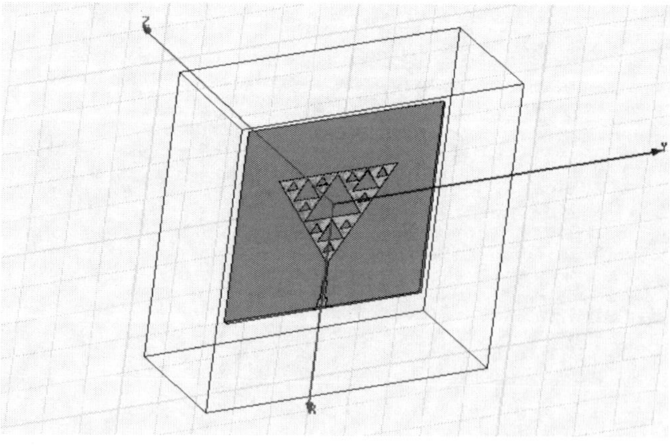

Figure 16. Iteration 3 of the antenna.

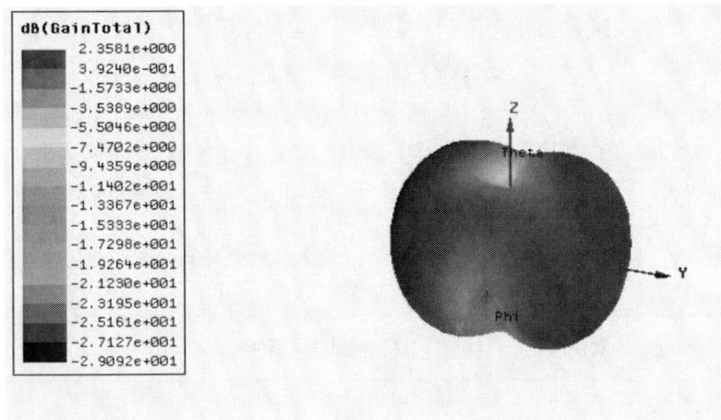

Figure 17. Gain for iteration 3 antenna.

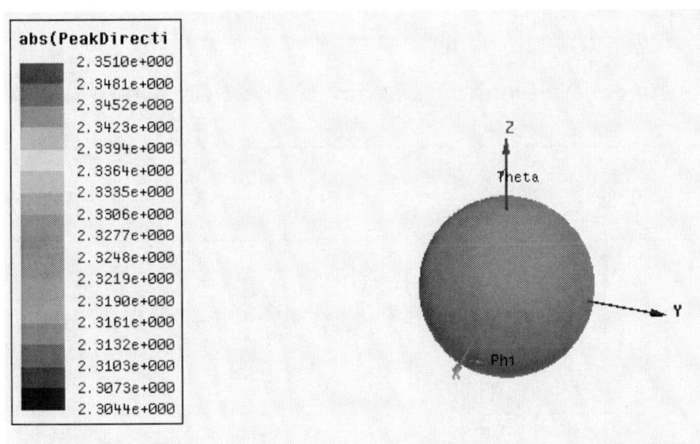

Figure 18. Directivity for iteration 3 antenna.

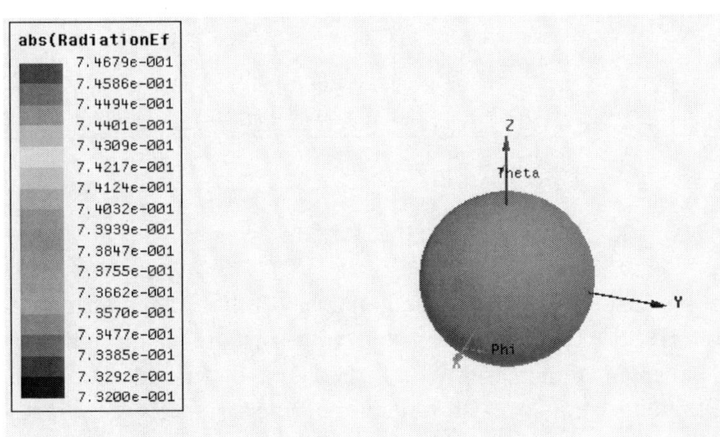

Figure 19. Rad. efficiency for iteration 3 antenna.

V. The Results of Different Iterations of Proposed Antennas

The following tables give the values of radiation efficiency, directivity, gain, resonant frequencies, bandwidth, return losses and VSWR for 0th, 1st, 2nd and 3rd iterations.

Table 1. The values of radiation efficiency, directivity and gain

Iterations	Radiation efficiency (%)		Directivity(dB)		Gain (dB)	
	Theoretical	*Experimental*	*Theoretical*	*Experimental*	*Theoretical*	*Experimental*
1	**74.675**	**75.665**	**2.2612**	**2.282**	**1.6718**	**1.692**
2	**74.39**	**74.958**	**2.2583**	**2.283**	**1.6634**	**1.684**
3	**74.679**	**75.692**	**2.251**	**2.291**	**2.3581**	**2.378**

Table 2. The values of resonant frequency, return loss, bandwidth and VSWR

Iterations	Resonant Frequency (GHz)		Return losses (dB)		Bandwidth (GHz)		VSWR	
	Theoretical	*Experimental*	*Theoretical*	*Experimental*	*Theoretical*	*Experimental*	*Theoretical*	*Experimental*
0	1.9	1.88	-6	-6.4	Very less	----	3	2.8
1	7.8	8.01	-10.7	-10.9	0.123	0.12	1.72	1.74
2	8.9 9.6	8.98 9.58	-16.2 -21	-16.26 -21.89	0.15 0.2	0.155 0.25	1.36 1.2	1.4 1.28
3	16.7 18.3 19.6 22.2	16.9 18.0 19.2 22.56	-28.2 -17 -25.2 -24.3	-28.26 -17.06 -25.28 -24.39	1.56 0.44 0.8 0.48	1.5 0.48 0.83 0.492	1.1 1.34 1.1 1.13	1.12 1.39 1.12 1.15

Conclusion

The present paper describes design and analysis of the triangular fractal antenna upto 3rd using the Ansoft HFSS 11,and observed that with increase in the number of iterations the bandwidth of the antenna increases & on second and third iterations the antenna starts showing the multiband behavior with radiation efficiency 74.679%, directivity 1.351dB and gain 2.3581dB. Increase in iterations also led to improvement is VSWR (1.15–2.8) & return losses (10.9-24.4 dB). The triangular fractal antenna is observed to possess multiband behavior similar to the Sierpinski gasket antenna. The simulated and experimental results are found to in good agreement. This new fractal antenna allows flexibility in matching multiband operations in which a larger frequency separation is required, according to these frequencies this antenna can be used as a multiband antenna in the wireless applications and RFID readers.

ACKNOWLEDGMENT

Authors are very thankful to Dr. S.K. Koul, Professor, CARE, Indian institute of Technology, Delhi and Dr. Ashok De, AIT, Delhi for providing the testing facility and valued suggestions for this work.

REFERENCES

[1] J. Bahl and P. Bhartia, "Microstrip Antennas", *Dedham, Ma, Artech House*, 1981.

[2] C. Puente, J. Romeu, R. Pous, X. Garcia, and F. Benitez, "Fractal multiband antenna based on the Sierpinski gasket," *Electron. Lett.*, vol. 32, no. 1, pp. 1-2, Jan. 1996.

[3] C. Puente, J. Romeu, R. Pous, and A. Cardama, "On the behavior of the Sierpinski multiband fractal antenna," *IEEE Trans. Antennas Propagat.*, vol. 46, pp. 517-524, Apr. 1998.

[4] S. N. Khan, J. Hu, J. Xiong, and S. He, "Circular fractal monopole antenna for low VSWR UWB applications", *Progress in Electromagnetic Research Letters,* Vol. 1, pp. 19-25, 2008.

[5] E. Lule, et al, "Koch island fractal ultra wideband dipole antenna", *IEEE, Antennas and Propagation Society International Symposium*, Vol. 3, pp. 2516-2519, June 2004.

[6] J. P. Gianviffwb and Y. Rahmat-Samii, "Fractal antennas: A novel antenna miniaturization technique and applications", Antennas Propagate. Mag. Vol. 44, pp. 20-36, 2002.

[7] P. W. Tang and P. F. Wahid, "Hexagonal Fractal Multiband Antenna", *IEEE Antennas and Wireless Propagation Letters*, Vol. 3, pp. 111-112, 2004.

[8] R. Kumar and P. Malathi, "Design of CPW – Fed Ultra wideband Fractal Antenna and Backscattering Reduction", *Journal of Microwaves, Optoelectronics and Electromagnetic Applications,* Vol. 9, No. 1, pp. 10-19, June 2010.

[9] J. S. Dahele and K. F. Lee, "On the resonant frequencies of the triangular patch antenna," *IEEE Trans. Antennas Propagate.*, vol. AP-35, pp. 100-101, Jan. 1987.

[10] R. Singh, A. De and R. S. Yadava, "comments on 'An improved formula for resonant frequency of the triangular microstrip patch antenna," *IEEE Trans. Antennas Propagate*, vol. AP-39, pp. 1443-1444, Sept. 1991.

[11] M. Polivka, M. Drahovzal, J. Rohan and P. Hazdra, "Multiband Patch Antenna with Perturbation Elements Generated By Genetic Algorithm", *CTU in Prague, FEE, Dept. of Electromagnetic Field Technicka* 2, 166 27 Prague.

[12] K. Verma and Nasimuddin, "Fast and accurate model for analysis of equilateral triangular patch antenna," *Journal of Microwaves and Optoelectronics*, Vol. 3, No. 4, April 2004.

[13] R. Garg and S. A. Long, "An improved formula for the resonant frequency of triangular microstrip patch antenna," *IEEE Trans. Antennas Propagate.*, vol. AP-36, p. 570, Apr.1988.

[14] K. Verma and Nasimuddin, "Resonance frequency and bandwidth of rectangular patch antenna on thick substrate," *IEEE Microwave and Wireless Components Letters*, vol. 12, No. 2, pp. 60-62, Feb. 2002.

[15] E. Hammerstad and O. Jensen, "Accurate models for microstrip computer Aided Design," *IEEE MTT-S, Int. Microwave Symp., Digest,* 1980, pp. 407-409.

In: Knowledge Management and Communication … ISBN: 978-1-62808-250-0
Editor: Rafiqul Islam

Chapter 11

PEDESTRIAN CRASH CHARACTERISTICS IN DIFFERENT AGE GROUPS: A CASE STUDY IN THE LAS VEGAS METROPOLITAN AREA

Ganesh J. Karkee *
The University of Texas at San Antonio
Department of Civil and Environmental Engineering
One UTSA Circle, San Antonio, TX, US

ABSTRACT

All the crash data were based on the reported police crash report. Crash data were obtained from Nevada Department of Transportation (NDOT). Those data were collected for five years, from 1996 to 2000. The obtained data did not provide more information about psychological aspects of the individuals for traffic accidents. The fatality rate of older people exceeds than those of the younger people. Approximately 45 percent of fatal crashes were related to either alcohol or drugs. Fatal and total pedestrian crashes reduced from 1996 to 2000 by 35 percent and 10 percent, respectively. The age segments 64 years and younger comprised of approximately 3 fatalities per 100,000 population comparatively lower than the older population segments of 65 years and older, which contributed approximately 7 fatalities per 100,000 population. On the other hand, the injury crash rate is declined along with decrease in age of pedestrians. Fatal pedestrian crashes are greater in elder population groups than in younger population groups. Injury crashes are greater in younger population groups than elder population groups.

1. INTRODUCTION

Pedestrians are more vulnerable in traffic crashes, in terms of number of people involved in traffic accident and fatality of crashes. Pedestrians differ in age groups from younger to older and some are physically challenged people. With the increase of people's age, their

* E-mail address: ganesh.karkee@gmail.com.

perception response (PR) time also increases (American Automobile Association, 1952). Research also shows that the younger population has less PR time than that of the older population. Similarly, fatigue and driving under the influence (DUI) road users elapse more time to respond to any cues they encountered.

There are several psychological processes involved in pedestrian's task in crossing a street. Some of them are as follows: recognition, identification, detection, and decision. There are various types of factors that affect pedestrian behavior and safety, environment (road type, intersections, and lighting), traffic (volume, moving and stationary vehicles, and communication), personal (physical, psychological, motivation, experience, and psychological state), and social (presence of others and purpose of journey).

1.1. Background

In the United States, pedestrian and motor vehicle crashes loss of 5,449 lives in 1996; the number of pedestrian fatalities reduced to 4,763 in 2000 (National Highway Traffic Safety Administration, 2010; Insurance Institute for Highway Safety, 2010). The number of injuries involving pedestrian is also falling, from 82,000 in 1996 to 78,000 in 2000 (United States of Department of Transportation, 2001; United States of Department of Transportation, 1997).

Along with new inventions in the medical field, better medical facilities, health consciousness, and physical exercise increase life expectancy of Americans. Roads are also aging due to fatigue. Making the country's roadways safe for all travelers requires constant monitoring and timely repairs. Some of the possible faults in road combinations that may lead to accidents are the following: intersections with poor lighting, faulty traffic signals, fallen or missing stop signs, roadways obstructed by overhanging tree branches, lanes that are too narrow, curves that are too sharp or entrance and exit ramps are too short.

In the entire population of American drivers', nearly 30 percent consists of 50 years and older. Starting in 2011, as the baby boomers (those born between 1946 and 1964) start turning age 65, the older population will begin to increase dramatically. The older population is projected to be 71.5 million by 2030 more than double than the older population in 2000 (Minnesota Department of Health, 2005; United States Department of Health and Human Services, 2004). Americans aged 85 and older population group already comprise the fastest growing segment of the population, this population segment is expected to increase from 4.7 million in 2003 to 9.6 million by 2030 (United States Department of Health and Human Services, 2004). As the proportion of senior's increases, government and concerning agencies need to focus on safety aspects of senior drivers and pedestrians. Basically, in driving tasks, there are four steps involved in decision-making process. This process is also termed as perception-response time. These four steps are as follows: perception, intellection, emotion, and volition.

The time required to make decision in driving task is not the same for all age groups. From the human factor perspective, senior people take longer time to make decisions for driving cues, due to poor eye sight, prolong duration to identify any event, etc. The perception-response time also varies with dexterity of the driving tasks. Recently licensed teenage drivers do not have the same dexterity as an experienced driver, in driving task. Familiarity of the road network is also an important factor. Senior people are also as vulnerable as pedestrians at intersection and mid block.

In 2000, there were 41,945 fatalities on US roads, or 1.53 fatalities per 100 million vehicle miles traveled (United States of Department of Transportation, 2001). There has not been a significant decrease in fatalities since 1992. In 2000, there were a total of 58,741 traffic crashes in Nevada. Property damage only crashes were 36,680. In the same way, 18,795 were injury and 266 were fatal crashes. Among the total crashes in 2000, the most crashes occurred in Clark County, which was 74.2 percent of total crashes in Nevada.

1.2. Data Base Characteristics

Crash data were obtained from the Nevada Department of Transportation (NDOT). Those data were based on the reported police crash report. These reported data were collected from 1996 to 2000 in the Las Vegas metropolitan area. The Las Vegas metropolitan area consists of the City of Las Vegas, the City of Henderson, the City of North Las Vegas, and unincorporated Clark County. Table 1 shows the total fatal and injured crashes in Las Vegas metropolitan area from 1996 to 2000.

Table 1. Total Crashes in the Las Vegas Metropolitan Area from 1996-2000

Year	Fatal	Injury	Total
1996	51	707	758
1997	45	683	728
1998	31	718	749
1999	47	743	790
2000	33	652	685

1.3. Population

Clark County is one of the fastest growing counties in the United States, its population almost doubled from 1990 to 2000 increased from 768,203 to 1,428, 690 (Clark County Nevada, 2010). Approximately 35 million tourists visited Las Vegas in 2000, as the traffic growth along with population continues.

Table 2 shows the total population of Clark County as of 2000 census (U.S. Census Bureau, 2010). Table 2 also depicts the different age distribution of the total population, such as age under 5 years, over 5 years to 9 years, 10 years to 14 years, etc. Same proportion of age distribution of Clark County population used to interpolate the population of different age groups in Las Vegas metropolitan area.

Over a billion dollar loss resulted from highway deaths and injuries in Nevada every year. The total estimated economic loss resulting from traffic crashes in Nevada for the year 2000 is $1,209,749,000 (Nevada Department of Transportation, 2000). This paper is aimed to analyze pedestrian crashes in different age groups in the Las Vegas metropolitan area.

Table 2. Age and Gender Distribution of Clark County Population (U.S. Census Bureau, Census 2000)

Total	1,464,653
Under 5 years	7.5%
5 to 9 years	7.6%
10 to 14 years	6.8%
15 to 19 years	6.2%
20 to 24 years	6.7%
25 to 34 years	16.2%
35 to 44 years	16%
45 to 54 years	12.9%
55 to 59 years	5.2%
60 to 64 years	4.2%
65 to 74 years	6.6%
75 to 84 years	3.4%
85 years and over	0.8%

2. Relevant Literature

Various articles related to traffic crashes were reviewed. In particular, researches on driving under influence, crash related to different age group and gender, crash related to different time of the day and days of the week, crash related to weather conditions, perception time of the road users were reviewed.

The field of vision is shrinking with age and also part of peripheral vision is lost which is important under driving conditions for detecting objects moving into the driving path. Lerman and Borkman (1976) found that overall transmittance between 20 years and 80 years drops from about 55% to 32% at 600 mm wavelength. Peripheral vision is important to detect objects that might move into the driving path. Owsley et al. (1998) have suggested that reduction in the field of view increases crash risk in older drivers.

Dewar et al. (1994) found that older drivers had poorer understanding than younger drivers on 39% of the symbols examined. Of the 85 standard symbols, 16 were understood by over 95% of drivers; however, 10 were understood by fewer than 40%. Drivers from Texas, Idaho, and Alberta, Canada participated in the study. They found out the improvements can be made in understandability of traffic sign symbols through design revisions, especially for those signs which are poorly understood by drivers. One of the remedies for enhancing driver understanding of poorly understood symbols is a campaign to educate drivers about the meanings of these messages. Their findings also suggest the need for public education campaigns to pay particular attention to older drivers.

A study on attention capability of young and older drivers was conducted with twenty subjects (Mourant et al., 2001). Ten old and 10 young drivers ranged from 58 to 76 years and 23 to 46 years, respectively. They were asked to drive on a curvy road by using fixed-based driving simulator. Older drivers were less accurate in obtaining information from the in-vehicle display. The average lane-position error of older drivers was greater; they also spent more time driving outside their lanes. Researchers found that the poor performance of older drivers with the in-vehicle display was due to vision-related changes, such as longer eye accommodation times, rather than cognitive processes.

Kim and Yamashita (2001) observed approximately 203,717 vehicle crashes from 1986 to 1995; 6,516 were fatigue related and 76.2 percent of those involved were men and the remaining 23.8 percent were women. As per their finding, most of the fatigue-related crashes take place on highways and major roadways. Fatigue-related crashes peak on the weekends (Saturday), whereas other types of crashes peak on Friday. Similarly, in Hawaii 58% of fatigue-related crashes occur between midnight and 6:00 a. m. On the other hand, non-fatigue-related crashes have large mid afternoon peak and a smaller peak mid morning, coinciding with peak travel periods.

Hingson et al. (2002) found that the earlier the age of respondents started drinking, the more likely they were to report driving after drinking too much and being a motor-vehicle crash because of their drinking. Those who started drinking in age group under 21 were significantly more likely than those who started drinking after age 21 to have ever and in the past year in motor-vehicle crash after drinking too much. People who drive after drinking are more likely to speed and are less likely to wear seat belts. Consumption of more alcohol may further impair the driver's judgment.

Ferguson et al. (2002) found that despite drinking and driving in the United States has declined significantly during the past two decades; this trend has not been seen among Hispanic drivers. Fewer than half of the respondents were aware about the blood alcohol concentration threshold at or above which it is illegal to drive in California law. Some identified patterns were failure to yield, inattentive driving, speed too fast, failure to reduce speed, following too close, and excessive speed not necessarily over the speed limit were found to be contributing factors.

Light transmittance of the human lens will also decrease with age. The overall transmittance decrease 55 percent to 32 percent from 20 to 80 years at 600 mm length. It takes longer time to visualize the object so does to process from eye to brain through neurons. Ultimately reduces the capability to resolve details and visual acuity. Legibility distance also decreases; one 20 years old person can read 50 cm letter height from 350 m, the same object, an another person of 60 years old can read from 270 m. (Adrian, 1998). Older people also tend to walk more than others because they have more leisure time. Walking is a good exercise and also an inexpensive mode of travel for short trips. The elderly are more law abiding than the general population and trusting of traffic signals and drivers when crossing the street (Dewar and Olson, 2002).

More than two-third of all pedestrian crashes occur in urban areas. Almost half of fatal crashes involving pedestrians occur between the hours of 6:00 p. m. and midnight (United States Department of Transportation, 2003).

3. METHODOLOGY

Collected crash data for five years from 1996 to 2000 in the state of Nevada were analyzed. These data were obtained from the Transportation Research Center at the University of Nevada, Las Vegas.

4. Data Analysis and Discussion

The obtained pedestrian crash data were analyzed based on the different age groups. The detailed analysis is shown in the following sections and discussed in detail.

4.1. Accident Ratios Based on Population

Population of Las Vegas Valley from 1996 to 2000 is shown in Table 3. The fatal pedestrian crashes in 1996 are 4.7 per 100,000 population. Similarly, injury and total crashes are 65.8 and 70.6 per 100,000 population, respectively. On the other hand, in 2000, fatal, injury, and total crashes are 2.4, 47.7, and 50.1 per 100,000 population, respectively. Total as well as fatal pedestrian crashes declined from 1996 to 2000. Fatal and total pedestrian crashes reduced from 1996 to 2000 by 35 percent and 10 percent, respectively.

Figure shows that the fatal, injury, and total pedestrian crashes per 100,000 population in the Las Vegas metropolitan area from 1996 to 2000. Injury and total pedestrian crashes declined from 1996 to 2000.

Table 3. Pedestrian Crashes per 100,000 Population in Las Vegas Valley

Las Vegas Valley	1996	1997	1998	1999	2000
Population	1,074,362	1,147,501	1,220,639	1,293,778	1,366,916
Fatal	4.7	3.9	2.5	3.6	2.4
Injury	65.8	59.5	58.8	57.4	47.7
Total	70.6	63.4	61.4	61.1	50.1

Note: Population from 1997-1999 was obtained by interpolating population of 1996 and 2000.

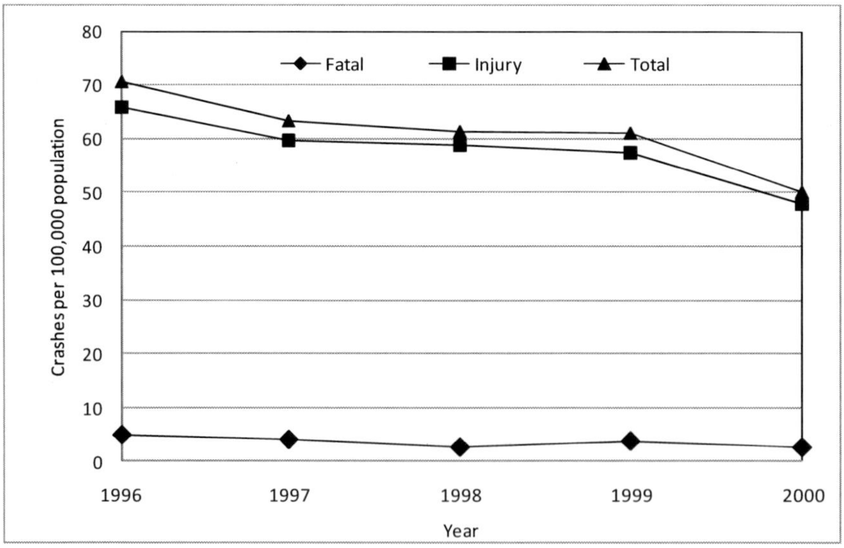

Figure 1. Pedestrian Crashes per 100,000 Population in Las Vegas Metropolitan Area from 1996-2000.

4.2. Pedestrian Crashes with Age Group

The age groups of all pedestrian crashes in 1996 are divided into 13 age brackets from under 5 years to over 85 years. Table 4 shows the percentage of age group of the total population and percentage of crashes. For age group under 5 years, this group consists of 7.5 percent of the total population, which contributes 3.9 percent fatal, 10.2 percent injury, and 9.76 percent of total crashes in 1996. Age group 25 to 34 year is comprised of the highest proportion of the total population, i.e., 16.2 percent, but it contributes only 13.7 of fatal and 16 percent of total crashes.

Figure 2 depicts the proportions of population and their share in fatal, injury, and total crashes in 1996. The age group 35 to 44 years shares the most in fatal, injury, and total crashes. More than 27 percent of fatal crashes are from this age bracket. In the same manner, 17 percent and 17.7 percent of injury and total crashes respectively, are also from this age group.

Older pedestrians are more likely to be fatal crash victim than that of injury crash victim. The fatal crash rate is increased with an increase in age of pedestrians as shown in Figure 3. The age group greater than 85 years contributed in an average 10 fatalities per 100,000 population from 1996-2000 (except in 1997). The younger population segments, 34 years and younger pedestrians, contributed approximately 4 or less fatalities per 100,000 population (with an exception of under 5 years age segment in 1999). The age segments 64 years and younger are comprised of approximately 3 fatalities per 100,000 population comparatively lower than the older population segments of 65 years and older, which contributed approximately 7 fatalities per 100,000 population. On the other hand, the injury crash rate is declined along with decrease in age of pedestrians; which is shown in Figure 4. In an average, injury rate is approximately 60 injuries per 100,000 population for age segments 64 years and younger and approximately 39 injuries per 100,000 population for 65 years and older pedestrians.

Table 4. Percentage of Pedestrian Crashes (1996) in Different Age Groups

Age Bracket	Percentage of Total Population (%)	1996		
		Fatal (%)	Injury (%)	Total Crash (%)
Under 5 years	7.5	3.9	10.2	9.8
5 to 9 years	7.6	0	7.6	7.1
10 to 14 years	6.8	0	9.2	8.6
15 to 19 years	6.2	2.0	7.9	7.56
20 to 24 years	6.7	2.0	6.4	6.1
25 to 34 years	16.2	13.7	16.1	16.0
35 to 44 years	16.0	27.5	17.0	17.7
45 to 54 years	12.9	19.6	10.9	11.5
55 to 59 years	5.2	3.9	3.3	3.35
60 to 64 years	4.2	7.8	3.4	3.7
65 to 74 years	6.6	9.8	4.8	5.1
75 to 84 years	3.4	7.8	2.7	3.0
> 85 years	0.8	2.0	0.6	0.7
Total	100	100	100	100

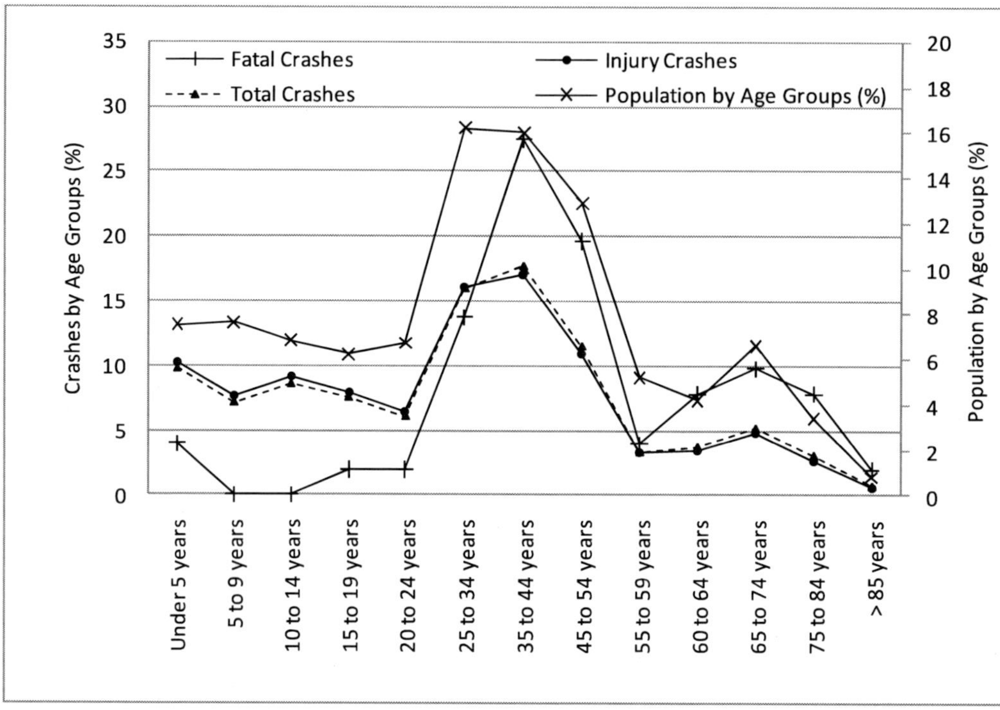

Figure 2. Pedestrian Crashes in Percentages by Different Age Groups in 1996.

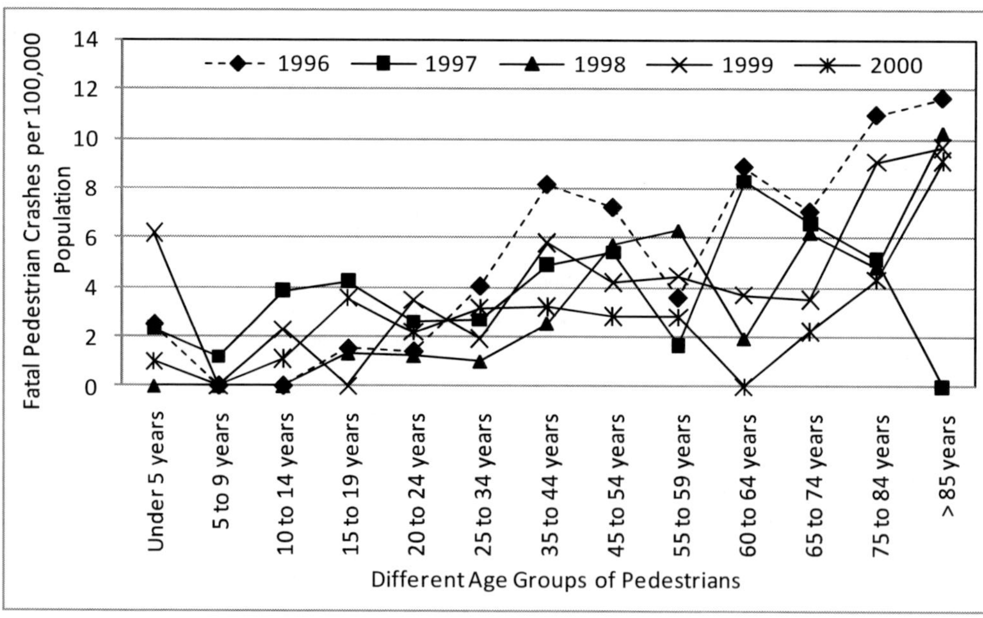

Figure 3. Fatal Pedestrian Crashes per 100,000 Population in Different Age Groups.

Some of possible causes of more elderly pedestrian fatalities are as follows: i) gap adjustment – misjudging the distances and intervals approaching vehicles; ii) attention - distracted when stepping off the side walk; iii) visual search - watching the traffic instead of traffic light; iv) expectation – misinterpreting the movement of vehicles and assuming that drivers will yield; v) impassion – impatiently crossing after waiting, and crossing midblock between parked cars.

Older drivers are easily distracted and tend to look down while crossing the street. Some of the physical challenges to the older pedestrians are as follows: walking more slowly because of unsure footing and chances of falling, poor balance and reduced ability to catch them if they slip, reduced agility, and difficult walking due to ailments (Wickens and Hollands, 2000).

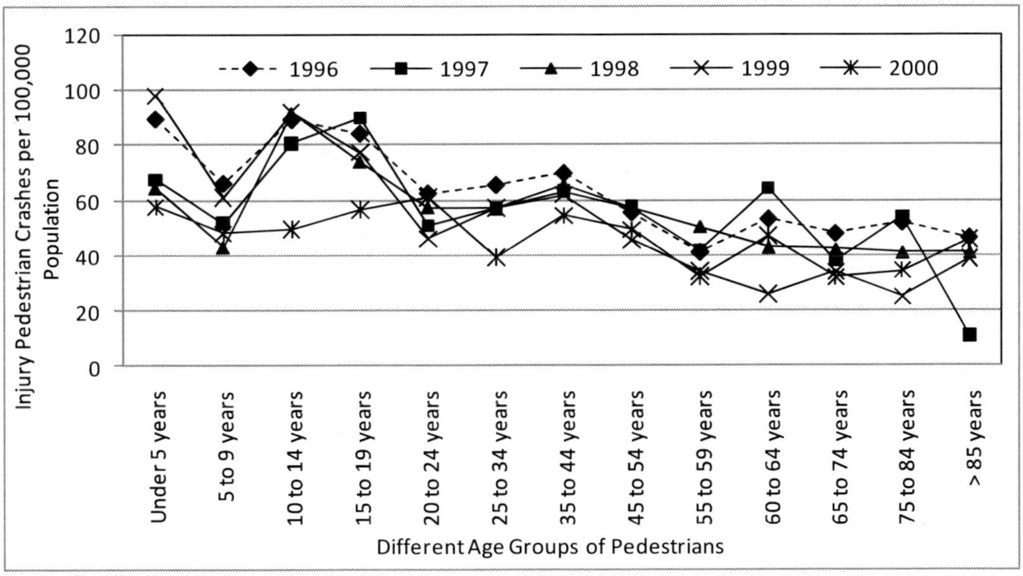

Figure 4. Injury Pedestrian Crashes per 100,000 Population in Different Age Groups.

Conclusion

Fatal as well as injury pedestrian crashes declined from 1996 to 2000. The fatal crash rate is increased with an increase in age of pedestrians. Injury crashes are greater in younger population groups than elder population groups. Older people have longer perception time, slower walking speed along the cross walk, may not be able to anticipate the distance of oncoming vehicle. Eye physiology also changes along with age. Limited vision and hearing, slower reaction time and decision making, lower level of attention and reduced walking speed make older population segment more vulnerable in the event of crash. These factors also make older people susceptible to more physical damage in the event of crash than that of younger counterpart.

REFERENCES

[1] American Automobile Association, Traffic Engineering and Safety Department. 1952. *Age and Complex Reaction Time*. Report No. 41.

[2] Adrian, W. 1999. *On the Influence of Age on Vision Under Driving Conditions*. In *Proceedings of the Progress in Automobile Lighting (PAL) September 28-29*, Department of Lighting Technology, Darmstadt University of Technology, Germany. pp. 1-8.

[3] Clark County Nevada. *Clark County Population 1980-2006*. 2010. Available at: http://www.accessclarkcounty.com/depts/assessor/pages/Census. Accessed November 12, 2010.

[4] Dewar, R. E., Kline, D. W. and Swanson, H. A. 1994. *Age Difference in the Comprehension of Traffic Sign Symbols*. Transportation Research Record, 1456:1-10.

[5] Dewar, R. E. and Olson, P. L. 2002. *Human Factors in Traffic Safety*. Lawyers & Judges Publishing Company, Inc. Tucson, AZ.

[6] Ferguson, S. A., Burns, M. M., Fiorentino, D., Williams, A. F. and Garcia, J. 2002. *Drinking and Driving among Mexican American and Non-Hispanic White Males in Long Beach, California*. Accident Analysis and Prevention 34(4):429-437.

[7] Hingson, R., Heeren, T., Levenson, S., Jamanka, A. and Voas, R. 2002. *Age of Drinking Onset, Driving after Drinking and Involvement in Alcohol Related Motor-Vehicle Crashes*. Accident Analysis and Prevention 34(1):85-92.

[8] Insurance Institute for Highway Safety. Fatality Facts 2005: Pedestrians. 2010. Available at: http://www.iihs.org/research/fatality_facts_2005/pedestrians.html. Accessed November 12, 2010.

[9] Kim, K. E. and Yamashita, E. Y. 2001. *Asleep at the Wheel Spatial and Temporal Patterns of Fatigue-Related Crashes in Honolulu*. Transportation Research Records, 1779:46-53.

[10] Lerman, S. and Borkman, R. 1976. *Spectroscopic Evaluation and Classification of the Normal, Aging, and Cataractous Lens. (With 1 Color Plate)*. Ophthalmic Research. Vol. 8(5):335-353.

[11] Minnesota Department of Health. *Demographics of an Aging Population*. 2005. Available at: www.health. Accessed November 12, 2010.

[12] Mourant, R. R., Tsai, F., Al-Shihabi, T. and Jaeger, B. K. 2001. *Measuring Divided-Attention Capability of Young and Older Drivers*. Transportation Research Records, 1779:40-45.

[13] National Highway Traffic Safety Administration. Ranking of State Pedestrian Fatality Rates – State: USA. 2010. Available at: http://www-fars.nhtsa.dot.gov/States/States Pedestrians.aspx. Accessed November 12, 2010.

[14] Nevada Department of Transportation. *Nevada Traffic Crashes 2000*. 2000. Available at: http://www.nevadadot.com/reports_pubs/nv_crashes/2000/. Accessed November 13, 2010.

[15] Owsley, C., Ball, K., McGwin, G., Sloane, M.E., Roenker, D.L., White, M.F. and Overley, E.T. 1998. *Visual Processing Impairment and Risk of Motor Vehicle Crash*

Among Older Adults. The Journal of the American Medical Association, Vol. 279(14):1083-1088.

[16] United States Department of Health and Human Services. *A Profile of Older Americans: 2004*. 2004. Available at: http://www.aoa.gov/AoAroot/Aging_ Statistics /Profile/2004/index.aspx. Accessed November 12, 2010.

[17] United States Department of Transportation. *Pedestrian Roadway Fatalities: DOT HS 809 456*. 2003. National Center for Statistics Analysis. Available at: www:nrd.nhtsa.dot.gov/Pubs/809-456.PDF. Accessed November 19, 2010.

[18] United States Department of Transportation. *Traffic Safety Facts 1996 – A Compilation of Motor Vehicle Crash Data from the Fatality Analysis Reporting System and the General Estimates System*. National Highway Traffic Safety Administration. Washington DC: 1997.

[19] United States Department of Transportation. *Traffic Safety Facts 2000 – A Compilation of Motor Vehicle Crash Data from the Fatality Analysis Reporting System and the General Estimates System*. National Highway Traffic Safety Administration. Washington DC: 2001.

[20] U.S. Census Bureau. Census 2000. Available at: http://censtats.census.gov/data/NV /05032003.pdf. Accessed November 12, 2010.

[21] Wickens, C. D. and Hollands, J. G. 2000. *Engineering Psychology and Human Performance*. Third Edition, Prentice Hall.

In: Knowledge Management and Communication …
Editor: Rafiqul Islam
ISBN: 978-1-62808-250-0

Chapter 12

PROGRESSIVE STUDY IN SUPERSTRATE COVERED CIRCULAR MICROSTRIP ANTENNAS

***R. K. Yadav*[1*] *and R. L. Yadava*[2†]**
[1]Department of Electronics & Communication Engineering
ITS Engg College, Gr. Noida, U.P. India
[2]Department of Electronics & Communication Engineering
Galgotias College of Engg & Tech.
Gr. Noida, U.P. India

ABSTRACT

This paper provides a case study of superstrate loading effects on circular patch antennas with emphasis on theoretical and practical aspects for last three decades. Several theoretical analysis methods have been summarized, including Transmission line, Method of Moment (MOM), Finite Difference Time Domain (FDTD) as well as Modified Wolf Method (MWM). In particular, effect of superstrate loading on co-polarisation and cross-polarisation with emphasis on radiation and directivity has also been studied. The paper aims those readers who involve R & D in the field of superstrate loaded circular patch antennas.

Keywords: Circular Patch Antenna, Dielectric Covers, Numerical Techniques, Frequency Alteration, Environmental Hazards

1. INTRODUCTION

The environmental effects (such as snow, raindrops, etc.) deteriorate the performance of antenna; particularly resonance frequency and reflection coefficient. This is the reason, superstrate (cover) dielectric layers are often used to protect printed circuit antennas (PCA's)

---

* E-mail address: ravipusad@gmail.com.
† E-mail address: rlyadava@rediffmail.com.

from external hazards, or may be naturally formed (e.g. ice layers) during flight or severe weather conditions. Whether a cover layer is naturally formed or imposed by design, it may affect adversely the antenna performance characteristics, such as gain, radiation resistance and efficiency. For this reason, it is important to analyze superstrate loading effects from a fundamental point of view, so that the performance may be understood better or a proper choice of cover parameters may be implemented. There are various shapes of patch antennas whose characteristics have been analyzed for dielectric covers. Such as rectangular/square, triangular, equilateral triangular, elliptical and circular patch antennas. However, the lowest profile and circular polarization characteristics of circular patch make it a dominant candidate for modern communications application. Therefore authors presents the progressive studies of dielectric loading on such antennas and describe loading effects on the various parameters; resonance frequency, radiation, bandwidth and beam-width etc.

II. Effect on Resonance Frequency

As the antenna is a resonance device, its resonance frequency is the most important parameter, because it also affects other parameters such as bandwidth, gain and efficiency etc. The available literatures on superstrate loading patch antenna reveals that, dielectric loading significantly changes the resonance frequency hence causes detuning. K. M. Luk et al. in 1989 first studied the characteristics of circular patch under various dielectric cover thicknesses (Figure 1(a)).

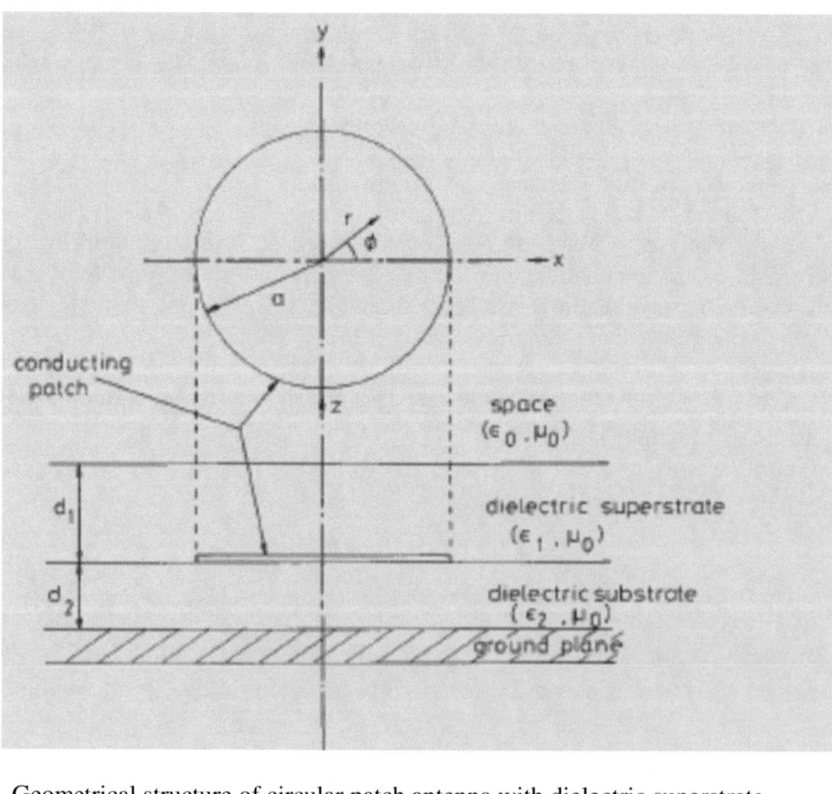

Figure 1(a). Geometrical structure of circular patch antenna with dielectric superstrate.

Using Hankel transform analysis method for TM_{11} mode, they found that resonant frequency decreases with increasing the thickness of superstrate, while half-power bandwidth changes slightly [1] as shown in Figure 1(b). While A. K. Bhattacharyya, in year 1991, derived the expressions for wall admittance, wall conductance and radiation efficiency of dielectric loaded circular patch antenna. In this study, he used principle of equivalence, which yields equivalent electric and magnetic surface currents for the outside fields, that is responsible for radiation. He concluded that for a small substrate thickness, the magnetic current model yields fairly accurate results, which deteriorates with an increase in the substrate thickness.

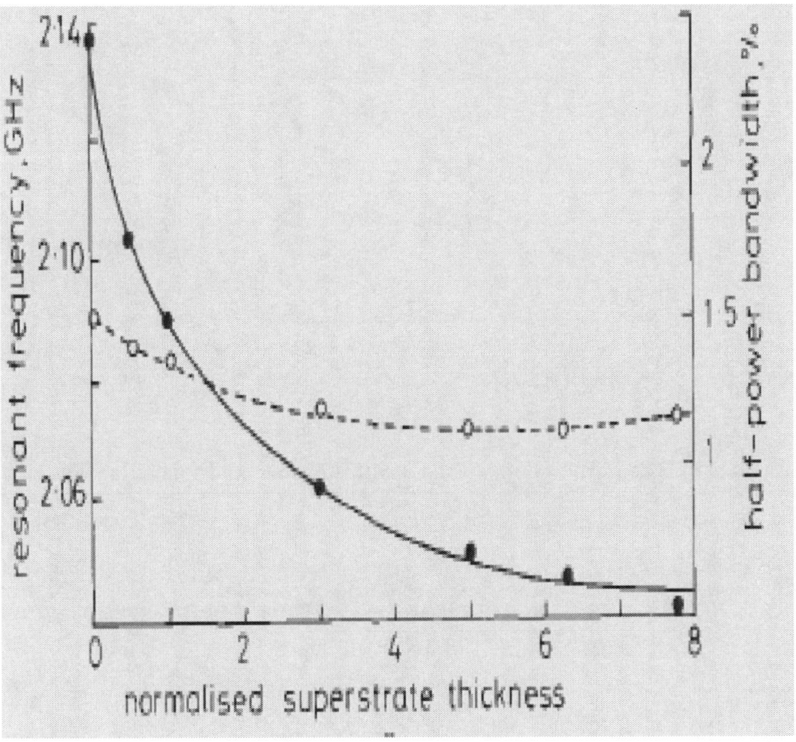

Figure 1(b). Variations of resonant frequency and bandwidth with superstrate thickness.

The radiation efficiency decreases initially and then increases with the thickness of the dielectric layer. The maximum value of the radiation efficiency is obtained in the case of an antenna with different substrate and superstrate materials than that of an antenna with substrate and superstrate of same dielectric. The wall admittance of the patch is representative of the total power output and this output power has two components; the space wall power *(P_{sp})* and surface wave power *(P_{su})*. The radiation efficiency η is defined as [2]

$$\eta = \frac{g_{sp}}{g_{sp} + g_{su}} \% . \tag{1}$$

Where g_{sp} and g_{su}, are the space and surface wave conductance's respectively. The variations of space and surface wave conductance's are shown in Figures 2(a) and 2(b), whereas radiation efficiency is computed and plotted against the normalized substrate/superstrate thickness is shown in Figure 3.

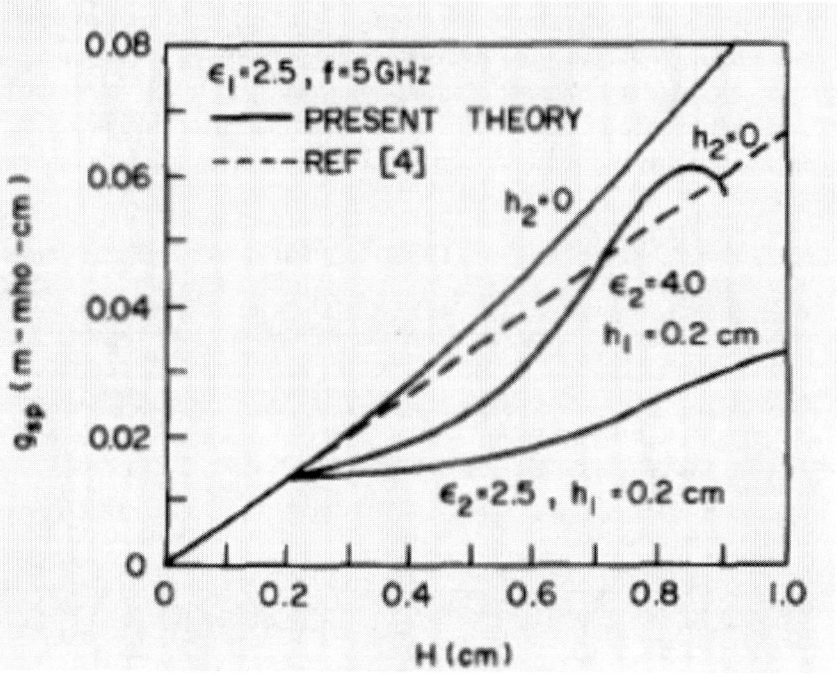

Figure 2(a). Variations of space wave conductance's with substrate and superstrate thickness (a = 1.1 cm, f = 5 GHz).

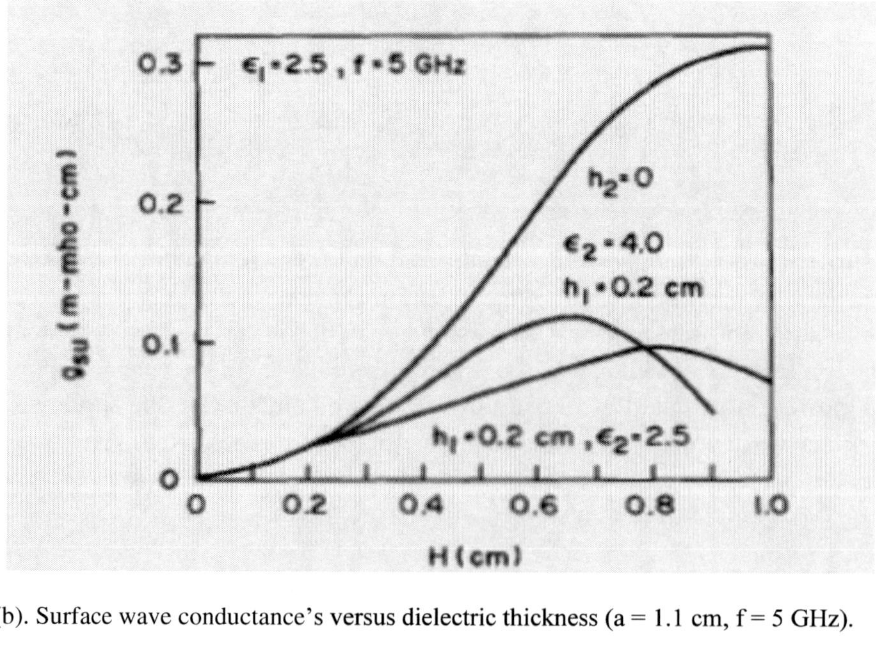

Figure 2(b). Surface wave conductance's versus dielectric thickness (a = 1.1 cm, f = 5 GHz).

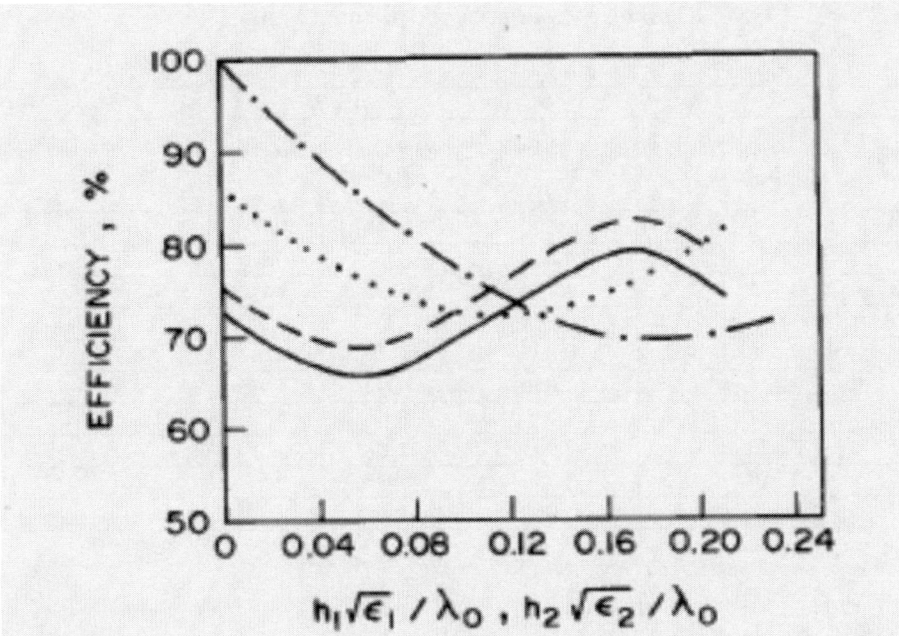

Figure 3. Radiation efficiency versus normalized Substrate superstrate thickness.

A new modified Wolff model (MWM) has also been proposed to calculate the resonance frequency of the circular covered and uncovered patch. And found that for uncovered patches the calculated frequency is within 0.5% of the measured value whereas for covered patch antennas the resonance frequency is within 1% of the results using from Spectral Domain Analysis (SDA).

The resonance frequency of circular covered and uncovered patch obtained;

$$f_r = \frac{v_o \cdot \alpha_{nm}}{2\pi . r_e \sqrt{\varepsilon_{dyn}}} . \tag{2}$$

Where for the fundamental TM_{11} model, $\alpha_{nm} = 1.841$. v_o is the velocity of light. Wolff and Knoppik have calculated effective radius $\boldsymbol{r_{ef}}$ by using the Kirchhoff approximate result which is suitable only for an air filled capacitor, (equations 2-7, [3]). The values of resonance frequencies calculated using various models are tabulated in Table 1.

Therefore the MWM is capable of giving highly accurate resonance frequency for both dielectrics covered and uncovered microstrip rectangular and circular patches [3]. Z. Shen and R. H. MacPhie proposed a waveguide modal analysis to calculate the input impedance of a circular patch antenna covered by a uniaxial anisotropic superstrate and fed by a centered coaxial line. The obtained results indicate that both the thickness and the dielectric constant of the superstrate have a significant influence on the antenna's performance. Finally, the effect of the anisotropy of the substrate and superstrate on the input impedance of a center-fed microstrip disk antenna was also examined for a circular patch of radius 15 mm on a substrate of thickness h = 1.6 mm and covered with a superstrate with the same parameters as the substrate.

Table 1. Resonance frequency results

				Wolff		*Modified Wolff*	
Radius r_0(cm)	***h_1 (mm)***	ε_{r1}	***f_r (Measured) GHz***	***Original GHz***	***Error* %***	***GHz***	***Error* %***
3.493	1.588	2.50	1.570	1.569	0.06	1.570	0.0
1.270	0.794	2.59	4.070	4.267	4.84	4.179	2.60
3.493	3.175	2.50	1.510	1.526	1.06	1.525	1.00
13.894	12.70	2.70	0.378	0.362	4.23	0.374	0.95
4.950	2.350	4.55	0.825	0.836	1.34	0.826	0.20
3.975	2.350	4.55	1.030	1.042	1.17	1.027	0.20
2.990	2.350	4.55	1.360	1.17	1.77	1.368	0.60
2.000	2.350	4.55	2.003	3.19	3.19	2.017	0.70
1.040	2.350	4.55	3.750	5.68	5.68	3.761	0.30
0.770	2.350	4.55	4.945	7.34	7.34	4.946	0.02

* Error in f_r, = (f_r measured $-f_r$ theoretical)/ (f, measured)

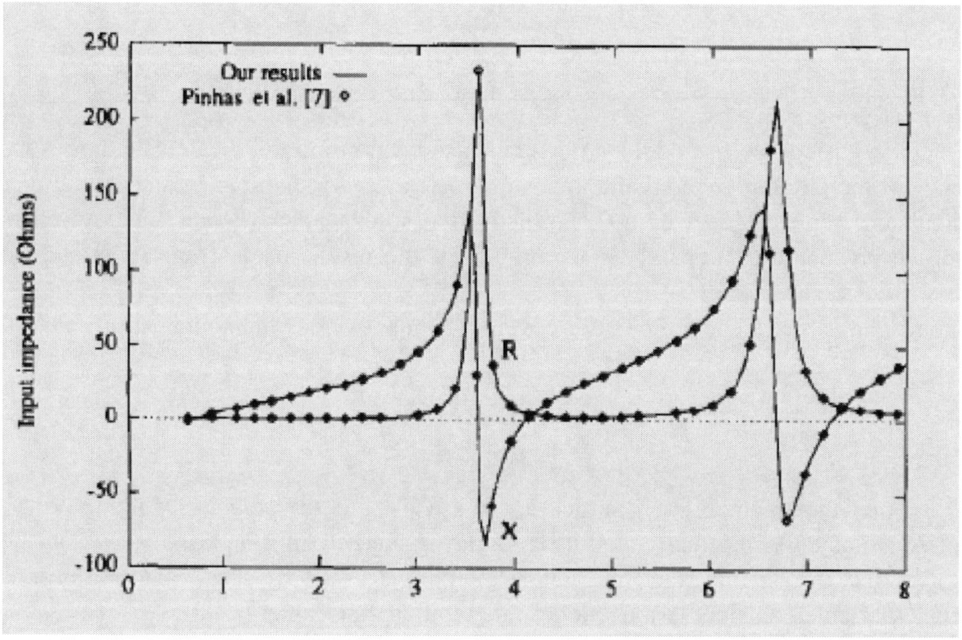

Figure 4. Input impedance of a circular patch antenna fed by a centred coaxial line.

It is also noted that the substrate anisotropy decreases the resonant frequency, because a negative uniaxial substrate equivalently produces a large effective dielectric constant of the substrate [4].

Debatosh Guha and Jawad Y. Siddiqui minimised the complexity of calculating the resonant frequencies of superstrate loaded circular microstrip antennas (Figure 5(a)), to both low and high dielectric constant superstrates [5].

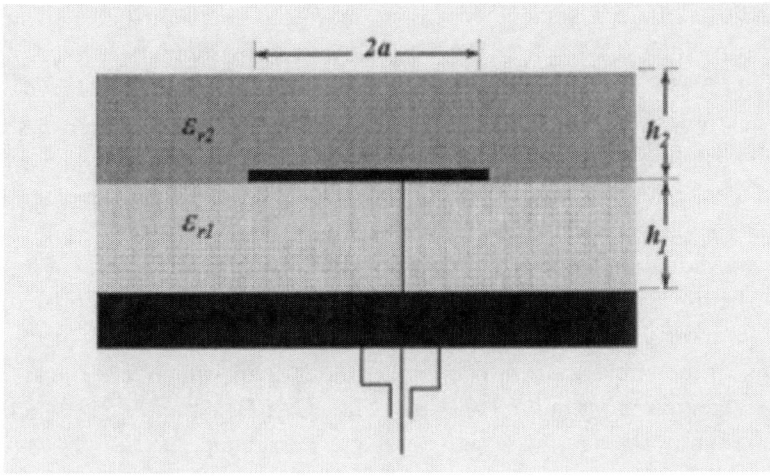

Figure 5(a). Coax-fed microstrip circular *patch* antenna covered with a dielectric superstrate.

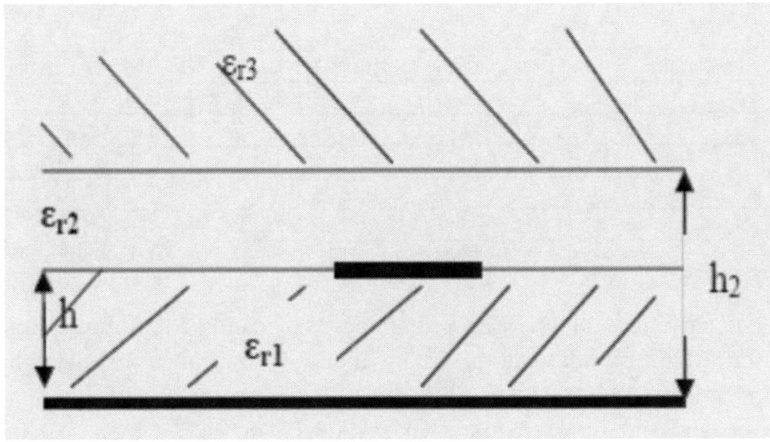

Figure 5(b). Circular Microstrip Patch with spaced Superstrate.

The proposed method is based on cavity model, and found to be closest approximations with the experiments with average percentage error as low as 0.52%.[Equations: 1-12, **5**]. In addition to superstrate loading (ICE) effects of air gap between substrate and superstrate on the performance of a circular patch antenna (Figure 5(b)) has also been studied [6]. The resonance frequency of the proposed antenna was estimated by:

$$f_{mn} = \frac{\alpha_{n.m}.c}{2\pi a_e(\varepsilon_e)^{1/2}}. \tag{3}$$

α_{nm} = m^{th} Zero of the derivative of Bessel's Function of order *n* [5].
c = Velocity of light.
a_e = Effective radius of circular patch.
ε_e = Effective relative permittivity.

Analysis indicates that % change in f_r , is more for small air-gap and tends towards Circular Microstrip patch (CMP) for very long air gap. In particular % fractional change is less than 0.4% and 0.75% for superstrates of $\varepsilon_r \cong 3.27$ respectively.

K. Guney has proposed a method to study the performance characteristics of a circular patch antenna loaded with the dielectric cover (Figure 6(a)). The method is based on adaptive network fuzzy interference system (ANFIS). The ANFIS used in this paper implements a first-order Sugeno fuzzy model which is the most widely applied one for its high interpretability and computational efficiency, and built-in optimal and adaptive techniques. For the ANFIS, the inputs are h_1/a, h_2/a, ε_{r1}, and ε_{r2}, and the output is the normalized resonant frequency f_r/f_{no}. f_r and f_{no} represent, respectively, the resonant frequencies of the TM_{11} mode (dominant mode) of the circular MSA with and without dielectric cover. The ANFIS utilized six different optimization algorithms; HL, LSQ, NM, GA, DEA, and PSO which are used to determine the optimum values of the design parameters and adapt the FIS [7].

The optimum design parameters of the ANFIS determined with the use of HL, LSQ, NM, GA, DEA, and PSO are given in Table 2.

Table 2. Training and test errors of ANFIS models optimized by different algorithms

Optimization algorithms	RMS training errors	RMS test errors
HL	0.0004860	0.0013710
LSQ	0.0003898	0.0002597
NM	0.0004809	0.0005790
GA	0.0004868	0.0009843
DEA	0.0004867	0.0009824
PSO	0.0004899	0.0009589

These RMS error values clearly show that the best results for training and testing are obtained from the ANFIS model optimized by the LSQ optimization algorithm. The test results of ANFIS optimized by the LSQ are compared with the results of the moment-method full-wave analysis and the CAD formula in (Figures 4-6, [7]) which are in very good agreement. In order to make a further validation, the results of ANFIS optimized by the LSQ are compared with the results of Galerkin's method in the Hankel transform domain in Figure 6(b).

It is seen from Figure 6 that the results of ANFIS are in good agreement with the results of Galerkin's method in the Hankel Transform domain. It is found that the result of ANFIS agrees better with the measured result than the results of [7]. In this paper, only the resonant frequency of the TM_{11} mode is calculated by using the ANFIS because this circular microstrip patch mode is widely used in MSA applications. However, the ANFIS can be easily adapted to compute the resonant frequencies of higher order modes of practical interest if the data sets for these modes are available. It must also be emphasized that the proposed method is not limited to the resonant frequency calculation of circular. The ANFIS has the advantages of easy implementation and good learning ability. The high-speed real-time computation feature of the ANFIS recommends its use in antenna CAD programs.

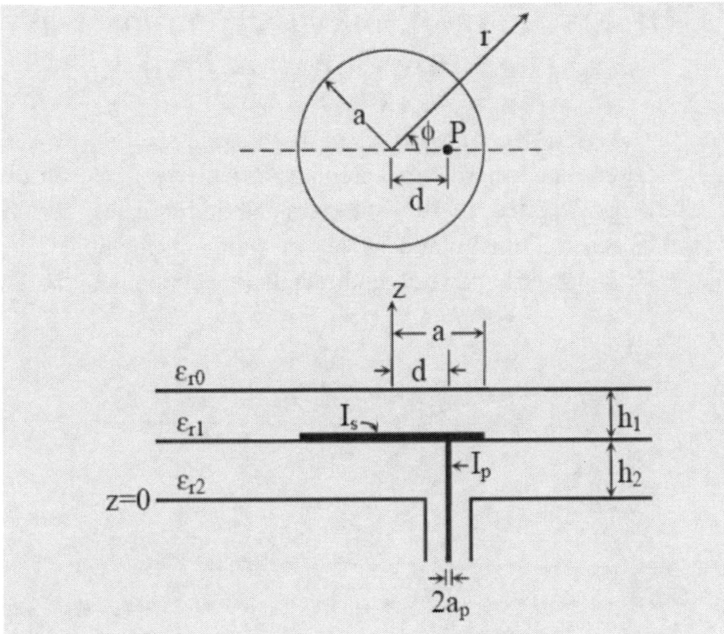

Figure 6(a). Geometry of the circular microstrip antenna with a dielectric cover.

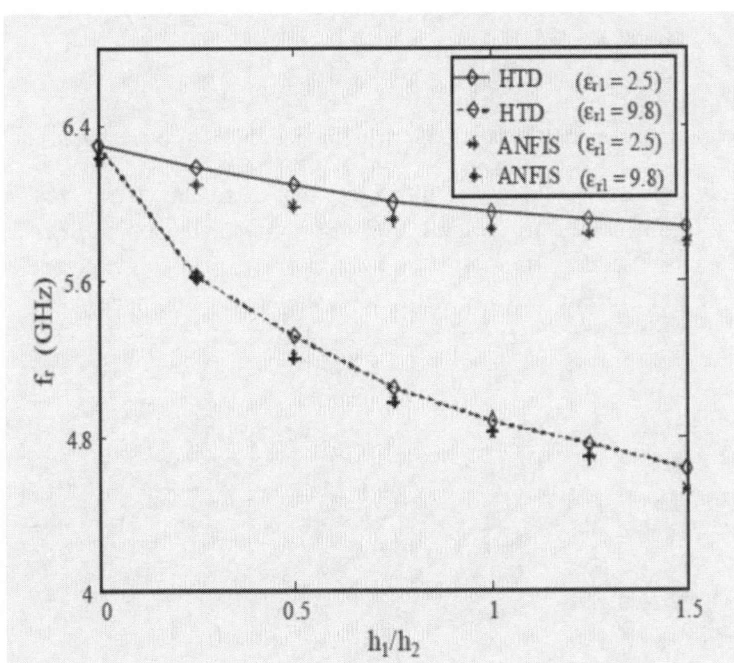

Figure 6(b). Comparison of the resonant frequency results of the ANFIS optimized by the LSQ and Galerkin's method in the Hankel transform domain (HTD) for h_2 = 1.5875 mm, a/h_2 = 5, and ε_{r2} = 2.5.

III. Effect on Radiation Pattern and Half Power Bandwidth

Kin-Lu Wong et al. described the effect of superstrate loading especially of different thickness on spherical-circular microstrip antenna particularly its resonance and radiation characteristics. They investigated these said parameters by using the Green's function formulation in the spectral domain and Galerkin moment method, and calculated the resonance frequency, half power bandwidth and radiated the antenna in the TM_{11} mode.

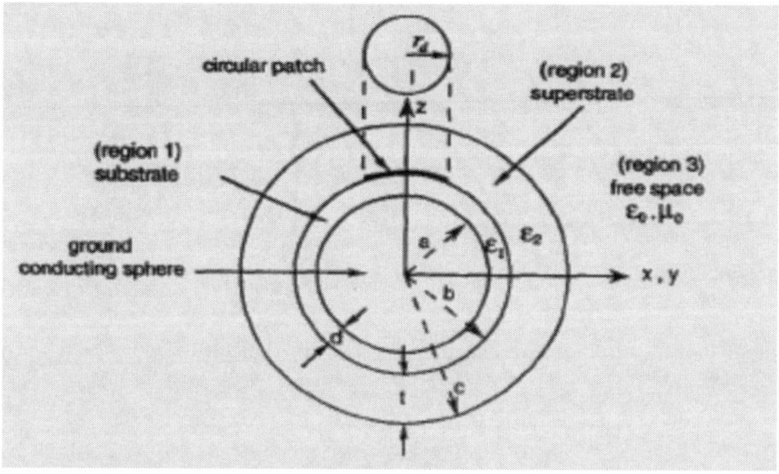

Figure 7. Geometry of a superstrate-loaded spherical-circular microstrip patch antenna.

The obtained results of analysis indicates that resonant frequency and half power bandwidth vary significantly for higher dielectric constants superstrates especially when dielectric constants of superstrate is greater than that of substrate whereas beam width of E-plane pattern is found to be decreasing with increasing cover thickness.

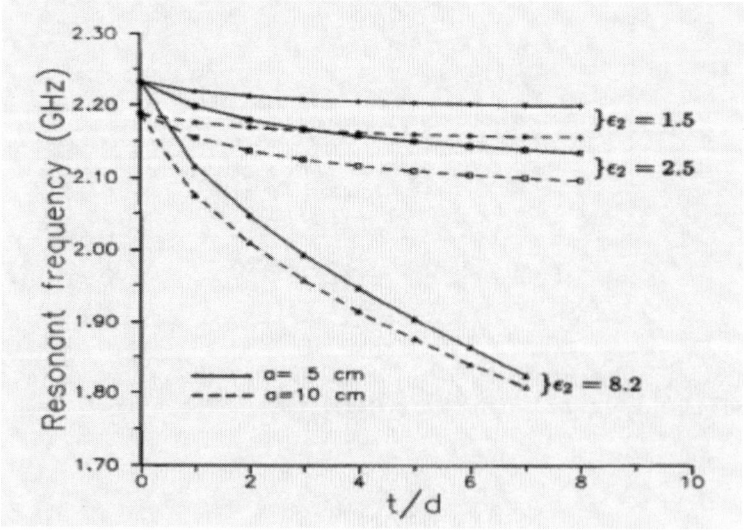

(a)

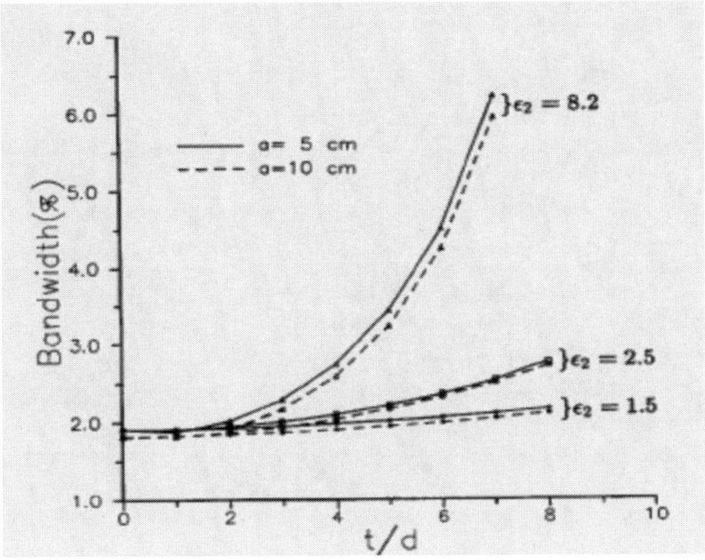

(b)

Figure 8. Dependence of (a) the resonant frequency and (b) the half-power bandwidth on the superstrate thickness for the cases of $\varepsilon_2 = 1.5, 2.5, 8.2$; $\varepsilon_1 = 2.5$, $r_d = 2.5$ cm, and d = 1.588 mm. Two different sizes of the sphere, with a = 5 and 10 cm, are shown.

But the H-plane radiation pattern is slightly affected by the superstrate loading [8]. It is also found that side lobe and back lobe levels for E-plane patterns increases with increasing the radius of sphere. The obtained results are shown in Figure 8 and Figure 9.

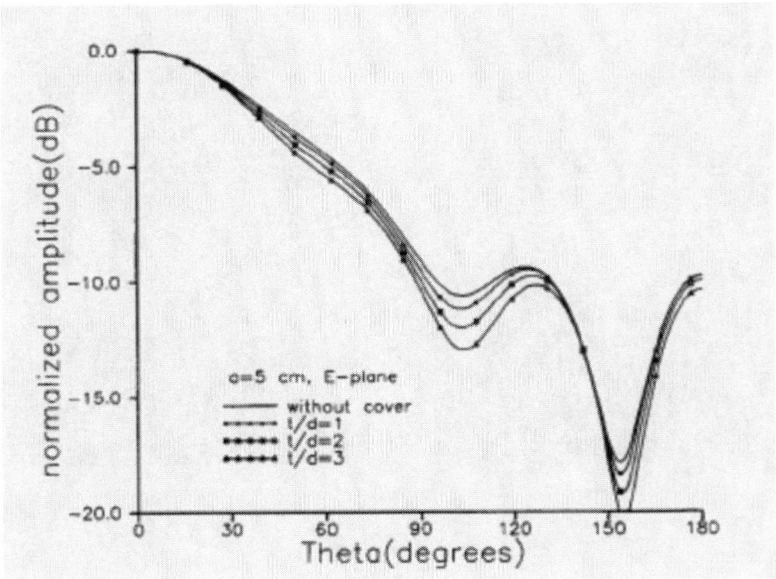

(a)

Figure 9. (Continued)

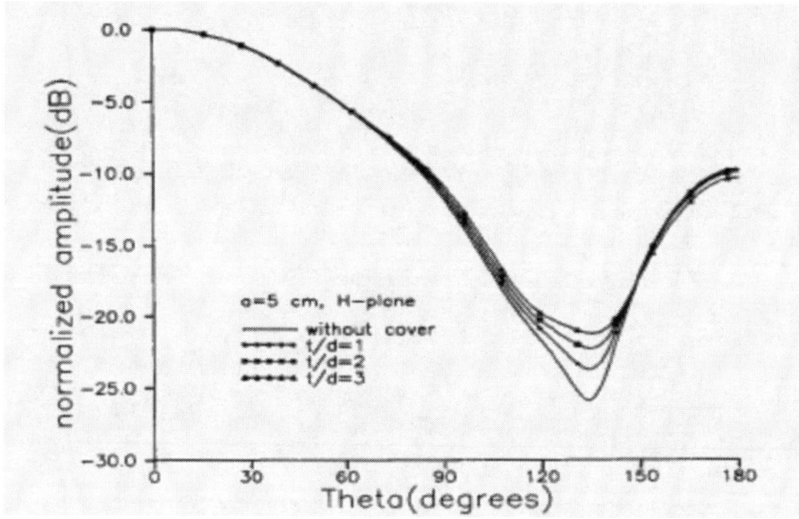

(b)

Figure 9. Far-field radiation patterns of the spherical-circular patch antenna with a = 5 Cm d = 1.6, rd = 1.88 Cm $\varepsilon_1 = 2.47$, $\varepsilon_2 = 2.32$, t / d = 0, 1, 2, and 3. (a) E-plane pattern, (b) H-plane pattern.

R. J. Allard, et al. extended the Reciprocity approach for the analysis of arbitrarily Y – shaped superstrate loaded microstrip antenna mounted on circularly- cylindrical platforms, and has been studied the effects of a superstrate in the far-field radiation patterns of the elliptical patch antennas [9]. The obtain results well agreed with expected results, Figure 10.

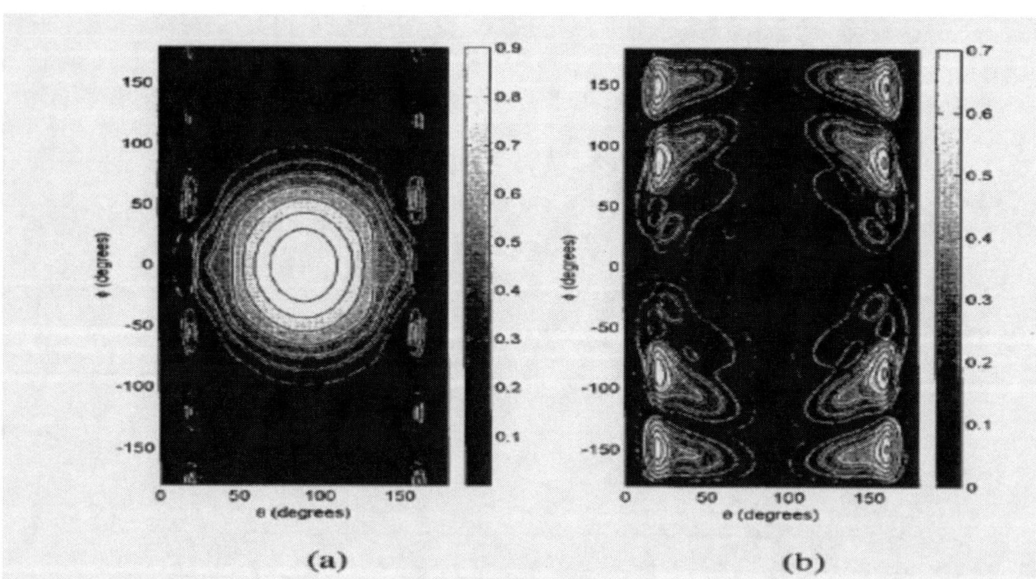

Figure 10. Contour plots of (a) E_θ and (b) $E_\varnothing$ for an elliptical patch with Duroid Substrate (**ε_{r1}** = 2.33) and Duroid Superstrate (**ε_{r2}** = 2.33).

IV. MULTILAYER LOADING

F. Zavoslt and J. T. Aberle described a cavity-backed multi-layers circular patch-antenna, and found that the antenna can provide improved performances; gain enhancement, polarization diversity, frequency agility, or increased bandwidth. In particular, E-plane co-pol. (co-polarization) pattern for varying superstrate permittivity indicates that the broadside gain of the antenna increases and side-lobe levels as well as bandwidth decreases, with increasing superstrate permittivity (Figure 11), which illustrates that the bandwidth decreases from about 2%, for the lowest-gain design, to less than 1%, for the highest-gain case [10].

In addition, Figure 12 shows that the resonant frequency of the patch, with superstrate of relative permittivity of 10 and 20, decreases by nearly 1.8 GHz and 2.3 GHz respectively.

Although substantial reduction in the operating frequency is achieved with a corresponding decrease in impedance bandwidth, it may also be noted that the presence of the cavity suppresses guided waves that are excited in the superstrate; hence, no reduction in radiation efficiency occurs. V. Losada et al have used Galerkin's method in the Hankel transform domain (HTD), for computing the resonant frequencies, quality factors, and radiation patterns of the resonant modes of circular microstrip patches embedded in multilayered dielectric substrates. The numerical results obtained via Galerkin's method in the HTD have been compared with previously published numerical results, measurements and numerical results obtained via the electromagnetic simulator "Ensemble". Good agreement has been found in all cases.

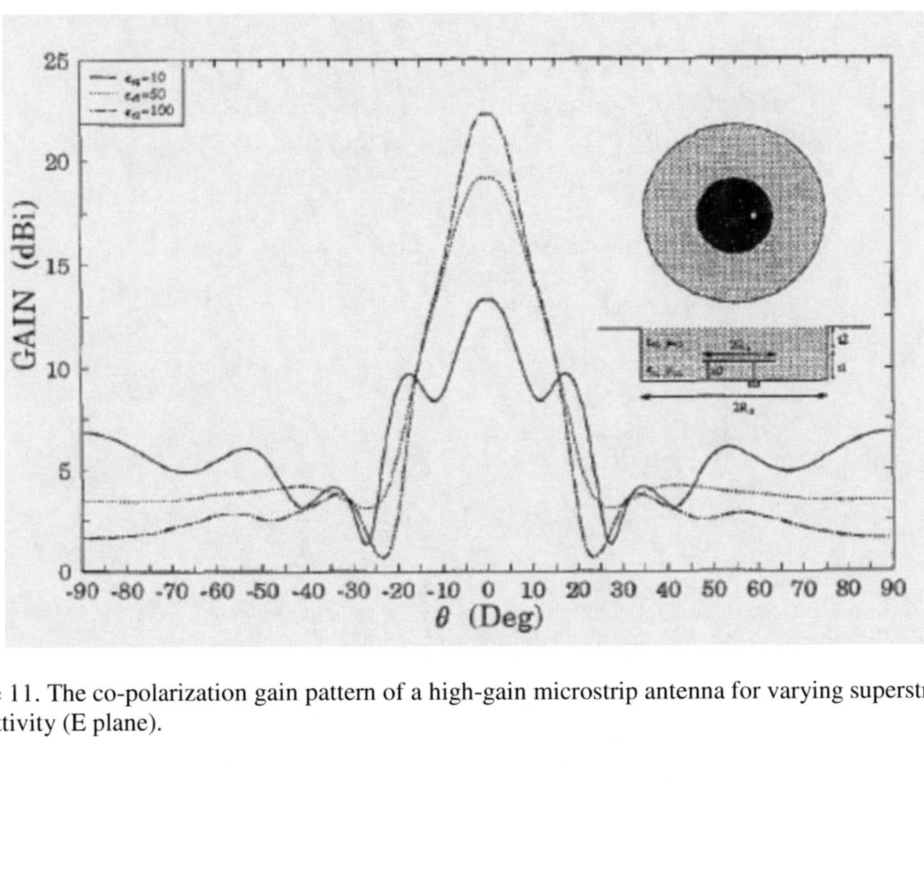

Figure 11. The co-polarization gain pattern of a high-gain microstrip antenna for varying superstrate permittivity (E plane).

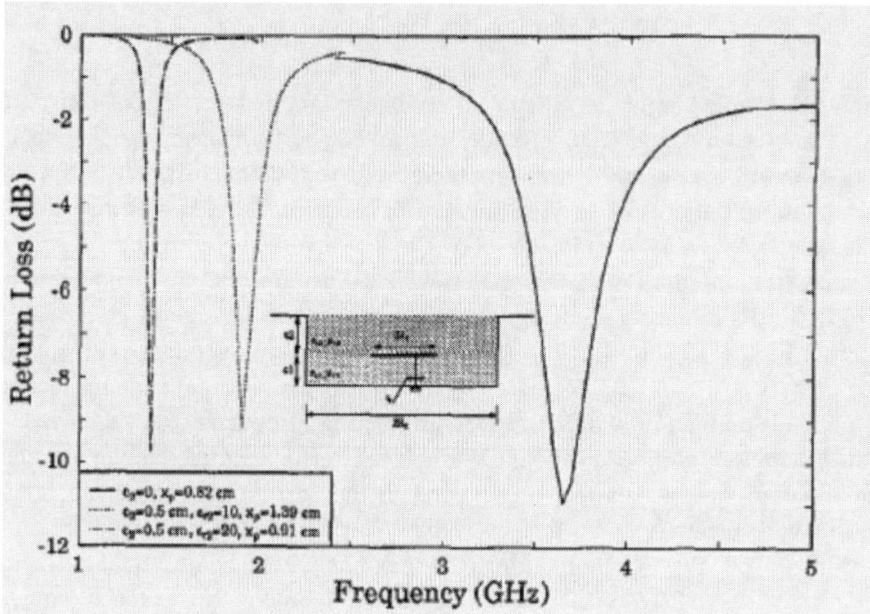

Figure 12. The return loss of a recessed cavity-backed patch antenna versus frequency for varying superstrate permittivity.

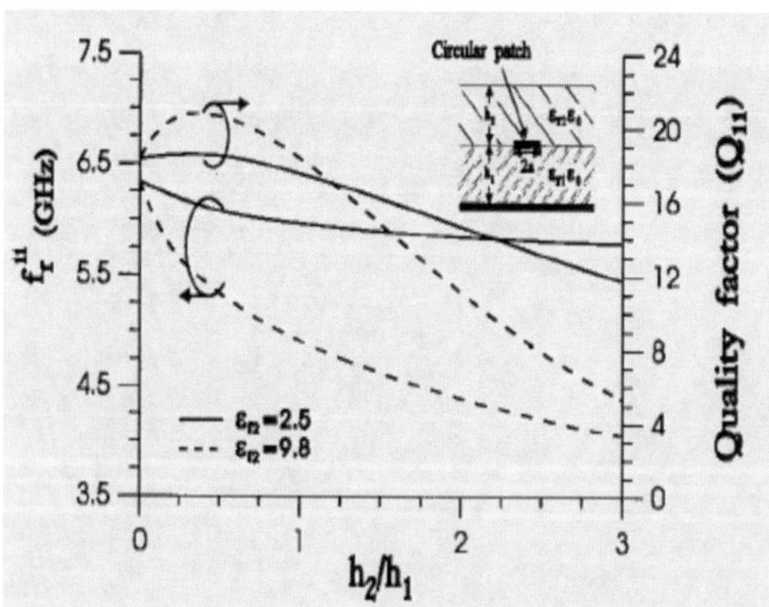

Figure 13. Resonant frequencies and quality factors of the fundamental resonant mode (m = 1; n = 1) of a circular microstrip patch printed on a dielectric substrate with two different superstrates of varying thickness.

The variation of resonance frequency and quality factor for superstrate thickness is shown in Figure 13, which shows that the effect of the superstrate on the resonant frequency and quality factor of the circular patch analyzed is stronger as both the thickness and permittivity of the superstrate increases [11]. Using FDTD method, Raj Kumar, et al have analyzed superstrate loading effects on multilayer antenna and design a four layered circular microstrip patch antenna calculated resonant frequency [12]. The average deviation for circular patch antenna with and without superstrate cover is found to be 0.4% and 0.56% respectively. However gain enhancement effect of doubling the layer on a cavity-backed slot antenna at 9.5 GHz has been studied using Transmission line model. And it is found that if the thickness and dielectric constant of each of the superstrates are properly chosen, gain up to 16.7 dB with 10 dB return loss bandwidth around 12% can be achieved.

The performance of the TM_{02} mode circular patch antenna with two superstrate layers has also been investigated by Jun Xia et al in year 2000. The study indicates that the antenna pattern exhibits minimum radiation in the zenith direction and is fairly symmetrical in azimuth, which is suited for wireless LAN application. It is also found that the gain is increased by about 1.6 dBi when the two superstrate layers are added. That is with the two superstrate layers, the maximum gain of the antenna increase, and by tuning the thickness of the superstrate layers, the elevation angle of the maximum gain can be changed to suit the various wireless LAN applications [13].

Table 3. Comparison of Resonant Frequency Calculated by Present Method and FDTD Method

$\varepsilon_{r2} = 2.43$, $\varepsilon_{r3} = 2.5515$, $\varepsilon_{r4} = 3.2$, $h_{12} = 0.49$ mm, $h_3 = 0.49$ mm, $a = 11.78$ mm, $h_{11} = 0$, $\varepsilon_{r2} = 1.0$				
***h*₄ (mm)**	***h*₄/*h*₃**	***fr* (FDTD)**	***Return loss***	***fr* (PM)**
0.2	0.408	4.40	-18.17	4.3923
0.5	1.022	4.39	-18.82	4.3701
1	2.04	4.36	-19.5	4.3496
2	4.08	4.34	-18.58	4.3229
5	10.2	4.29	-23.19	4.2636

Resonant frequency of circular microstrip patch $f_r = 4.5611$ GHz.
Resonant frequency of covered microstrip patch $f_r = 4.5276$ GHz for $h_3/h_{12} = 1$.

V. POLARISATION

Hong-Twu Chen et al. analyzed a section-spherical dielectric loaded circular microstrip patch antenna and found that there are two resonant peaks in the resistance locus. The first peak, around 1.97 GHz, is due to the presence of the section spherical dielectric cover; the second peak, around 2.1 GHz, is due to the circular microstrip patch. The E-plane and H-plane radiation patterns for the antenna operated at 2.1 GHz are also determined and shown in Figure 14. It is also seen that co-polarization patterns are broadside, the cross-polarization radiations for both E- and H- planes are below -20 dBi, and the antenna gain is increased up to **8** dBi [14]. In addition it was also found that, L. Economou et al has analyzed circular patches printed on RT Duroid, with glass laminated superstrates and found that the glass improved the bandwidth of the antenna significantly, from under 2% to about 7'%, due to the thick high-permittivity glass superstrate. However, variations in the dimensions of the glass in production are significant, up to 15%. The effect of such dimensional tolerances was to change the resonant frequency by about *3%,* and the bandwidth also varied by about 1.7%. Figure 15 shows the measured input impedances for a 16 mm diameter disc before, and after, it was attached to the glass laminate and found that the bandwidth for the patch alone was initially 1.8%, and this increased to 6.8% when attached to the glass; the input impedance fell from 58Ω to 47Ω respectively. This is a useful improvement in bandwidth; particularly in communication systems often need bandwidths exceeding *5%* [15].

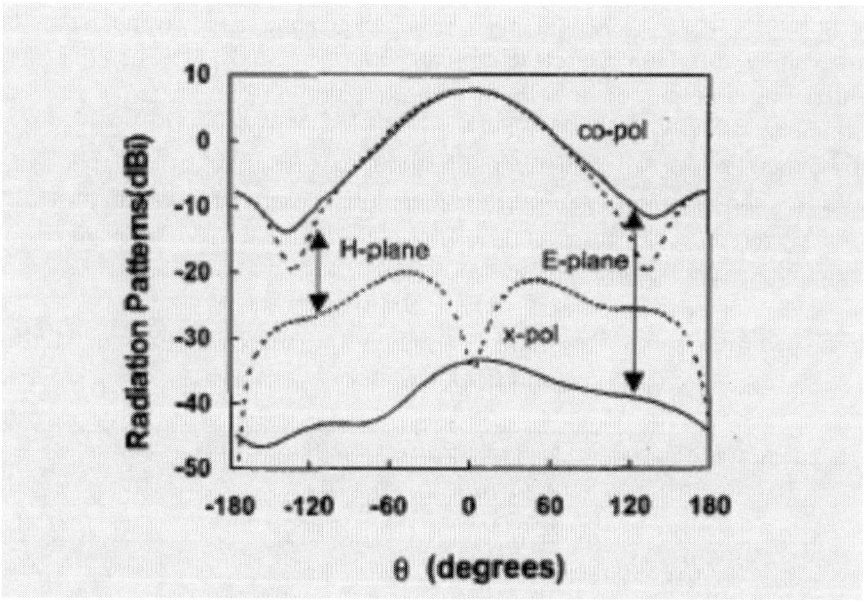

Figure 14. Simulated radiation patterns for the section-spherical dielectric loaded circular microstrip patch antenna.

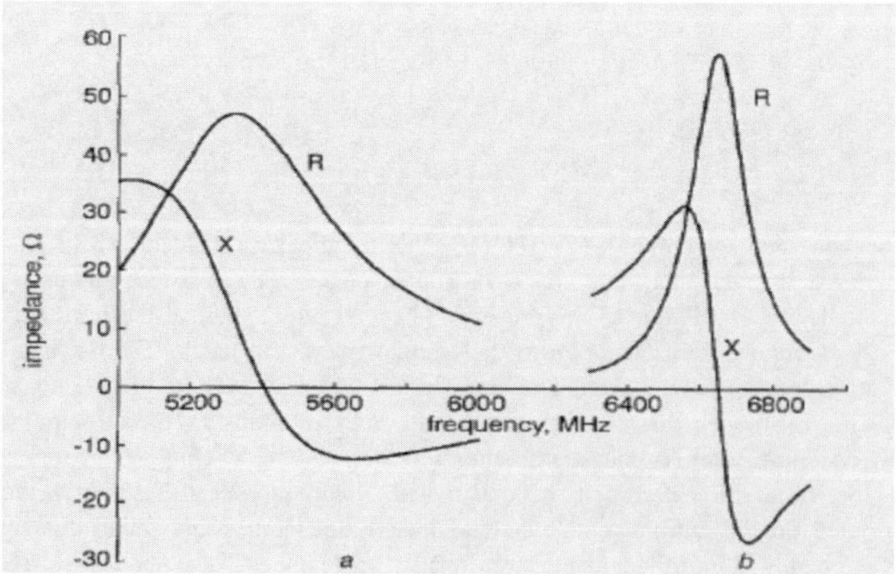

Figure 15. Measured input impedance of circular patch antenna 16nm in diameter (a) with and (b) without glass superstrate attached.

Later D. D. Krishna et al. [16] proposed a superstrate loaded slotted circular dual frequency microstrip antenna shown in Figure 16 and found that the superstrate, not only acts as a random, but also improves the operating bandwidth of the antenna.

The overall size of the antenna reduces to about 60% than that of a un slotted patch along with good efficiency, gain and bandwidth.

The resonance frequency f_r of a circular patch with air as substrate ($\varepsilon_r = 1$) and a thin superstrate (ε_{r1}, h_1) is approximated as follows [17]

$$f_{r1} = \frac{1.841.c}{P_e.\sqrt{\varepsilon_{eff}}}. \tag{4}$$

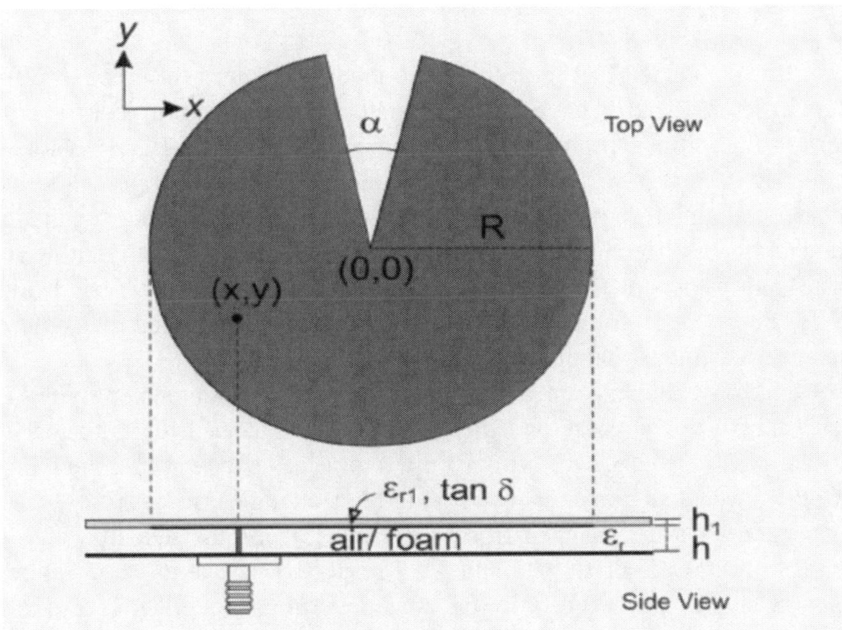

Figure 16. Geometry of the microstrip-fed sector slotted circular patch antenna with a random.

Where

ε_{eff} = The effective permittivity of a two layered patch antenna.

P_e= the effective circumference of the equivalent circular patches C_1 or C_2.

$$P_e = P_i\left[1 + \frac{2h}{\pi.R.\varepsilon_{re}}\left(ln\frac{\pi R}{2h}\right) + 1.7726\right]^{1/2}. \tag{5}$$

Where

$$i = \begin{cases} 1, P_1 = Circumference\ of\ C_1 = R(2\pi - \alpha) + 2R \\ 2, P_2 = Circumference\ of\ C_2 = R(2\pi - \alpha) \end{cases}$$

R is the radius and α is the slot angle of the proposed antenna.

The main findings of their studies are: by varying the slot angles from (10^0 to 90^0).The ratio of the two resonances can be changed from 1.402 to 1.51.The radiation efficiency and bandwidth can also improved up to 98% and 3.9% respectively.

CONCLUSION

Therefore, it is concluded that the superstrate loading changes the resonance frequency, gain, radiation efficiency and bandwidth of the patch antenna, and amount of effects vary with thickness of dielectric, types of dielectrics, impedance matching and polarization. The resonant frequency decreases with increasing the thickness of superstrate, while half-power bandwidth changes slightly. The maximum value of the radiation efficiency is obtained in the case of an antenna with different substrate and superstrate materials than that of an antenna with substrate and superstrate of same dielectric. The half power bandwidth varies significantly for higher dielectric constants superstrates especially when dielectric constants of superstrate are greater than that of substrate.

In case of multi-layer superstrate loading, it is found that the antenna provides improved performances, and if the thickness and dielectric constant of each of the superstrates are properly chosen, gain up to 16.7 dB with 10 dB return loss bandwidth around 12% can be achieved. For the covered slot antenna, by varying the slot angles from (*10^0 to 90^0*), the ratio of the two resonances can be changed from 1.402 to 1.51 and the radiation efficiency and bandwidth can also improved up to 98% and 3.9% respectively. Therefore, in future research works all above studies are needed to be considered for the effectiveness of modern technology with emphasis on compensation of described effects.

REFERENCES

[1] K. M. Luk et al., “Analysis of circular microstrip antennas with superstrate. *Proceedings of the Institute of Electrical Engineering,* Vol. 136, Pt. H No. 3, pp. 261-262, June 1989.

[2] A Bhattacharyya, “Characteristics of circular patch on thick substrate and superstrate,” IEEE Trans. Antennas Propagation, Vol. 39, No. 7, pp. 1038-1041, July 1991.

[3] K. Verma and Z. Rostamy, “Modified Wolff model for the determination of resonant frequency of dielectric covered circular microstrip patch antenna,” *Electron Letters* vol. 27, No. 24, pp. 2234-2236, Nov. 21, 1991.

[4] Z. Shen and R. H. MacPhie “Input impedance of a centre- fed circular patch antenna covered by a uniaxial anisotropic superstrate”, *Microwave and optical technology letter*, Vol. 13, No. 2, pp. 77-81, October 5, 1996.

[5] D. Guha and J. Y. Siddiqui, "Resonant frequency of circular microstrip antenna covered with dielectric superstrate," *IEEE Trans. Antennas and Prop*agation. Vol. 51, No. 7, pp. 1649-1652, July 2003.

[6] Raj Kumar, P. Malathi and J. P. Shinde, "Design of Circular Microstrip Patch with Spaced Dielectric Superstrates ", *IEEE*, pp. 315-317, 2007.

[7] K. Guney, N. Sarikaya," Resonant frequency calculation for circular microstrip antennas with a dielectric cover using adaptive network –based fuzzy inference system optimised by various algorithms "*Progress In Electromagnetics Research,* PIER, pp. 72, 279-306, 2007.

[8] K. L.Wong, S. F.Sun-Fu Hsiao and H. T. Hong -Twu Chen, "Resonance and radiation of a superstrate-loaded spherical-circular microstrip patch antenna," *IEEE Trans. Antennas Propagation.*, Vol. 41, No. 5, pp. 686-690, May 1993.

[9] R. J. Allard, D. H. Werner and J. S. ZmysloA, "Domain- decomposition Reciprocity technique for the analysis of arbitrarily - shaped dielectric- loaded microstrip antennas as mounted on circularly- cylindrical platforms," *IEEE*, pp. 2-5, 2000.

[10] Frank Zavoslt and Janies T. Aberle "Improving the Performance of Microstrip-Patch Antennas", *IEEE Antennas and Propagation Magazine,* Vol. 38, No. 4, pp. 7-10, August 1996.

[11] V. Losada, R. R. Boix, and M. Horno, "Resonant modes of circular microstrip patches in multilayered substrates", *IEEE Trans. Microwave Theory Tech.,* Vol. 47, pp. 488-498, April 1999.

[12] Raj Kumar, P. Malathi and Yogesh B. Thakere, "On the Design of Four layered Circular Microstrip Patch Antenna", *Microwave Optical Technology Letters*, Vol. 50, No. 12, pp. 3206-3212, December 2008.

[13] Jun Xia, Soon Hie Tan and Jeffrey S Fu, "Analysis of TM_{20}, -mode circular patch antenna with two superstrate layers for wireless LAN's", *IEEE,* pp. 1264-1267, 2000.

[14] Hong-Twu Chen, Yuan-Tung Cheng, Shyh-Yeong Ke "Section-Spherical Dielectric Loaded Circular Microstrip Patch Antennas", *IEEE*, pp. 786-789,2000.

[15] L.Economou et al, "Circular microstrip patches antennas on glass for vehicle applications", *IEE Proc. Microwave* Antennas *Propagation*, Vol. 145, pp. 416-420, 1998.

[16] D. D. Krishna et al, "Compact dual band slot-loaded circular microstrip antenna with a superstrate," *Progress in Electromagnetic Research*, Vol. 83, pp. 245-255, 2008.

[17] R. Garg et al., "Microstrip Antenna Design," *Handbook, Artech House*, 1995.

In: Knowledge Management and Communication ...
Editor: Rafiqul Islam
ISBN: 978-1-62808-250-0

Chapter 13

A Non Linear Optimization for Microwave Active Circuits Using the Retreat Function

S. Aidel*, N. Boucenna, A. Said and S. Belaidi
LABCOM, Department of Electrical Engineering
Centre Universitaire de Bordj Bou Arreridj, Algeria

Abstract

The main objective of this paper is to show one way of optimization, analysis and design of microwave active circuits, based on the transformation of the reflection coefficient for an amplifier or a microwave oscillator. This improved method is based on a non linear optimization of the circuit. This way consists on introducing a transformation invariant which allows to test the appropriate transistor for a given frequency bandwidth.

The transformation invariant is called coefficient N, and it determines very quickly, the possible modifications of the active multiport network by external circuits connected to it. This procedure is applied by looking at the microwave transistor as a three- port or as an equivalent two-port. The present method appears to be accurate and fast in the design and optimization of a microwave oscillator or an amplifier based on H.F transistors.

Keywords: Microwave oscillator, Optimization, Computer aided design, H.F transistor

Nomenclature

N - Coefficient of Level
M - Coefficient of Modification
W - Load Coefficient
Γ_A - Coefficient of Reflection of Reference Plane
Γ_R - Coefficient of Reflection of Tuning Plane

* E-mail: aidel_salih@yahoo.fr.

Γ_{A0} Coefficient of Reflection for Small Signals
I_c - Current Source
Ps - Power Source
f - Frequency
T_A - Reference Plane
T_M - Modification Plane
$R(P_S)$ - Retreat Function
θ_{A0} - Angle Phase for Small Signals
θ_R - Reflection Angle Phase
$\phi(R(P_S))$ - Retreat Function Angle Phase
S_{AA},S_{MM} - Scattering Parameter Coefficients
S_{AM},S_{MA} - Scattering Parameter Coefficients

I. Introduction

The last four decades have been marked by rapid developments in RF and microwave circuits and by their wide application in a variety of commercial and military systems. Examples of systems in which such integrated circuits have been applied range from cellular telephones and microwave links on the commercial side to missile and electronic warfare systems on the military side, with many dual-use types of applications such as in radar and navigation systems. During the past three decades, several robust optimization techniques have been developed. These techniques supplied designers with strong and reliable tools necessary for the complex and demanding needs of modern circuit design. They utilize the circuit responses and possibly derivative information in the optimization loop.

Choosing the appropriate optimization techniques highly depends on the physics of the structure to be optimized, the required accuracy and available resources. In fact without comprehension of the nature of the problem the design process will be inefficient. Actually no algorithm for optimizing general functions exists that will always find the global optimum for a general minimization problem in a reasonable amount of time.

Recently, commercial software packages, for example, have been developed that does solve Maxwell's equations for circuits of arbitrary geometrical shapes. Such simulators are denoted as Electromagnetic methods. The active elements are represented by non linear one-port devices such as diodes and feedback transistors [1].

For nearly a half a century computer-aided design (CAD) of electronic circuits have evolved from a set of special purpose, rudimentary simulators and techniques to a variety of highly flexible and interactive, general purpose software systems, with amazing visualization capabilities [2].

The big progress in the microwave semiconductor components technology allows more easily the realization of microwave sources with some transistors that are easily available.

By taking into consideration the elementary circuits theory and by applying the condition of quasi-linear oscillation, we can design tuned electrically oscillators.

The choice of a transistor is a very important parameter. In this paper we suggest a new technique that consists in optimizing and choosing the most appropriate high parameter for the large band microwave oscillator circuit [3].

The method consists for example the optimization of microwave sources.

It is based on the introduction of a parameter of homographic transformation that we call it coefficient of level *N*.

This criterion is simple and very quick. It is based on the coefficient of level *N*, on the modification *M* and on the load coefficient *W*. The *N* coefficient is a transformation invariant by any active multi-port [4].

Some of the components of the design optimization loop such as update strategy are common in any design practice. In contrary, some aspects of the optimization exclusively apply to the optimization in electromagnetics domain. Among all this microwave-specific components, defining the objective function is not only distinct for microwave domain (in comparison to other optimization problems) but also it varies among different microwave design problems. Thus the derivation of the objective function in microwave CAD will be covered in more detail. Numerical analysis of microwave structures is another aspect which is distinct for microwave engineering, although these methods will be deployed for different applications beforehand.

II. Necessity of the Non Linear Optimization

When we analyze an active circuit, many natural limitations and some other technological constraints make the practical realization very difficult. This is due the fact that analysis is always subject to compromise between conditions sometimes contradictory.

We can classify these conditions as follows:

1. It is almost always difficult to obtain values of reflection coefficients greater than one when the frequency increases.
2. If we adjust correctly the varactor, we can modify the variation of the area of the reflection coefficient phase, by coupling more or less the varactor and the circuit that is connected to it.
3. We observe a transistor reflectance decrease when we increase the frequency, although the variation of the variation of the diapason increases with the bandwidth.
4. When the coupling coefficient increases, we observe some losses in the reflection coefficient. In almost all cases, the absolute value of this coefficient decreases to stop the oscillation process or the amplification process.

III. Active Circuit Representation

The most important component of the microwave amplifier or the microwave oscillator is the active circuit. It can be designed by using the elementary circuits theory and by applying the quasi-linear oscillation condition. A simple representation of the active circuit is given in Figure 1 [5]. The active circuit is the H.F transistor. In this circuit and at the level of reference plane that is chosen, we have a port by which the transistor decreases the losses of the connected circuit. The reflection coefficient Γ_A has to be greater than one.

It is a non linear coefficient [6] that depends on:

1. The source *(V_{cb}, I_c)*.
2. The generated power by the circuit *(Ps)*
3. The frequency *f*.
4. The value of Γ for small signals denoted Γ_{A0}.

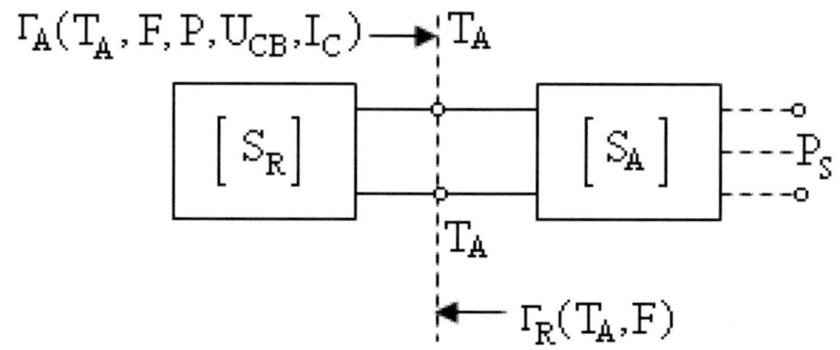

Figure 1. Circuit of the electrically tuned active circuit.

The adjustment circuit is defined by its coefficient Γ_R that depends on the frequency. The oscillation condition can be written as follows:

$$\Gamma_A(V_{CB}, I_C, T_A, P, f).\Gamma_R(T_A, f) = 1 \tag{1}$$

By introducing the retreat function *R (P_S)* we can establish the oscillation as follows:

$$|\Gamma_A|.|\Gamma_R| = \frac{1}{|R(P_S)|} \tag{2}$$

Introducing the phase's condition:

$$\theta_{A0} + \theta_R = -\varphi(R(P_S)) \tag{3}$$

We have:

$$R(P_S) = \frac{\Gamma_A}{\Gamma_{A0}} \tag{4}$$

$$R(P_S) = |R(P_S)|.e^{j\phi_R} \tag{5}$$

$$\Gamma_R = |\Gamma_R|.e^{j\theta_R} \tag{6}$$

$$\Gamma_{A0} = |\Gamma_{A0}|.e^{j\theta_{A0}} \tag{7}$$

P_S is the output power.

These conditions are always verified at the stable point of oscillation [7],[8],[9].

The stability factor should be less than unity for any possibility of oscillation. If this condition is not satisfied, either the common terminal should be changed or positive feedback should be added. Next, the passive terminations must be added to resonate the input and output ports at the frequency of oscillation.

The microwave transistors used germanium and the silicon. The choice of transistors is between the silicon BJT and the gallium arsenide MESFET. A comparison of these devices permits to say that Silicon has another important advantage, a lower flicker corner frequency, which is important for oscillator applications [10].

IV. Scale Coefficient and Active Circuit Modification

The relation that shows this modification and that links the two reflection coefficient depends on the homographic transformation between the two parameters newly introduced and that are the modification coefficient and the load coefficient [11].

The modification is the variation of a circuit parameter in a chosen direction. To understand this, we compare many values of level coefficients that are obtained for either the same structure of the circuit, or for many structure based on different transistor *(FET* or *Bipolar*). We consider the active circuit as a two-port between the T_A and T_M planes. T_A is the connection plane showed in Figure 2.

The relation between Γ_A and Γ_M:

$$\Gamma_A = S_{AA} + \frac{S_{AM}.S_{MA}\Gamma_A}{1 - S_{MM}\Gamma_M} \tag{8}$$

Another possible written form is:

$$\Gamma_A = S_{AA}.(1 + N.M) \tag{9}$$

where:

$$N = \frac{S_{AM}.S_{MA}}{S_{AA}.S_{MM}} \tag{10}$$

is the level coefficient and

$$M = \frac{W}{W - 1} \tag{11}$$

is the load coefficient and

$$W = S_{MM}.\Gamma_M \tag{12}$$

is the modification coefficient.

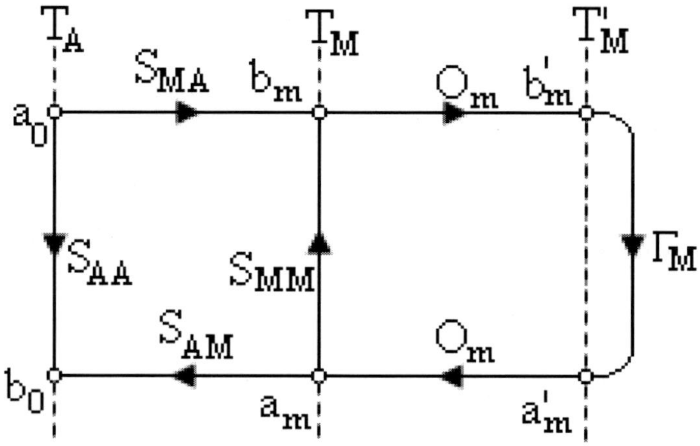

Figure 2. Fluency graph issued from the transformation of Γ_M to Γ_A.

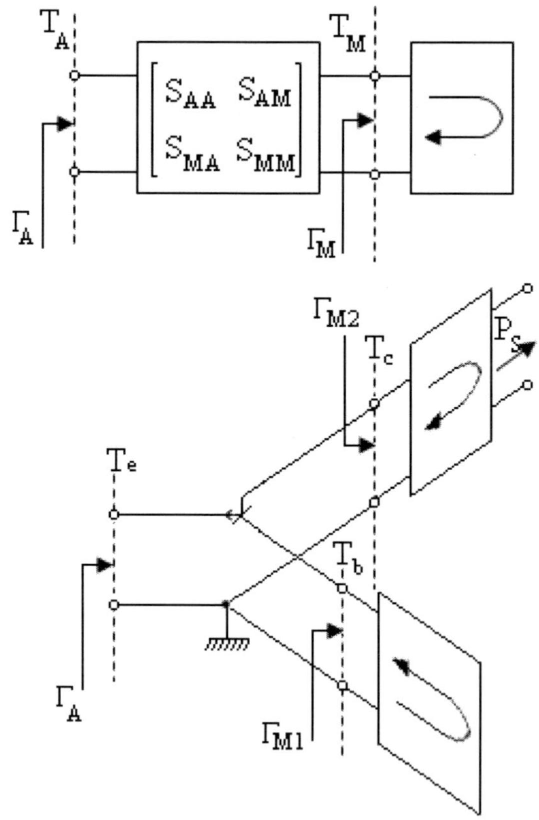

Figure 3. Active circuit modification by external circuits.

This technique uses the transistor as the three-port or an equivalent two-port and using the modification theory represented in Figure 3. [12], [13].

The main principle of this technique is to consider the transistor as an equivalent two or three-port. And by using the modification theory based on the introduction of the level's coefficient *N*, we can show that some configurations are more appropriate than some others. We present also the Smith chart that shows the transformation *M* and the modification and reflection coefficients shape in the active plane of the oscillator and this over a large wide band of frequencies.

To understand the modification of the transistor, we compare the three measured scale coefficients N_{eb}, N_{ec}, N_{bc} for the same transistor taken as a three-port *(HP 35820A).*

The three-port matrices are useful for converting CE to CB and CC configurations. This is most easily done using the three-port *Y* parameters, since the three-port and two-port *Y* parameters are measured to be the same. If we arbitrarily label the emitter port 1, the collector port 2, and the emitter port 3, we may easily show the common-emitter.

In Figure 4, the best possibility to modify this transistor is the one taken between the base and the emitter.

By comparing the bipolar transistor reflectance taken as an equivalent two-port, we remark that the biggest value of this reflectance is issued from the B-C configuration in Figure 5.

An additional advantage of the transfer matrix is revealed in cascaded structures, where the matrix of the cascaded connection can be determined by simply multiplying the transfer matrices of its components. [14], [15].

The disadvantage of using wave matrices is that there is no information about input structure and processes because the matrix describes characteristics of outside parameters only.

A frequency resonance condition simply requires the circuit imaginary term be zero.

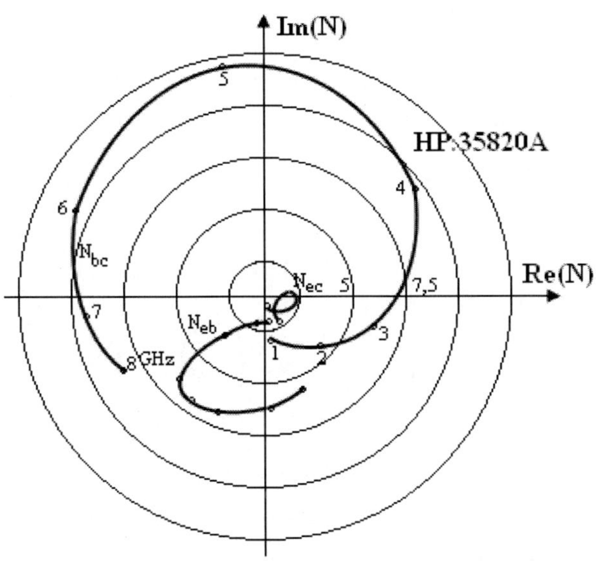

Figure 4. Comparison between 3 level coefficients of transistor HP 35830A.

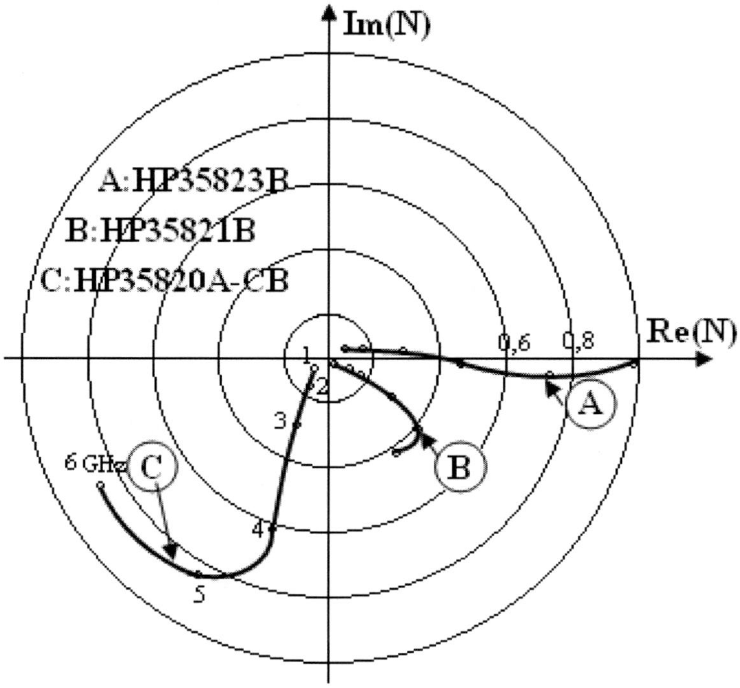

Figure 5. Comparison between bipolar scale coefficients.

If the impedance resonance is on the left-hand real axis, this is a series resonance; that is, at frequencies above resonance the impedance is inductive and below resonance the impedance is capacitive. If the impedance resonance is on the right-hand real axis, the resonance is a parallel resonance; that is, at frequencies above resonance the impedance is capacitive and below resonance the impedance is inductive.

These circuits are investigated in the steady state, and a frequency domain is applied. However, some questions for the best configuration are better answered by a time domain analysis. The investigated circuit is then regarded as being quasi-linear because the time domain investigation of non linear circuits cannot normally be carried out with analytical functions. [16], [17], [18].

V. MEASURMENT AND TEST

The elements of a scattering matrix are relatively easy to measure experimentally. *S*-parameters are related to the power incident on and waves reflected from network ports. In the microwave range, besides frequency we can measure only power and reflection coefficients (module and phase) [19].

These measurements are equivalent to measurements of incident and reflected waves. The *S* parameters related to the terminal voltages and currents cannot be measured because appropriate microwave test equipment does not exist, and also because broadband short and open circuits are very difficult to achieve [20], [21].

The principle of conservation of energy for a given device is easily expressed through the unitary condition of the scattering matrix; it is not easy to verify this condition when *Z* and *Y* matrices are used. When reference planes of networks are changed, only the phases of scattering matrix elements are altered. By contrast, Z and Y elements change both in module and phase. An additional advantage of the transfer matrix is revealed in cascaded structures, where the matrix of the cascaded connection can be determined by simply multiplying the transfer matrices of its components.

The disadvantage of using wave matrices is that there is no information about input structure and processes because the matrix describes characteristics of outside parameters only. Extending the foregoing small-signal model for nonlinear operation can be done in a variety of ways. The most obvious is to base a new circuit model on large-signal S parameters. This can easily be accomplished by measuring device S parameters at elevated power levels. The power level is usually chosen to correspond to the level encountered in the final circuit application. However, there are some negative aspects of using large-signal S parameters. First, as we have learned, S parameters are defined in a linear n-port system with constant load impedance. Under large-signal conditions, the microwave n-port is not linear and large-signal S parameters cannot predict device performance for load impedances other than the one used during measurement [22], [23]. Also, the value obtained for $S22$ is not the conjugate of the optimum load impedance. Having established the oscillation conditions, we now apply the equivalent-circuit elements and the nonlinear equations from the dc (I–V) simulation to determine the oscillator properties under steady-state oscillation conditions for the permissible range of load terminating conditions. The result of the nonlinear analysis is shown in Figure 4.

Shown are closed constant-output power contours (in dBm) as a function of load (drain) terminating conditions. Also shown are intersecting loci of constant-frequency contours. For example, the 7.5-GHz contour shows the predicted power output at various terminating admittance levels. The power levels indicated are in the range obtained experimentally, as shown by the measured data [25],[26].

To alleviate the limitations noted above, the active elements must be characterized as a function of terminating impedance as well as a function of drive level. This technique, commonly called “load pull,” has the advantage that the FET is operated. It is evident from the discussion above that a numerical model with the efficiency of a frequency-domain linear solution is required for efficient interactive nonlinear design. Several such modeling approaches, commonly referred to as “harmonic balance” methods, are solved in the frequency domain but employ time-domain descriptions for the active-element nonlinearities. Thus nonlinear behavior of the total microwave circuit is obtained. Development of a CAD model of this type is relatively straightforward and begins with the linear FET model described previously [27], [28], [29].

Two-port oscillator design may be summarized as follows:

1. Select a microwave transistor with level coefficient, sufficient gain and output power capability for the frequency of operation. This may be based on oscillator data sheets, amplifier performance, or S-parameter calculation.
2. Select a topology that gives a correct N at the operating frequency.
3. Select an output load matching circuit that gives $|S11|>1$ over the desired frequency range. In the simplest case this could be a 50 Ω load.
4. Resonate the input port with a lossless termination so that $\Gamma_A = 1$. The value of $S22$ will be greater than unity with the input properly resonated.

In all cases the transistor delivers power to a load and the input of the transistor [30], [31]. Practical considerations of realizability and dc biasing will determine the best design.

For both bipolar and FET oscillators, a common topology is common base or common gate, since a common-lead inductance can be used to raise *S*22 to a large value, usually greater than unity even with a 50 Ω generator resistor. However, it is not necessary for the transistor *S*22 to be greater than unity, since the 50 Ω generator is not present in the oscillator design. The requirement for oscillation is $|S11| < 1$; then resonating the input with a lossless termination will provide that $|S22| > 1$. [32], [33].

A simple example will clarify the design procedure. A common-base bipolar transistor (H35830A) was selected to design a fixed-tuned oscillator at 5 GHz. [34], [35].

The measurement results are shown in Figures (4 and 5) where the curves A, B, C represent the reflectance of the transistor with the base short-circuited, and the reflectance of the active circuit charged by the output circuit with the resonator. The change in reflectance of the transistor as a function of HP35820A collector current *Ic* for the voltage *Vcb* = 10V shows that the circuit becomes active currents above 3 *mA*. The value of the module is this reflection coefficient *2.05* for the value of 10 *mA* collector current. Curves A and B show the variation of the modulus of the reflectance.

See in function of frequency in the band (24 GHz), for *current Bias Ic=5 mA* and 10 *mA*. Note the sharp drop modulus of the reflection coefficient for frequencies higher than *2.75 GHz*. This parasitic effect is reduced by the operation of the circuit output, which is presented by curve C. The fact of not having obtained a practically large enough value of modulus of the reflectance for the higher frequencies is due to the cutoff frequency of the transistor which equals to *4.5 GHz*. These Figure show the topography of the generator provided to around *2 GHz*. The shape of the power measured with or without resonator, this source as a function of frequency is shown in figures (4 and 5).

The results of tests undertaking in this work show that this technique allow us to choose the appropriate H.F transistor, so, to optimize the structures of microwaves transistors, because the choice of the required transistor of the microwave amplifier or the microwave oscillator for our research is essentially based on the retreat function and the configuration of different coefficients of level *N* of different microwave transistors.

CONCLUSION

Another critical point in the algorithm is the choice of the first guess to choose *N*. A well chosen first guess of the best value of *N* will considerably ease the convergence of the algorithm to the correct solution. If the circuit is mildly nonlinear, the linear solution, obtained for a low-level input, will probably be a good first guess for a given *N* and *M*. If the circuit is driven into strong non-linearity, a continuation method will probably be the best approach. The level coefficient of the input signal is first reduced to a quasi-linear excitation and a mildly nonlinear analysis is performed; then, the input level is increased stepwise, using the result of the previous step as a first guess. In most cases the intermediate results will also be of practical interest, as in the case of a power amplifier or an oscillator driven from small-signal level into compression. Most commercially available CAD software automatically enforces this method when convergence becomes difficult or when it is not reached at all. The tests results show that this technique permits to choose the appropriate HF transistor and to be an efficient selection criterion to optimize active microwave structures, because the choice of the required transistor is based essentially on the analysis and on comparison of the level coefficients of different useful transistors. The methods produce comparable results but are implemented quite differently.

The presented circuits may thus be regarded as examples which can further be utilized for the treatment of many similar circuits.

An interesting comparison can be made between our optimization method for nonlinear optimization using the retreat function, and convex optimization or stochastic methods such as Genetic Algorithms (GA) and Design of Experiments (DoE), indirect optimization methods such as Neural Networks (NN) and Gradient Optimization (GO). Since differentiability of the objective and constraint functions is the only requirement for most local optimization methods, formulating a practical problem as a non-linear optimization problem is relatively straightforward. The art in local optimization of microwave active circuits is in solving the problem (in the weakened sense of finding a locally optimal point), once it is formulated. In convex optimization these are reversed: The art and challenge is in problem formulation; once a problem is formulated as a non linear optimization problem using the retreat function, it is relatively straightforward to solve it.

REFERENCES

[1] M. Golio, J. Golio, *RF and Microwave Applications and Systems*,CRC Press Taylor and Francis Group, 2008, I1-I20.

[2] T. Bercelli, *Nonlinear Active Microwave Circuits* (New York: Elsevier, 1987, 69-71).

[3] G.D. Vandelin, *Design of amplifiers and oscillators by the S parameters method* (New York: John Wiley and sons, 1982, 132-135).

[4] C.F. Scheilbold, An approach to realizing multioctave performance in GaAsFET Yig-tunedd oscillators, *IEEE,* MIT-Digest, chapJ-2, 1985.

[5] R. Gilmore, L. Besser, Practical R.F Circuit Design for Wireless systems, *Active Circuits and Systems,* Vol. 2: Artech House microwave library, 2003, 121-123).

[6] F. Giannini, G. Leuzzi, *Nonlinear Microwave Circuit Design* (New York, John Wiley and Sons, 2004, 172-178).

[7] G.D. Vandelin, A.M. Pavio, U.L. Rohde, *Microwave Circuit Design Using Linear and NonLinear Techniques,* John Wiley and sons, New Jersey, 2005, 520-525.

[8] S.Aidel, A.Szymczak, A. Said, Non-Linear Analysis of the electrical properties of the active circuits of high frequency oscillators, Engineering Institute of Canada, *Canadian Conference on Electrical and Computer Engineering*, Vo. 1 and 2, 4P, SEE N-94-22101 05-33

[9] S.Aidel, A.Szymczak, A. Said, *Conception optimale de circuits micro-ondes*, COMCONEL 90, CAIRO 1990, pp.174-177.

[10] S.Rosloniek, *Algorithms for Computer-aided Design of Linear Microwave Circuits* (Boston: Artech House, 1990, 159-164.

[11] G. Gonzales, *Foundations of Oscillator Circuit Design*: Artech House, 2007, 76-89).

[12] K.Kurokawa, *An Introduction of the Theory of Microwave Circuits*, New York, 1969, Academic Press,380-397.
[13] L.Besser,G.Gilmore, *Practical RF Circuits Design for Modern Wireless Systems Boston*, Artech House, 2003, 286-289.
[14] R.C.Booton, *Computational Methods for Electromagnetics and Microwaves*, New York J.Wiley, 1998, 150-160.
[15] David.M.Pzar, *Microwave Engineering*, J. Wiley, New York, 1998, 635-648.
[16] E.Fooks, R.Zakreviclus, *Microwave Engineering Using Microstrip circuits*, 1990, London, Prentice Hall,28-35.
[17] U. L. Rohde, *Digital PLL Frequency Synthesizers Theory and design*, Prentice-Hall, Englewood Cliffs, N.J., 1983.
[18] D. B. Leeson, "A Simple Model of Feedback Oscillator Noise Spectrum," *Proceedings of the IEEE,* Vol. 54, February 1966, pp. 329–330.
[19] K. Kurokawa, "Noise in Synchronized Oscillators," *IEEE Transactions on Microwave Theory and Techniques*, Vol. MTT-16, April 1968, pp. 234–240.
[20] S. Hamilton, "FM and AM Noise in Microwave Oscillators," *Microwave Journal*, June 1978, pp. 105–109.
[21] K. L. Kotzebue and W. J. Parrish, "The Use of Large Signal S Parameters in Microwave Oscillator Design," *Proceedings of the International IEEE Microwave Symposium on Circuits and Systems*, 1975.
[22] K. M. Johnson, "Large Signal GaAs MESFET Oscillator Design," *IEEE Transactions on Microwave Theory and Techniques,* Vol. MTT-27, March 1979, pp. 217–227.
[23] R. J. Gilmor e and F. J. Rosenbaum, "An Analytic Approach to Optimum Oscillator Design Using S-Parameters," *IEEE Transactions on Microwave Theory and Techniques,* Vol. MTT-31, August 1983, pp. 633–639.
[24] Y. Tajima and B. Wrona, "GaAs FET Large Signal Model and Design Applications," *Technical Digest of IEDM*, 1980, pp. 122–125.
[25] Y. Tajima, B. Wrona, and K. Mishima, "GaAs FET Large Signal Model and Its Application to Circuit Designs," *IEEE Transactions on Electron Devices*, Vol. ED-28, February 1981, pp. 171–175.
[26] Bahl and P. Bhartia, *Microwave Solid State Circuit Design*, Wiley, New York, 1988, Chapters 3 and 9.
[27] P. S. Khanna and Y. Garault, "Determination of Loaded, Unloaded, and External Quality Factors of a Dielectric Resonator Coupled to a Microstrip Line," *IEEE Transactions on Microwave Theory and Techniques*, Vol. MTT-31, March 1983, pp. 261–264.
[28] D. Kajfez and P. Guillon, Dielectric Resonators, Artech House, Dedham, Mass., 1986.
[29] N. K. Osbrink, "YIG-Tuned Oscillator Fundamentals," *Microwave Systems News*, Vol. 13, November 1983, pp. 207–225.
[30] G. D. Vendelin, W. C. Mueller, A. P. S. Khanna, and R. Soohoo, "A 4 GHz DRO," *Microwave Journal*, June 1986, pp. 151–152
[31] Kipnis, "Silicon Bipolar MMIC for Frequency-Conversion Applications up to 20 GHz," *IEEE MTT-S Digest*, 1987, pp. 855–858.
[32] J. Obregon and A. P. S. Khanna, "Exact Derivation of the Non-Linear Negative Resistance Oscillator Pulling Figure," *IEEE Transactions on Microwave Theory and Techniques*, Vol. MTT-30, July 1982, pp. 1109–1111.

[33] K. Kurokawa, "Injection Locking of Microwave Solid State Oscillators," *Proceedings of the IEEE,* Vol. 61, 1973, p. 1386.

[34] J. A. Barnes, A. R. Chie, and L. S. Cutter, "Characterization of Frequency Stability," *IEEE Transactions on Instrumentation and Measurement*, Vol. IM-20, No. 2, May 1971, pp. 105–120.

[35] D. J. Healey III, "Flicker of Frequency and Phase and White Frequency and Phase Fluctuations in Frequent Sources," *Proceedings of the 26th Annual Symposium on Frequency Control,* Fort Monmouth, N.J., June 1972, pp. 43–49.

[36] P. Lesage and C. Audoin, "Characterization of Frequency Stability: Uncertainty Due to the Finite Number of Measurements," *IEEE Transactions on Instrumentation and Measurement,* Vol. IM-22, No. 2, June 1973, pp. 157–161.

In: Knowledge Management and Communication … ISBN: 978-1-62808-250-0
Editor: Rafiqul Islam

Chapter 14

APPLICATION OF THE PILOT TONES AND UNUSED CARRIERS (PTUNC) ALGORITHM FOR THE FOURTH GENERATION (4G) MOBILE CELLULAR SYSTEMS

Djamel Slimani* [*] *and Younes Bahri
Instrumentation Scientific Laboratory (PNR RM4G) Electronic Department
Faculty of Technology, University of Ferhat Abbas Sétif, Algeria

ABSTRACT

Increasing demand for wireless data communication and services has introduced new technologies capable of handling higher data rate. 4th Generation wireless mobile systems are widely expected to cope with such demands. Orthogonal Frequency Division Multiplexing (OFDM) is proving to be as the multiple access technology and numerical modulation scheme in 4G. This paper investigates the Peak Average Power Ratio (PAPR) reduction in the OFDM-4G transmitter signal by using the pilot tones and unused carriers (PTUNC) algorithm. A dynamic adapting for a kind of 4G that can be selected depends on the chosen design parameters. The simulation results are shown for different OFDM-4G parameters.

1. INTRODUCTION

The technology of the Orthogonal Frequency Division Multiplexing (OFDM) is considered as the best leading candidate for the fourth generation (4G) mobile cellular systems [1]. Because of an OFDM signal is produced by an Inverse Fast Fourier Transform (IFFT) operation followed by a digital-to-analog converter [2].This simple implementation has attracted a lot of attention to OFDM systems as a simultaneous numerical modulation and multiplexing technology for wireless communication systems. This is because, today Fast Fourier Transform (FFT) and IFFT processes can be implemented on fast Digital Signal

[*] E-mail: slimani.djamel@ymail.com.

Processing (DSP) or on Field Programmable Gate Access (FPGA) modules and they make OFDM an efficient technology for the new generation of network mobile 4G systems. The huge amount of information interchange is increasing rapidly from day to day because of complex graphics and video required by anticipated future applications [3]. 4G systems are also expected to fulfill the requirements of next generation internet, featuring quality of service (QOS) and mobile IP.

Furthermore, wireless local area networks (WLANs) can be connected to the 4G systems. Smart antenna technologies can be performance and reduce interferences. However one of the major problems posed by OFDM is its high Peak-to-Average Power Ratio (PAPR), which seriously limits the power efficiency of the transmitter's High Power Amplifier (HPA). This is because PAPR forces the HPA to operate beyond its linear range with a consequent nonlinear distortion in the transmitted signal. This distortion is viewed as a major impediment to progress by the RF system design community. Due to importance of this problem several techniques have been proposed [4, 5]. In this paper we reconsider the PTUNC algorithm developed by Carol [6] with our proper allocation manner of the pilot tones. A rigorous work is carried out on the two types of 4G which are distinguished by their design parameters. The key link parameters of OFDM-4G are applied for a potential 4G cellular network using PTUNC algorithm in order to reduce the peak average power ratio (PAPR) of the OFDM-4G transmitted signals. In section 2 below, we explain PTUNC algorithm. In section 3, we describe the simulation process and the pilot tones allocation. The obtained results and discussion are presented in section 4 and our concluding remarks are in section 5.

2. Application of the Pilot Tones and Unused Carriers (PTUNC) Algorithm on 4G

In practical OFDM based systems, not all subcarriers are used to transmit the information data, in fact that can be employed to reduce the peak average power ratio (PAPR) described as the PTUNC algorithm which is explained here and illustrated in figure 1 [4]:

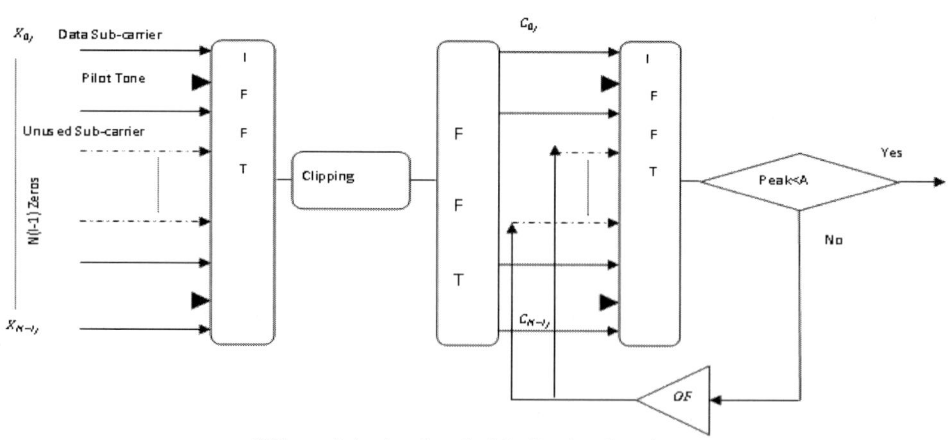

Where A is the threshold clipping level.

Figure 1. The PTUNC algorithm.

- The time domain signal is hard-clipped.
- Following an FFT transform, the data carriers and the magnitude of the pilot tones are restored to their original values while a small, pre-selected number of unused tones together with the phases of the pilot tones retain their clipped values.
- The signal is then converted back into the time domain to check for the presence of peaks. The unused tones may be scaled in the frequency domain to increase their absorption of the clipped effects bringing the time domain peak value down further. An optimal factor (OF) can be found after few iterations.

3. SIMULATION

The simulation process is carried out under MATLAB environment on the two types of 4G. Since 4G systems will be used in wide range of environments, a fixed set of parameters is not capable of achieving high performance in all causes. Consequently 4G systems will be very capable of dynamically adapting one or more of the following design parameters may offer superior performance [7]

Table 1. OFDM design parameters for 4G

Parameter	Type A of 4G	Type B of 4G
Operating Frequency	2GHz	2GHz
Bandwidth ($\boldsymbol{B}$)	4096kHz	4096kHz
Useful Symbol Duration ($\boldsymbol{T}$)	62.5 $\boldsymbol{\mu s}$	125 $\boldsymbol{\mu s}$
Guard Interval Duration ($\boldsymbol{T_g}$)	15.625 $\boldsymbol{\mu s}$ ($\boldsymbol{T/4}$)	31.25 $\boldsymbol{\mu s}$ ($\boldsymbol{T/4}$)
Total Symbol Duration ($\boldsymbol{T_{symbol}}$)	78.125 $\boldsymbol{\mu s}$ (with GI=$\boldsymbol{T/4}$)	156.25 $\boldsymbol{\mu s}$ (with GI=$\boldsymbol{T/4}$)
Inner Channel Coding	Punctured ½ rate convolution code, Constraint length 7.$\{\mathbf{133, 171}\}_{octal}$	Punctured ½ rate convolution code, Constraint length 7.$\{\mathbf{133, 171}\}_{octal}$
FFT Size	256	512
Number of data Sub-carriers ($\boldsymbol{N_D}$)	216	432
Sub-carrier spacing(Δ_f)	16 kHz	8 kHz

In an OFDM system with N sub-carriers, the baseband signal can be written as

$$x(t) = \frac{1}{\sqrt{N}}\sum_{k=0}^{N-1} x_k e^{j(2\pi k \Delta f t)}, 0 \le t \le T \tag{1}$$

where x_k is the symbol carried by the Kth sub-carrier, Δf is the frequency difference between sub-carriers, and T is the OFDM symbol duration. In the transmitter, the temporal signal is produced by Inverse Fast Fourier Transform (IFFT) of the N-point $\{x_k\}$ sequence, and at the receiver, the Fast Fourier Transform (FFT) is employed to restore the signal. Since the signal which passes through the power amplifier is in the continuous time domain, and may have an instantaneous value significantly higher than the discrete time estimate, it is necessary to oversample the signal by a factor of at least 4 to enable accurate peak detection. For an over sampling factor of I, the input signal to the IFFT is extended by including $N(I-1)$ zeros in the center of the signal. However, a major drawback of OFDM system is the manner in which the phases can align in the frequency domain causing high peaks to result in the time domain. So it is preferable to reduce the PAPR of the signal before it enters the power amplifier in order to avoid its saturation.

The PAPR signal of the OFDM signal in the time domain can be expressed as

$$PAPR(x)_{dB} = 10\log_{10}\left(\frac{max|x(t)|^2}{E[|x(t)|^2]}\right) \tag{2}$$

where $E[.]$ denotes the expectation operation. As more sub-carriers are added, higher peak values may occur, hence the PAPR increases proportionally with the number of sub-carriers.

In the 4G type A there are 208 data carriers, 8 pilot tones and 40 unused tones, while in the 4G type B there are 416 data carriers, 16 pilot tones and 80 unused tones.

The adopted allocation manner of the pilot tones in this work is done in the following way:

Pilot tones are transmitted on individual sub-carriers during the transmission period as shown in figure 2. Such a strategy will be advantageous in moderately frequency-selective and high Doppler channels.

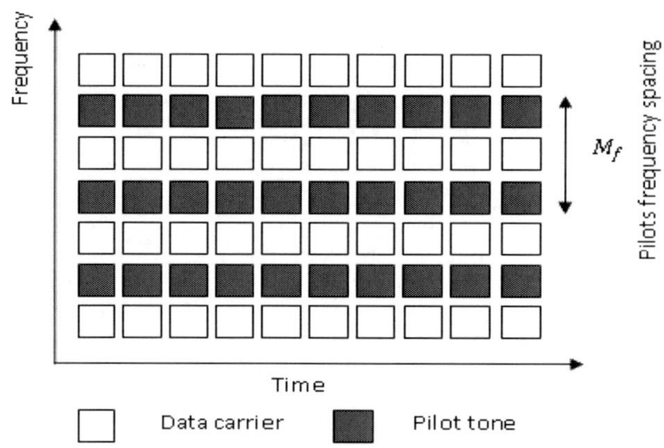

Figure 2. Pilot Tones Allocation.

Note that frequency spacing M_f between the pilots should be reasonable enough to fully estimate the channel. We note that for ideal reconstruction, the frequency spacing M_f between the pilots should fulfill the requirements of the 2D sampling theorem as

$$M_f < \frac{N}{2B\tau_{max}} \tag{3}$$

Where

M_f: is the pilot frequency spacing

N: is the number of sub-carriers in the OFDM system

B: is the bandwidth of the OFDM system
τ_{max} : is the maximum excess delay of the channel

So, in our case

$$M_f < \frac{256}{2*4096*10^3[Hz]1*10^{-6}[s]} = 31.25 \tag{4}$$

For 4G type A, and

$$M_f < \frac{512}{2*4096*10^3[Hz]1*10^{-6}[s]} = 62.5 \tag{5}$$

For 4G type B

So, the allocation of pilot tones can be found by dividing the number of data sub-carriers N_D, by the number of pilot tones. In our case 27 is chosen as the pilot tones spacing in both types that satisfies the above conditions (4) and (5).

4. Performance Results

In this paper performance results for the two types of 4G are presented for the different modulation schemes (16-QAM, 64-QAM and 256-QAM). Figures 3 and 4 show the power spectral density for 4G types A and B respectively for different signals as indicated on the figures shown below. While figures 5 and 6 show the Bit Error Rate (BER) versus the ratio of bit transmission energy over the noise power ($\frac{Eb}{N0}$) for the types A and B of 4G respectively.

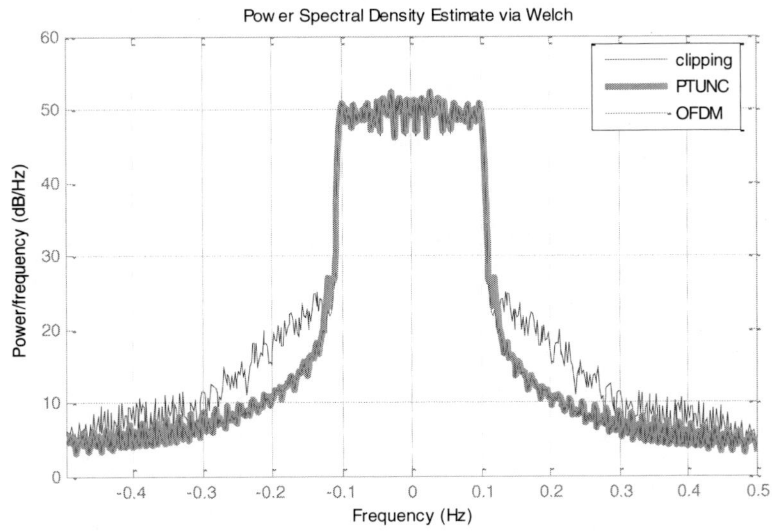

Figure 3. Power spectral density for 4G type A.

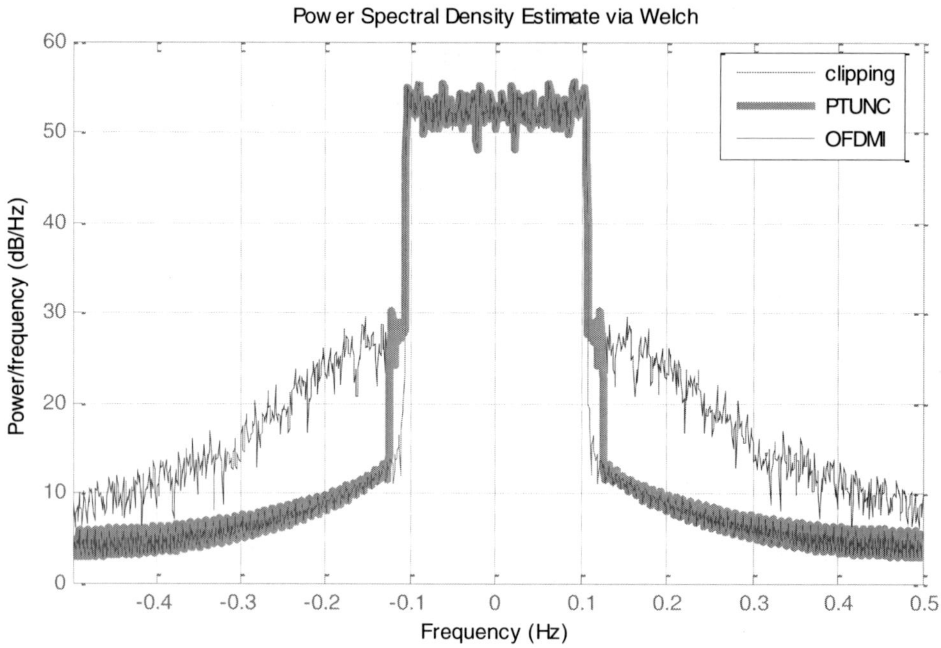

Figure 4. Power spectral density for 4G type B.

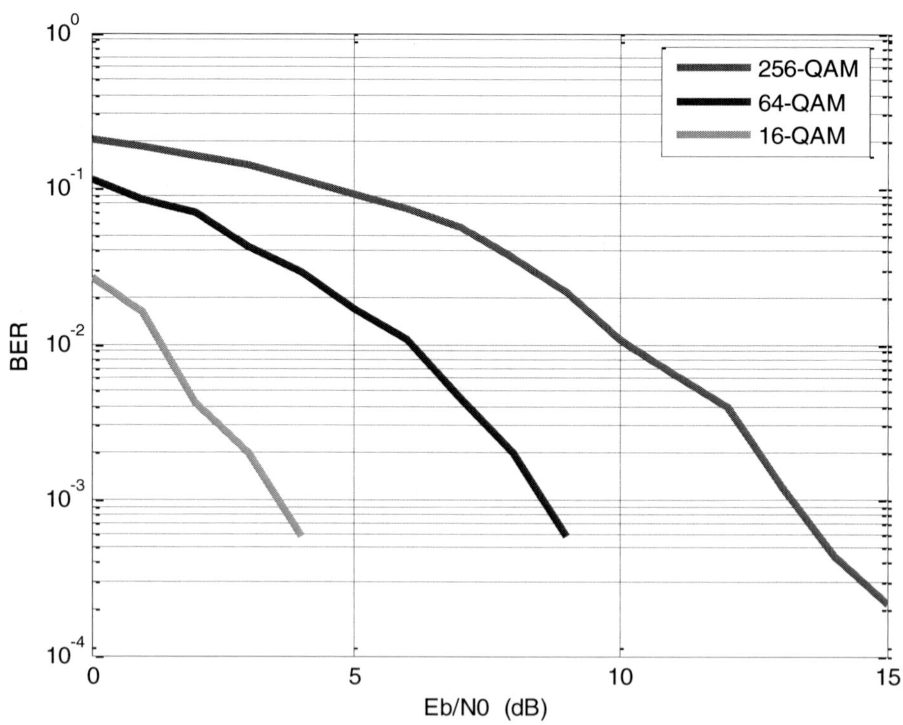

Figure 5. BER versus $\frac{Eb}{N0}$ for 4G type A for PAPR= 2 dB.

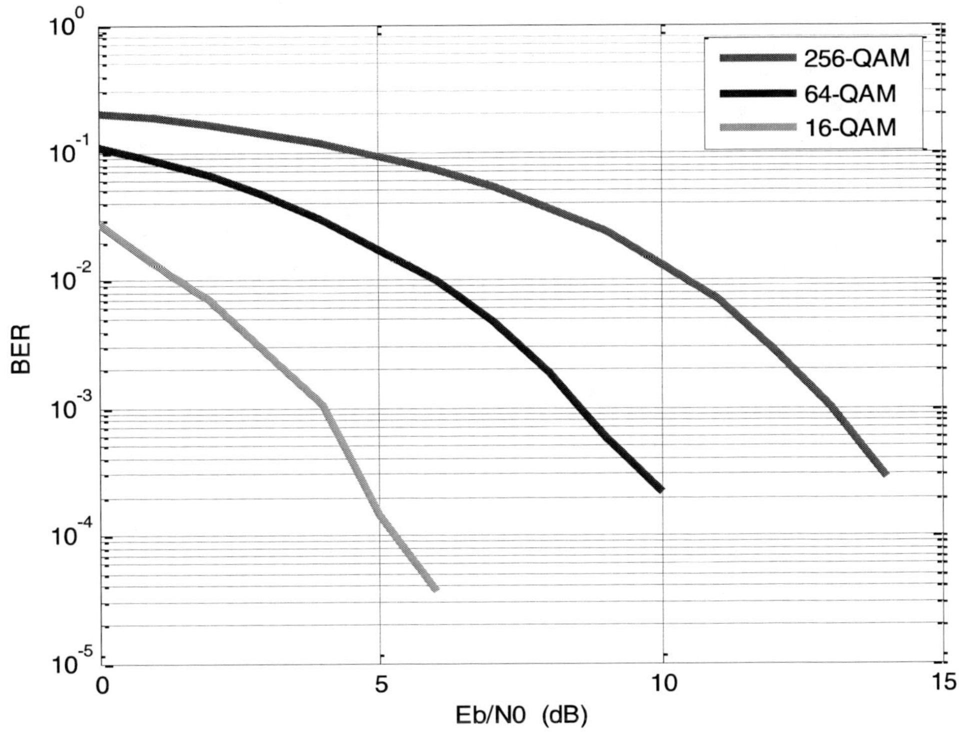

Figure 6. BER versus $\frac{Eb}{N0}$ for 4G type B for PAPR=3 dB.

From the obtained results, we can see that PAPR can be reduced by 2 dB for 4G type A and by 3 dB for 4G type B using PTUNC algorithm. From figures 5 and 6, we can conclude that to achieve high data rate we must increase the order of modulation but we have to pay in other side for the bit transmission energy Eb (dB).

CONCLUSION

We tested the PTUNC algorithm in both types of the 4G using OFDM-4G link design parameters. A variable set of parameters can give a dynamic adapting for type of 4G systems that can be used.

REFERENCES

[1] Jihad Qaddour, "*High Peak To Average Ratio Solution in OFDM of 4G Mobile Systems*" IWCMC'06 July 3-6, (2006) Vancouver, Canada.

[2] R.Van Nee and R. Prasaed, "*OFDM for wireless multimedia communications*", Artech House, Norwood, MA (2000).

[3] N.Nakajima,y.yamao, "Development for 4 th Generation Mobile Communications", *Wireless Communications and Mobile Computing*, Vol. 1, No. 1. Jan-Mar(2001).

[4] Hyun-Bae Jeon and al ' Bit-Based SLM for PAPR Reduction in QAM Modulated OFDM Signals ', *IEEE Transactions on broadcasting*, Vol. 55, September (2009).

[5] Byung Moo Lee and Rui J.P. de Figueiredo ' MIMO-OFDM PAPR reduction by selected mapping using side information power allocation', *Digital Signal Processing* 20 pp 462-471 (2010)

[6] Carole A.Devlin, Anding Zhu, and Thomas J. Brazil "*Peak to Average power Ration Reduction Technique for OFDM Using Pilot Tones and Unused Carriers*",1-4244-1463-6/08/© (2008) IEEE.

[7] Angela Doufexi, Simon Armour, Andrew Nix and Mark Beach,"Design Consideration and Initial Physical Layer Performance Results for a space Time Coded OFDM 4G Cellular Network", 07803-7589-0/02/© (2002) *IEEE.*

In: Knowledge Management and Communication …
Editor: Rafiqul Islam
ISBN: 978-1-62808-250-0

Chapter 15

FAULT DETECTION BY K-NN ALGORITHM AND MLP NEURAL NETWORKS IN A DISTILLATION COLUMN: COMPARATIVE STUDY

***I. Manssouri**[*1]**, M. Manssouri**[2] **and B. El Kihel**[2]*

[1]Laboratoire de Modélisation Mathématiques pour l'Analyse et la Prise de Décision, École Nationale Supérieure d'Arts et Métiers, Université Moulay Ismail, Meknès, Maroc

[2]Laboratoire de Génie Industriel et Production Mécanique, ENSA, Oujda, Maroc

ABSTRACT

This work, is placed within the framework of Detection by Pattern Recognition (DPR), we describe an approach with the detection of defects applied to a continuous distillation column which contains a binary mixture of toluene methylcyclohexane. Initially, by using the algorithm K-Nearest-Neighbors (KNN) and secondly, by using the Neural Networks of the type MLP (Multi Layer Perceptron) like classifiers of defects. The training base of our classifiers consists of two modes: normal mode and abnormal mode which is caused by an increase in the distillation to 100% of its capacity. In continuation we work out a base of a test in order to evaluate the quality of the various classifiers. To this end, we proceed in two different ways: Firstly, the base of the test consists of samples in normal mode and samples in an abnormal mode which do not resemble the abnormal mode used during the training. We take for example the dysfunction of feed rate to 100% of its capacity. Secondly, the base of the test consists of samples in normal mode and samples in abnormal mode, but this time the abnormal mode is similar to the abnormal mode used during the phase of training.

Keywords: Faults Detection, Classification, K-NN algorithm, Neural Networks-MLP, Distillation Process

[*] Corresponding author: E-mail: imade_mansouri@yahoo.fr.

1. Introduction

A surveillance system consists of two basic surveillance functions: detection and diagnosis. Surveillance methodologies are generally divided into two groups: Surveillance methodologies with model and others without model. The first category is based on the existence of a formal model of equipment, and generally uses the techniques of automatic. The second category of methodologies is more interesting when a model of equipment is non-existent or difficult to obtain. In this case, we use the tools of statistic and artificial intelligence. The surveillance function is then seen as an application of pattern recognition. The patterns represent an input vector composed of the different data of equipment and the classes represent the different functioning modes.

There are several detection and diagnosis methods to analyze functioning defects in the process; we only cite the proposed applications in the field of defects detection and diagnosis that use the neural networks, K-NN algorithm, and SVM algorithm.

Mamar et al. [1] use K-NN algorithm, neural networks RBF and SVM algorithm for the detection and diagnosis of defects of the wear away of the guiding system of a tramway on pneumatic. The execution of these methods of classification gives very satisfactory results, in particular the SVM classifier. It shows a big robustness and a good generalisation capacity with a learning base of weak dimension. The other classifiers, in particular, RBF networks are very sensitive to the space dimension of the representation.

Venkatasubramanian et al, [1-2] examined in detail many aspects of faulty classification techniques based on feed forward neural Networks (FANN). It is shown as well that the training of feed forward Neural Networks, serving to identify faults coming from a simple cause, can be used to detect multiple faults.

Similar results were presented for a simulated case study of column of engine-distillation. Kavuri and Venkatasubramanian [3] [4] Hoskins and Himmelblau [5] employ the simulated factory and bring back a very good diagnosis of defect. Patton et al. [6] present a short examination of several techniques of detection and diagnosis of defect such as the residual analysis, this model of recognition, the evaluation of parameter and fuzzy logic.

Chen et al. [7] present the results of the research on the detection and diagnosis of defect method for the circuit of the rail way. The proposed method employs hybrid quantitative/qualitative technique known under the name of neuro-fuzzy system. A hybrid system of detection and diagnosis of defects combing the advantages of the fuzzy logic and neural networks that is to say the capacity to deal with the imprecision of the system and learn through the formation process of neural networks. Chen et al. show that the proposed method detects and diagnosis correctly the most frequent defects of the railway circuit in a test installation in laboratory.

Jian–Da et al.[8] present a diagnosis technique of defects in internal combustion engine. The proposed technique is based on the analysis of WPT (Wavelet Packet Transform) combined with the Artificial Neural Networks (ANN) to detect and classify the functioning of defects in an internal combustion engine in various regimes. The proposed technique of WPT has many advantages with regard to CWT (Continuous Wavelet Transform) and DWT (Discrete Wavelet Transform), to check the efficiency of GRNN (Generalized Regression Neural Network) Jian –Da Wu et al. compare it to a network BPN (Back Propagation

Networks).The study shows that GRNN network was effective to diagnose and classify the functioning in a internal combustion engine in various regimes.

Kowalski et al. [9] deal with problems of defect diagnosis of rolling of induction engines in the case of rotor and stator. Kowalski et al. proposed two types of neural networks: The Multi-Layer Networks and Kohonen's auto organiser networks. The neural networks were formed and tested on a stator data base (Vibration mechanic spectre and currents). The efficiency of the neural detection was evaluated. The Multi-Layer Networks used for the detection of every kind of defect gave satisfactory results which are very important in the practice application. By contrast, Kohonen's auto organiser networks showed that they could be employed for the initial classification of engine defects as an introduction stage before the suitable neural detector.

Manssouri et al.[10] discuss the separation of two modes of functioning normal and abnormal by using the application of radial basis function (RBF) neural network to a continuous distillation column system of methylcyclohexane from a binary mixture toluene/methylcyclohexane.

Kun Mo et al.[11] worked on the application of neuron network of DNNA (Dynamic Neural Network Aggregation) for the detection, classification and prevision of cuts in nuclear centrals. The proposed system in this study uses a two level architecture of classifier with a model of DNNA instead of the conventional neuron network for defect diagnosis. DNNA model gives satisfaction especially on the level of memorisation capacity of the cut and detection of its position and proved to be the best comparison method.

Cau et al. [12] studied the conception problem of a Non-Destructive system for default detection in the canalization by means of artificial neural network models (MLP). Numerical techniques were employed to simulate the propagation of guided waves in the canalization. In particular, finished element method was employed to model the different defect types and to obtain return echo containing defect information. Firstly the obtained signals were dealt with in order to reduce the dimension of the data and extract pertinent information by the principal analysis component (PCA). Secondly, Cau et al. use the neural networks MLP for the size detection and the place of the different cut in the canalization.

Emmanouilidis et al. [13] present the development of flexible software for the watching and diagnosing of the machines, which can easily be adapted to the customer's needs to a wide range of supervision watching scenarios. Its principal elements are a certain member of modules of independent software such as Defection trees. The fuzzy classification module and the probabilistic neural networks (PNN) as under system of diagnosis. The software application is valid on two applications, namely the watching of machine-tool and the watching of gear box.

Simani et al. [14] give a methodology of fault diagnosis and identification (FDI) composed of two stages. In the first stage, the fault is detected on the basis of product residuals of kalman's filter bank, whereas, in the second stage, the fault identification is obtained from recognition technique of forms put in application by neural networks of the MLP type or the RBF type. The proposed tool of defect diagnosis was examined on an industrial model of a gas turbine. Satisfactory results are obtained with the use of MLP neural networks by contrast; RBF neural networks give bad results if the latter are formed by a wrong data base.

D'arazio et al.[15] deal with inspection problem and automatic detection of the abnormalities of composite material combining ultrasonic techniques and the neural network techniques, to do so, D'orazio et al. take into consideration two stages to deal with the automatic detection problems. The first stage consists of a pre-treatment of the different signals drawn from the different composite structure, these latter serve as a learning base of neural network. The second stage is the use of neural classifier for abnormalities detection existing in composite materials.

In the framework of surveillance, our study is about the detection of functioning defects (faults) and diagnosis failures. In this way, we proposed a new architecture of neural network of the MLP types and K-NN algorithm applied for the surveillance of a continuous distillation column of methycyclohexane from a binary mixture toluene\ methycyclohexane.

2. K-Nearest Neighbors Algorithm

The K – Nearest Neighbors method [16] is a non parametric approach which does not require any hypothesis on the division of the different classes or the nature of dividing surfaces which proved its efficiency facing data processing. The volume of "r" area around an xi form is chosen in a way to contain exactly k points of the observed sample. We generally choose regular volume centered in xi (hyper cube- hyper spheres) the method performances depend on the k value the number of the nearest neighbors, it consists in searching a new vector to classify the subset of k nearest vectors of the learning set in the sense of a distance (Euclidian distance, adaptive metric, Mahalanobis distance) which determine the neighboring type, then to apply the class greatly represented in the subset. This method is simple to implement without any parameter to fix, however, it is too much time consuming in calculating since it is necessary to calculate, with every new application to all xi distances to every learning example.

3. Multi Layer Perceptron Artificial Neural Network Models

The various models of network of artificial neurons used in this work were developed by using the software Matlab (Mathworks, Inc.).

The artificial Networks are composed of input layer of a hidden layer and an output layer. The independent variables inputs $X=(x_1, x_2, \ldots\ldots\ldots\ldots, x_N)$ are standardized between 0.1 and 0.9 and presented to the input layer of the network containing N neurals. They are first multiplied by the IW weights added to the IB bias existing between the input layer and the hidden layer. The neurons of the hidden layer receive the weighed signals. After the addition, they transform them by using a sigmoid non-linear function S (.) given by the following equation:

$$S(n) = \frac{1}{1+\exp(-n)}$$

The following mathematical model S(IW*X+IB) is introduced in the input of the output layer. This model will be multiplied by LW the weights and then added to the LB bias Existing between the hidden layer and the output layer, transformed then by a linear function (Means pure layer) L (.) stated by the equation:

$L(n)=n$

The mathematical model of artificial neural obtained is stated as follows:

$L\{LW^* S(IW^*X+IB)+LB\}.$

Different algorithms of training have been examined. The algorithm of back propagation [17-18] has been used to form artificial neural network in a robust and fast way.

The analysis was limited to networks which contain only one hidden layer since this architecture is able to classify all the data.

4. Experimental Operation

To have a D mixture rich in methylcyclohexane from a toluene/methylcyclohexane F mixture, the mass composition of which we defined 23 % in methylcyclohexane, we use the distillation technique. This latter allows the partial separation of the most volatile compound methylcyclohexane contained in a liquid mixture (figure 1).

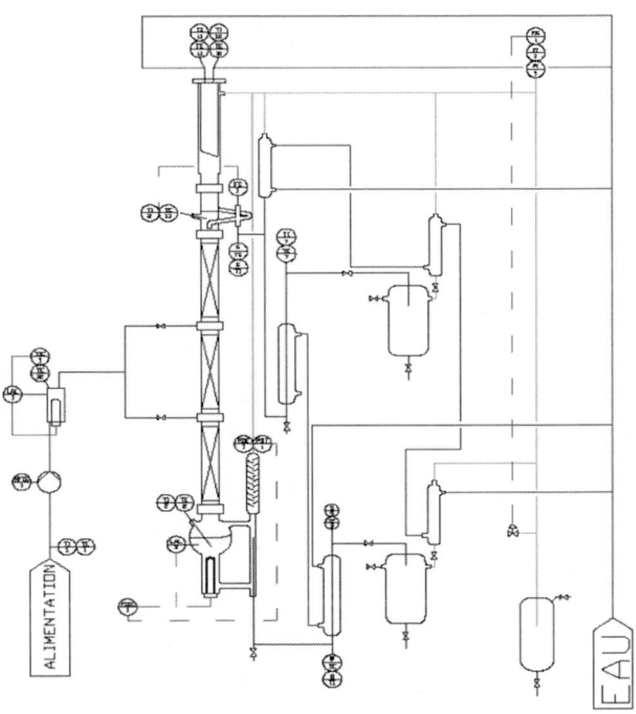

Figure 1. Diagram of installation of continuous distillation.

It consists of introducing a mixture flow of "toluene/methylcyclohexane" heated in a re-boiler under Q_B power where constant boiling pressure occurs. The vapor escaping from the re-boiler is condensed under Q_C power, which leads as an immediate effect the direct recovery of the condensate. The vapor stemming from the mixture is always richer in most volatile composite, so that the title X of the condensate is constantly superior to that of the title Z_F of the initial mixture. Continuing the operation gradually impoverishes constituting the most volatile liquid of the re-boiler. It results in a continuous decrease of the condensate title.

To work out a database serving, on the one hand as a basis of training and as test basis on the other hand, (the two formed of the normal mode and abnormal mode) we will define the inputs and the outputs of the distillation system.

5. Definition of the Relevant I/O

Before starting the study of inputs and relevant outputs behavior (figure 2), it is necessary to determine a working interval. For that, we base on the fact that we want to obtain a rich mixture in methylcyclohexane. Thus, starting from the isobar diagram liquid vapor of the mixture toluene/ methylcyclohexane, we will determine the temperature range (of title) by using the Wilson model for mixture modeling. The selected interval is [0.48; 0.84] taking into account the fact that in an industrial environment, the purification of this binary mixture is carried out in this range. The higher limit of this interval cannot reach a final purity equal to 1, and this because of the very great difficulty of separation of these two products at low temperature. This choice determines a range for each input of the distillation system so as to have a constant temperature at the output; but the flow of cooling remains fixed a raison that will not be taken into account in the database.

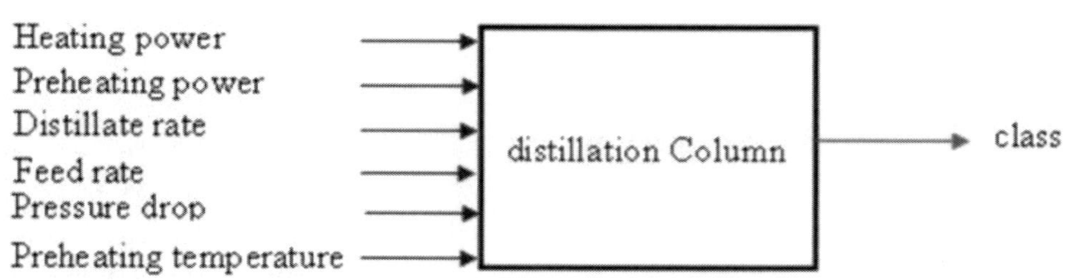

Figure 2. Determination of the inputs /outputs of the distillation column.

6. Identification of the Different Modes of Functioning

6.1. Normal Mode

The distillation admits two areas of operation, a first area, considered certain, where all the parameters of operation follow the normal operational mode and the second one, considered dangerous area, where the same parameters deviate from their initial dynamics.

All input variations that keep temperature at the column head equal to 101.5°C the equivalence of 84 % title X of the distillate defines the normal regime of the distillation column (figure 3). Several experiments have been elaborated for data collection reflecting the behaviour of each input for a normal functioning. This data will serve to the elaboration of a learning basis rich in frequency so that of the whole spectrum of the dynamic of the process is present.

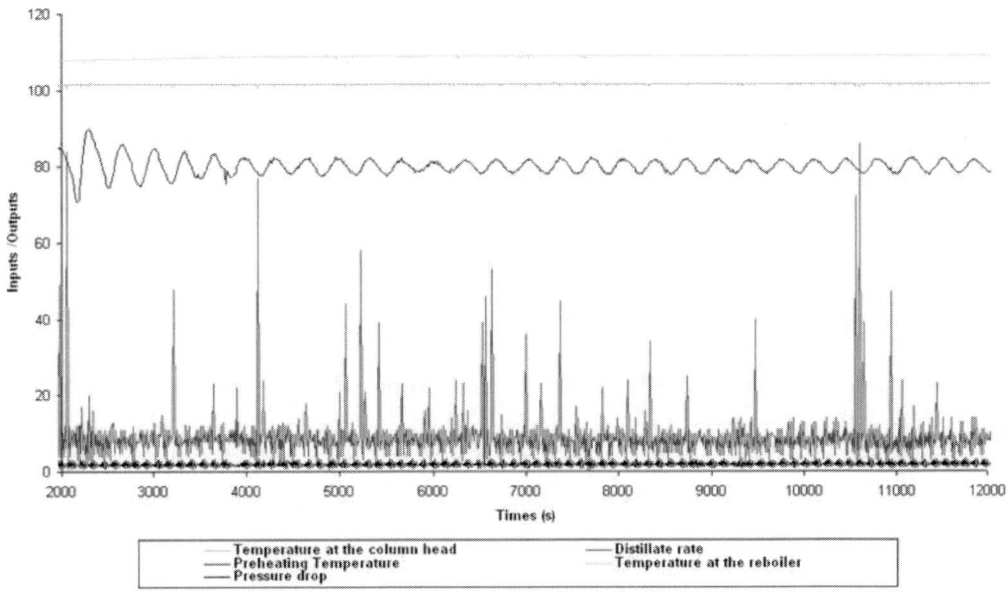

Figure 3. Normal mode in continuous distillation column.

6.2. Failing Functioning Modes

The test and learning data of the different classifiers is composed of two functioning modes, namely the normal mode and the abnormal mode. That is why we had recourse to some malfunctioning on the pertinent inputs in order to observe the reaction of output parameters.

On the test data we use at the beginning of a failing mode a malfunctions which is provocated by the increase of the flow rate by 100 % of its capacity. Figure 4 represents only the outputs which were perturbed by the malfunctioning. From what has been said before we can see that the sudden increase in the flow rate provokes an increase in the heat in the head of the column since the timer is by 100% lets flow all the distillate not only in the column as before but directly in the collected formula, which leads to the increase in the heat in the head of the column.

In the second time of a malfunctioning provoked by the increase of the feed rate output by 100% of it is capacity. Figure 5 represent only the outputs which were disturbed by this malfunctioning.

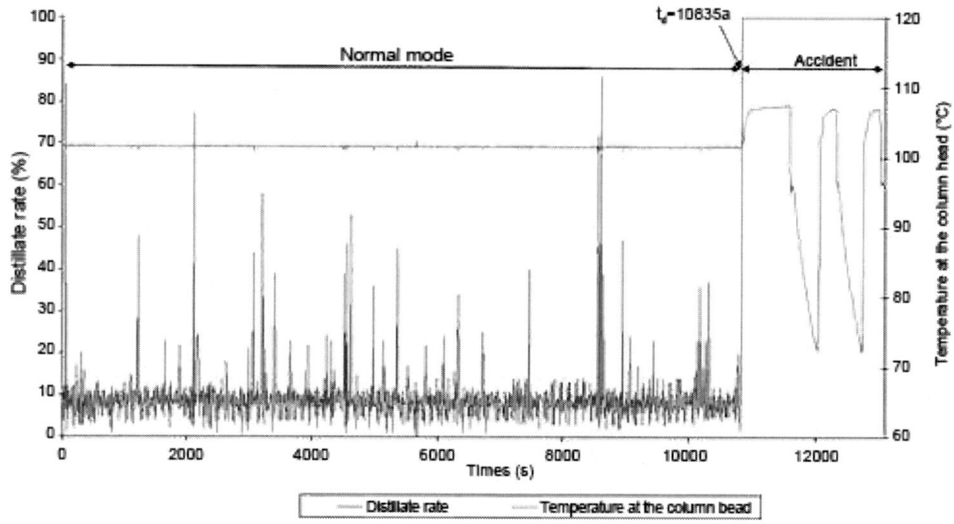

Figure 4. Dysfunction on of the distillate rate in accordance with time.

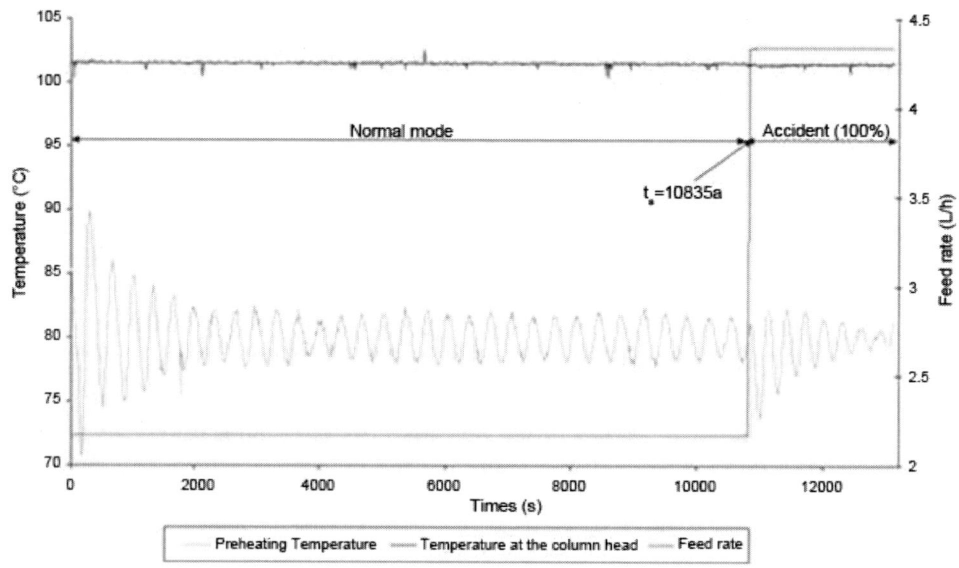

Figure 5. Dysfunction on of the feed rate in accordance with time.

7. Results and Discussions

7.1. K-Nearest Neighbors Algorithm

The technique of K-NN is a method of supervised classification; it thus requires the knowledge of the partition of the data in classes. The established procedure consists of two phases:

First of all, a regrouping of the base of training made up of 240 samples in a certain number of classes (normal class and failing class of rate of backward flow) by the use of the Principal Components Analysis PCA.

The Principal Components Analysis (PCA) is a mathematical technique making it possible to reduce a complex system of correlation in a smaller number of dimensions. Geometrically, the process of the factorization amounts placing axes in a sphere. The PCA state in general applies to a rectangular table of measurements (table of the data) in which the columns represent the variables with numerical values and the lines correspond to the individuals on which these variables are measured. In fact, it is a technique of analysis which mainly makes it possible to obtain charts between the lines and the columns of the table of figures, making it possible to visually represent the proximities between the individuals and the variables on plans.

The use of the analysis in principal components enables us to visualize the distribution of two classes: normal and failing (dysfunction of rate of backward flow at 100%) (Figure 6). With the first two principal components 78.872% of the information contained in the base is represented, from where the need for a three-dimensional projection.

According to this representation, we notice that the information added by the third principal axis (17.034%), with the first three principal components 95.906% of the information contained in the base is represented.

Then, once the training carried out, it is essential to evaluate the quality of classifier K-NN. The idea is to apply a phase of test and to vary the parameter K (the number of the nearest neighbors) to estimate the quality of classification (error rate). The error rate is estimated by the number of failures on the total number of vectors in the base of test.

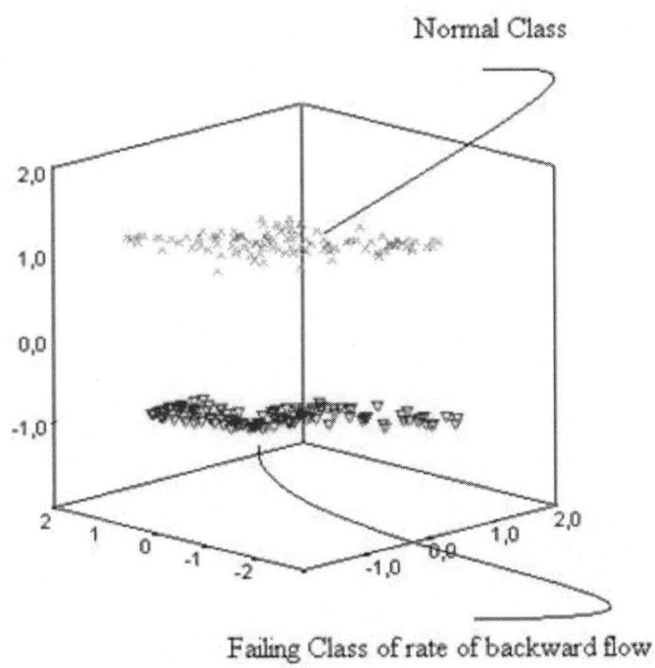

Figure 6. PCA of the two classes of learning.

For construction of the base of test one proceeds in two different ways: Firstly the base of test consists of 40 samples in normal mode and 40 samples in failing mode in this case the failing mode do not resemble the failing mode used during the training, one takes for example the dysfunction of rate of feed at 100% to supplement the base of test.

Once the base of test is elaborate; one fact of varying the value of K between 3 and 10 and one evaluates the percentage of error for each value of K, the error resulting being estimated at 35% for the value from K = 3 and 32.5% for different the values from K to be started from 4.In our case the value finally selected is k= 6.

Technique of 6-NN is applied to a base of test made up of 80 samples (40 samples of the normal class and 40 samples of the failing class of rate of feed at 100%). One notices that the majority of the samples of the failing class are not classified well.

Secondly, the base of test is always made up of 40 samples in normal mode and but 40 samples in failing mode in the same way standard than the failing mode used during the training (dysfunction of rate of backward flow at 100%).

Once the base of test is elaborate; one fact of varying the value of K between 2et 8 and one evaluates the percentage of error for each value of K, the resulting error being estimated at 0% for different values from K. In our case the value finally selected is k= 4.

Technique of 4-NN is applied to a base of test made up of 80 samples. The results obtained are presented on figure 7. It is noticed that the samples assigned to each of the two classes are perfectly well classified; the success rate is of 100%.

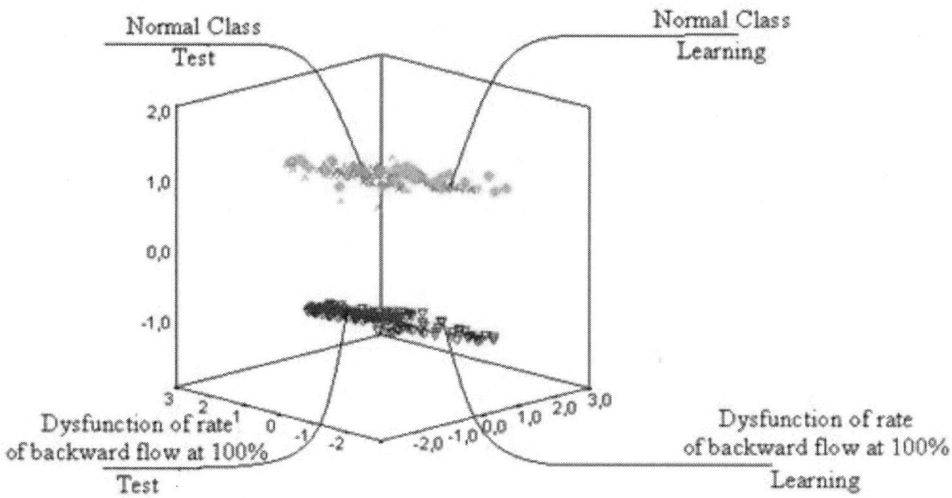

Figure 7. PCA State of the two classes of training and of the two classes test.

7.2. Neuronal Classifier of Type MLP (Multi Layer Perceptron)

7.2.1. Choice of Architecture Neuronal

In the optimization of a neural network trial and error method to find the best neural network architecture has been used. Early stopping is used to optimize learning epoch size and avoid over fitting to facilitate the interpretation, the results obtained in the quantisation of the samples in the training and test sets are expressed as root mean square errors (RMSE),

Maximum Average Percentage Error (MAPE), Akaike Information Criteria (AIC) and Bayesian Information Criteria (BIC). These tests are given by the following equations:

$$RMSE = \sqrt{\frac{\sum_{i=1}^{i=N}\left(E_{pi} - E_{ai}\right)^2}{N}}$$

$$MAPE = \frac{1}{N}\sum_{i=1}^{i=N}\left|\frac{\left(E_{pi} - E_{ai}\right)^2}{E_{ai}}\right| * 100$$

$$AIC = \ln\left(\frac{N}{2}L_F\right) + \frac{2N_W}{N}$$

$$BIC = \ln\left(\frac{N}{2}L_F\right) + \frac{N_W \ln(N)}{N}$$

with L_F as the loss function (or mean squared error).

$$L_F = \frac{1}{N}\sum_{i=1}^{i=N}(E_{pi} - E_{ai})^2$$

where Ea and Ep are the actual and predicted values from output neurons, respectively, N the numbers of compounds in the testing set, Nw represent the weights in a neural network. In order to determine the optimal number of hidden layer nodes, neural networks with different numbers of hidden nodes were trained. The numbers of hidden nodes vary between 1 and 25 to train the networks. The figures (8-11) of RMSE, MAPE, AIC and BIC - after 4500 iterations for the different ones numbers neurons of the hidden layer give the possibility of choosing two neurons for the hidden layer. We chose architecture 6-2-1 like "better" configuration of the network of neurons considering its good capacity of classification.

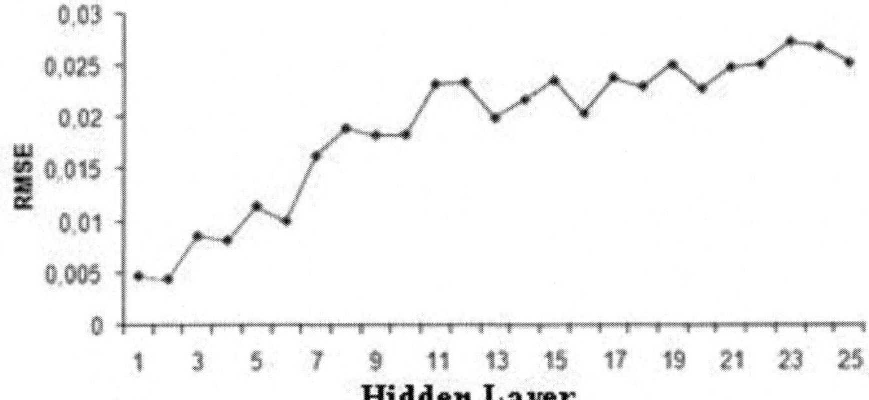

Figure 8. Test of robustness: Root Mean Square Error.

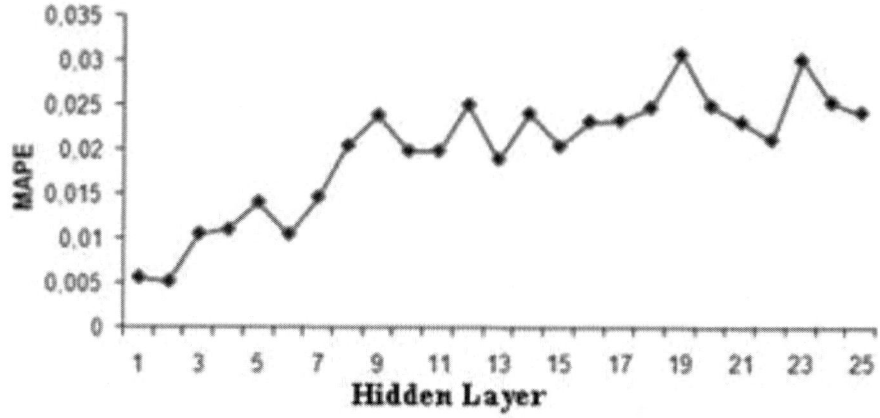

Figure 9. Test of robustness: Maximum Average Percentage Error.

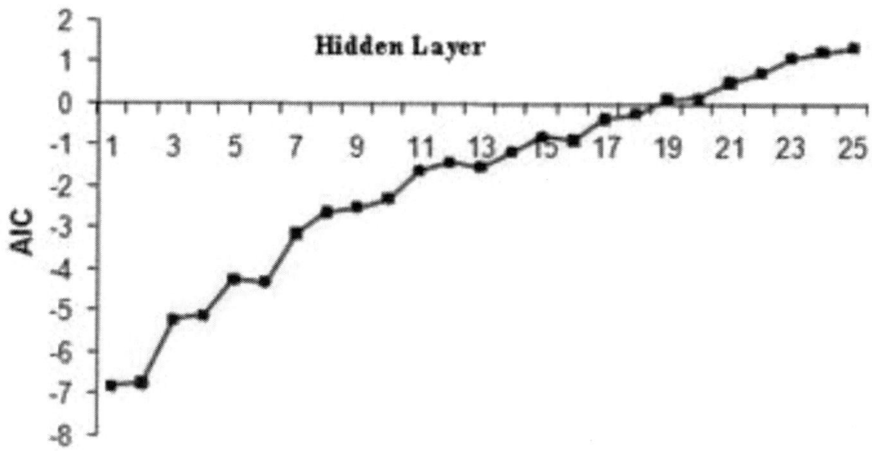

Figure 10. Test de robustesse: Akaike Information Criteria.

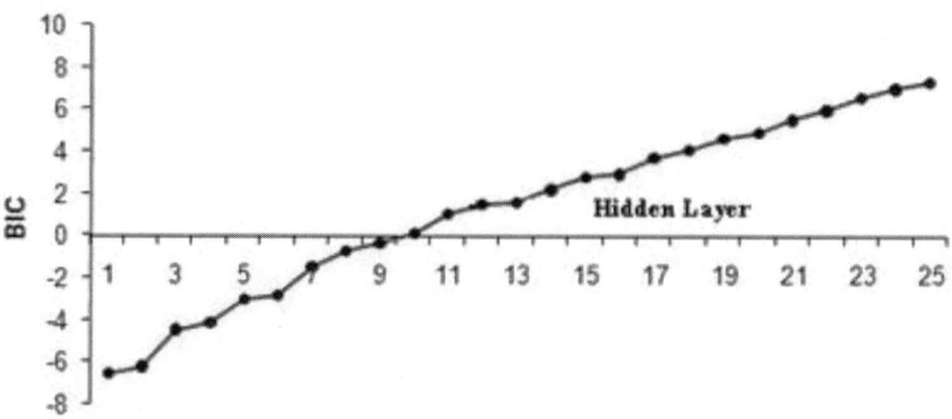

Figure 11. Test of robustness: Bayesian Criteria Information.

7.2.2. Training and Validation

The artificial network of neurons of type MLP "Multi Layer Perceptron" is composed of a layer of entry "input to bush-hammer" containing six neurons, a hidden layer "Hidden Layer" containing two neurons and a layer of exit "output to bush-hammer" containing only one neuron.

The base of training is composed of six vectors X_1, X_2, ..., X_6, independent and standardized between 0,1et 0,9 which are : respectively the power of heating, the power of preheats, the rate of backward flow, the rate of feed, the pressure losses and the temperature of preheats.

The base of training of the neural network consists of 120 samples in normal circumstances and 120 samples in abnormal mode obtained by the dysfunction caused by the increase in the rate of backward flow to 100% of its capacity. The weights and skews of the network were readjusted using the algorithms of back propagation.

Once the architecture, the weights and skews of the neural network were fixed, it should be known if this neuronal model is likely to be generalized.

The validation of neuronal architecture 6-2-1 thus consists in judging its capacity of classification by using the weights and the skews calculated during the phase of training, to apply them to two data bases tests:

The first base of test consists of 40 samples in normal mode and 40 samples in failing mode caused by the dysfunction of rate of feed to 100% of its capacity; this dysfunction does not resemble the failing mode used during the training. One uses first base of test to evaluate the quality of classification of architecture neuronal 6-2-1 by the success rate. The rate success is estimated by the number of individuals classified on the total number of vectors in the base of test. In our case the success rate is of 70%.

The second base of test consists of 40 samples in normal mode and of 40 samples in failing mode but this time the failing mode is in the same way standard than the failing mode used during the phase of training. Figure 12 translated the results of this test of classification; it is remarkable according to this figure that the data tests which compose the two modes "normal and abnormal" are perfectly well classified, the success rate is of 100%.

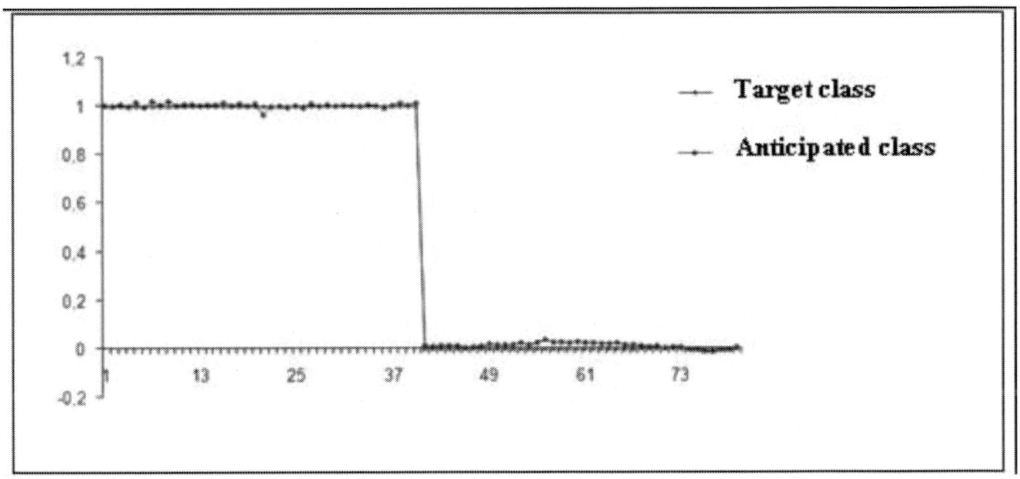

Figure 12. Result of test of classification of the neuronal model for the second base of test.

CONCLUSION

On one hand, this study made it possible to highlight the capacity of classification of the two modes (normal and abnormal) of a system of distillation of methylcyclohexane starting from a mixture toluene \ methylcyclohexane of the two classifiers (algorithm of K-statement and the network of neurons of the type MLP) for one base test which contains samples of comparable nature that the samples which forms the base of training.

On the other hand, two methods of classification give less satisfaction for the base of test which does not resemble the base of training, one finds a success rate of 70% for the network of neurons of the type MLP and 67.5% for algorithm K-statement could conclude that these two methods of classification are very simple to implement and lead to completely satisfactory results with the proviso of having the same type of failure presented in the two bases of training and test, if not, a base of training containing the various operating modes from the system of distillation (normal and failing).

REFERENCES

[1] Hamou Mamar, Z. Chainais P.et Aussem. A. 2006;"Diagnostic de l'usure des galets du système de guidage d'un tramway sur pneumatiques". *Journées d'Etudes Reconnaissance des Formes: quelles méthodes pour quelles applications?*, ENSTA Paris, France (actes).

[2] Venkatasubramanian, V. and Chan, K. 1989; "A neural network methodology for process fault detection", *AIChE Journal*, 35(12), pp:1993-2002.

[3] Kavuri, S.N. and Venkatasubramanian, V.1994; "Neural network decomposition strategies for largescale fault detection", *Int. J. Control*, 59(3), pp:767-792.

[4] Kavuri, S.N. and Venkatasubramanian, V. 1992; "Combining pattern classification and assumptionbased techniques for process fault detection", *Comput. and Chem. Eng*, 16(4), pp: 299-312.

[5] Hoskins, J.C. and Himmelblau, D.M.1988; "Artificial neural network models of knowledge representation in chemical engineering", *Comput. and Chem. Eng*,12(9), pp: 881-890.

[6] Patton, R.J., Chen, J. and Nielsen, S.B. 1995; "Modelbased methods for fault diagnosis", Some guidelines, *Trans. Inst. Measurement and Control*, 17, no. 2, pp : 73-83.

[7] Chen, J. Roberts, C. and Weston, P, 2008; "Fault detection and diagnosis for railway track circuits using neuro-fuzzy systems". *Control Engineering Practice*, Volume 16, Issue 5, pp: 585-596.

[8] Jian-Da Wu, C.H. 2008; "An expert system for fault diagnosis in internal combustion engines using wavelet packet transform and neural network". *Expert Systems with Applications,* In Press, Corrected Proof.

[9] Kowalski, C.T. Kowalska, O.T. 2003, "Neural networks application for induction motor faults diagnosis". *Mathematics and Computers in Simulation*, Volume 63, Issues 3-5, pp: 435-448.

[10] Manssouri, I., Chetouani, Y and El Kihel, B. (2008) "Using neural networks for fault detection in a distillation column", *Int. J. Computer Applications in Technology*, Vol. 32, No. 3, pp.181-186.

[11] Mo,K. Lee.S.J. and Seong,P.H. 2007; "A dynamic neural network aggregation model for transient diagnosis in nuclear power plants". *Progress in Nuclear Energy*, Volume 49, Issue 3, pp: 262-272.

[12] Cau, F. Fanni, A. Montisci, A. Testoni, P. Usai, M. 2006; "A signal-processing tool for non-destructive testing of inaccessible pipes" *Engineering Applications of Artificial Intelligence,* Volume 19, Issue 7, pp: 753-760.

[13] Emmanouilidis, C. Jantunen, E. MacIntyre, J. 2006; "Flexible software for condition monitoring, incorporating novelty detection and diagnostics". Computers *in Industry,* Volume 57, Issue 6, pp: 516-527.

[14] Simani, S. Fantuzzi, C. 2000; "Fault diagnosis in power plant using neural networks" *Information Sciences*, Volume 127, Issues 3-4, pp:125-136.

[15] D'Orazio, T. Leo, M. Distante, A. Guaragnella, C. Pianese, V. Cavaccini, G. 2008; "Automatic ultrasonic inspection for internal defect detection in composite materials". *NDT and E International*, Volume 41, Issue 2, pp: 145-154.

[16] MacQueen, J. B : "Some Methods for classification and Analysis of Multivariate Observations, Proceedings of 5-th Berkeley Symposium on Mathematical Statistics and Probability", Berkeley, University of California Press, 1:281-297. 1967.

[17] Rumelhart, D.E and McClelland, J.L. *Parallel distributed processing: Explorations in The microstructure of cognition* vol 1.: Foundations, MIT press,Cambridge,1986.

[18] Bishop, C. *Neural networks for pattern recognition*, oxford university press, New York, USA, 1995.

In: Knowledge Management and Communication … ISBN: 978-1-62808-250-0
Editor: Rafiqul Islam

Chapter 16

THE BEST COMBINATION SCHEME FOR ASYNCHRONOUS MULTI-CARRIER CDMA SYSTEM IN A RAYLEIGH FADING CHANNEL WITH MAI INTERFERENCE

Ismahene Ikhlef*[*1] and Abdellatif Said[2]
[1]University of Constantine (UoC), Faculty of Engineering, Department of Electronics, Constantine, Algeria
[2]United Arab Emirates (UAE) University, UGRU- Math/IT. Al Ain; Abu Dhabi, UAE

ABSTRACT

In this paper, we study the multi – carrier direct – sequence code division multiple – access system when the received signal is divided into predetermined spreading sequence, these sub - bands modulates M carriers. Because of the no – knowledge of the phase of the received signal in each sub - band, the receiver cannot coherently combine outputs of the correlators. We consider in this paper two non - coherent combining schemes; the equal gain combining and the selecting combining. The calculation of the probabilities of detection and false alarm for both multi - carrier and single carrier systems in a Fading Rayleigh channel demonstrate that the performance of the multi - carrier system is better than the performance of the single carrier in equal gain combining, while the performance of the single- carrier system in the selecting combining is better than the performance of the multi- carrier. The MC-DS-CDMA system using the EGC is robust against the MAI interference in comparison with that using the selecting combining (SC).

Keywords: CDMA, multi - carrier DS-CDMA, equal gain combining, selecting combining, non- coherent combing scheme

---

* Email: r_messas@yahoo.fr.

1. INTRODUCTION

If an interference or multipath is present (which is our case), degradation in system performance can result in both FDMA (Frequency Division Multiple Access) and TDMA (Time Division Multiple Access) systems [1]. For these reasons the Code Division Multiple Access (CDMA) has become a competitive multiple access scheme in certain situations [1].

We propose the multi-carrier direct sequence system transmitting in a Fading Rayleigh channel because of the anti-interference capability with is achieved by correlating the received signal with predetermined spreading sequence, thus allowing the inherent processing gain of the system to attenuate the interference [2]-[3]. For this the use of the chip waveform filter is necessary to reject the narrow band interference or self interference.

The serial acquisition performance of a multi – carrier direct sequence CDMA (MC-DS-CDMA) system has been investigated in [4] where each sub-carrier signal was subject to frequency-selective Fading.

The performance of the parallel code acquisition of the MC-DS-CDMA system is studied in [5] when each sub-carrier signal was subject to frequency-nonselective and modeled as Fading or non-Fading (with or without partial band interference by computing the probability of error).

However, in the MC-DS-CDMA system considered here, we study the performance of the MC-DS-CDMA and the single carrier CDMA (SC-CDMA) system in the presence of an additive white Gaussian noise (AWGN) and multiple access interference (MAI) in Fading Rayleigh channel for the equal gain combining (EGC) and the selecting combining by computing the probability of detection and false alarm. This paper is organized as follows:

Section 2 introduces the system model for the proposed multi-carrier DS-CDMA system. The statistics of the outputs of the non - coherent correlators and combiners for Fading Rayleigh channel are presented in sections 3 and 4 as well as the calculation of the probability of detection and false alarm. In section 5, we compare the numerical results for both multi-carrier and single carrier systems for the equal gain combining and the selecting combining. Finally, we provide our conclusions in section 6.

2. THE SYSTEM DESCRIPTION

A. The Transmitter

The block chain of the *k-th* active user's transmitter using CDMA scheme is depicted in Figure 1, where the data is generated as a random binary sequence, where $c_n^{(k)}$ is a pseudo – random (also named as Pseudo Noise-PN) spreading signature sequence with $n=\{1,...N_D\}$, N_D means the spreading gain. We assume that each signature has N_D chips of duration T_c per symbol which represent also the spreading gain code; the sequence is modulated by an impulse train with energy per chips equal to E_c .

The transmitted signal at the output of the chip wave – shaping filter modulates M subcarriers. The energy per-chip of the modulating signal over each carrier is $\frac{E_c}{M}$.

Finally, the spectral spreading is imposed on the complex signal by multiplying it with a spreading code. Therefore, the transmitted signal is given by [5]-[6]-[7] and has the following expression:

$$S_k(t) = \sqrt{\frac{2E_c}{M}} \sum_{n=-\infty}^{+\infty} d_{\left(n/N_D\right)}^{(k)} c_n^{(k)} h(t - nMT_c) \sum_{m=1}^{M} \cos(\omega_m + \theta_{k,m}) \tag{1}$$

where is the impulse response of the chip wave- shaping filter, and is a random phase uniformly distributed over [0, 2].

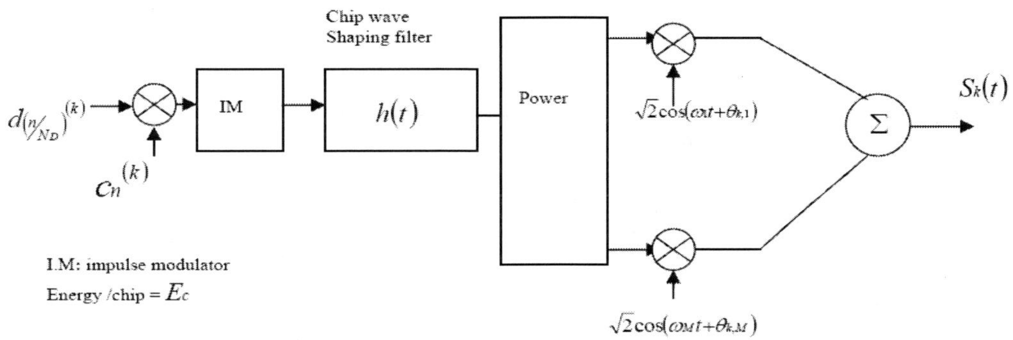

Figure 1. The schematic blocs of the MC-DS-CDMA transmitter.

B. The Channel

The channel is assumed to be a slowly varying frequency selective Rayleigh channel with the transfer function of the frequency band of the user $\alpha_{k,m}\exp(j\beta_{k,m})$, where and are independent and identically distributed (i.i.d) Rayleigh random variables with a unit second moment, and uniform random variables over [0, 2π], the received signal is given by [5]:

$$r(t) = \sum_{k=1}^{K} \sqrt{\frac{2E_c}{M}} \{ \sum_{n=-\infty}^{+\infty} d_{(n+D_k)/N_D}^{(k)} c_{(n+D_k)}^{(k)} h(t - nMT_c - \zeta_k)$$
$$\sum_{m=1}^{M} \alpha_{k,m} \cos(\omega_m + \theta''_{k,m}) \} + n_w(t) \tag{2}$$

We suppose in our case no-partial band interference (no-PBI).

C. The Receiver

The receiver is shown in Figure 2a, b[5].

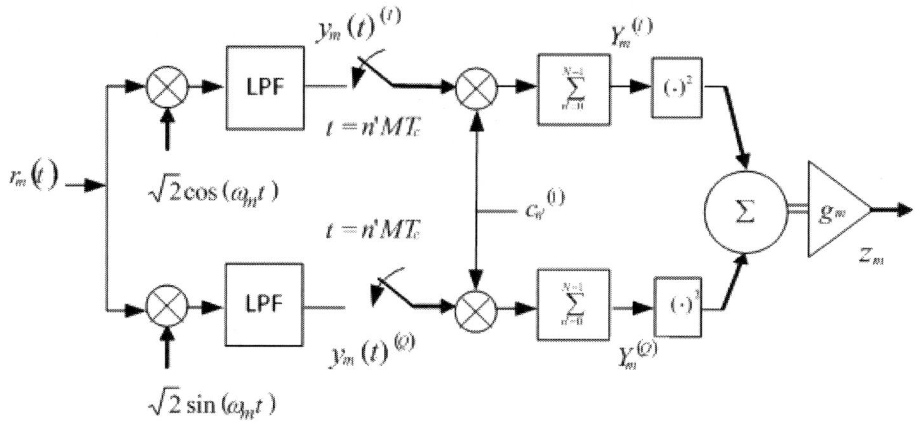

Figure 2a.The non-coherent correlator (I,Q) for the each m-th subcarrier.

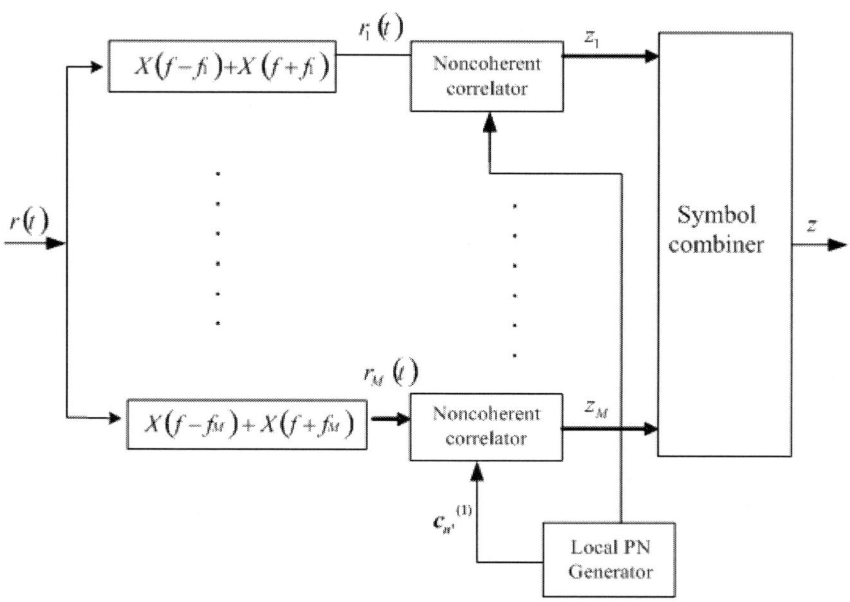

Figure 2b. The receiver of the multi-carrier DS – CDMA system.

We assume the chip wave-shaping filter is a raised-cosine filter as shown in (3) [5].

$$X(f)=\begin{cases}\frac{1}{W'} & |f|\le\frac{W'}{2}(1-\alpha)\\ \frac{1}{2W'}\left\{1-\sin\left[\frac{1}{2\alpha}\left(\frac{2\pi|f|}{W'}-\pi\right)\right]\right\} & \frac{W'}{2}(1-\alpha)\le|f|\le\frac{W'}{2}(1+\alpha)\\ 0 & elsewhere\end{cases} \tag{3}$$

3. The Statistics of the Outputs of the Non-Coherent Correlator and Combiner

A. The Statistics of the Non-Coherent Correlator

We assume during the acquisition process no data are sent during the acquisition process no data are sent

i.e. $d_{(n+D_1)/N_D}{}^{(1)} = 1$.

The outputs of the non-coherent correlator of the branch are given by [5]:

$$z_m = g_m^{-1}\left\{\left(Y_m^{(I)}\right)^2 + \left(Y_m^{(Q)}\right)^2\right\} \tag{4}$$

where g_m^{-1} is an adaptive gain control (AGC)

For the Rayleigh Fading channel, the conditional probability density function (p.d.f) on the hypothesis H_i of is given by [5]:

$$p\left(z_m / H_i\right) = \frac{1}{2v'_{m,i}} \exp\left(-\frac{z_m}{2v'_{m,i}}\right) \tag{5}$$
$$where \quad v'_{m,i} = v_{m,i} + \frac{m'^{\,2}_{m,i}}{2}$$

With z_m is the variable of decision and represent the output of the non-coherent correlator of the branch

H_i: represent hypothesis of decision, H_1 : represent de decision when the cellule is correct and H_0 : represent de decision when the cellule is incorrect.

B. The Outputs of the Non-Coherent Combiner

In this paper, we consider tow kinds of the non-coherent combiner scheme.

Equal Gain Combining

In this kind the output of the symbol combining is given by [5]-[9]:

$$z = \sum_{m=1}^{M} z_m \tag{6}$$

The probability density function of z is equal to:

$$p\left(z/H_i\right)=\frac{z^{M-1}\exp\left(-\frac{z}{2v'_{m,i}}\right)}{\left(2v'_{m,i}\right)^M Fact\left(M-1\right)} \tag{7}$$

Selecting Combining

We assume that all z_m are independent and identically distributed random variables. In the selecting combining, the phase corresponding to the largest test is assumed to be the correct phase (i.e. $z=\max(z_1,z_2..........z_M)$.

For this, we define the conditional joint distribution function as [9]:

$$F\left(z/H_i\right)=p\left\{\left(z_1\leq z, z_2\leq z.........z_M\leq z/H_i\right)\right\} \tag{8}$$

The conditioned (p.d.f) of is given by:

$$p\left(z/H_i\right)=\frac{M}{2v'_{m,i}}\exp\left(-\frac{z}{2v'_{m,i}}\right)\left[1-\exp\left(\frac{z}{2v'_{m,i}}\right)\right]^{M-1} \tag{9}$$

4. Calculation of the Probability of Detection and False Alarm for the Rayleigh Fading Channel for the Equal Gain Combining and the Selecting Combining

A. The Probabilities of Detection and False Alarm for the Equal Gain Combining

1) The Probability of Detection

We integrate the probability density function of see (7) over the normalized threshold γ^* over the correlation period N, we obtain:

$$p_{Dfe}=\sum_{m=0}^{M-1}\left(\frac{\gamma^*}{(B+a)}\right)\frac{\exp\left(\frac{\gamma^*}{(B+a)}\right)}{Fact(m)} \tag{10}$$

With $2v'_{m;i}=N(B+a)$

Where B and a are given by: $B=\frac{\left(1+(K-1)v\left(1-\alpha/4\right)\right)}{\left(1+Kv\left(1-\alpha/4\right)\right)}$

$$a = \frac{Nv}{\left(1 + Kv\left(1 - \alpha/4\right)\right)}$$

$$v = \frac{E_{c1}}{\eta_0} \; with \; E_{c1} = \frac{E_c}{M}$$

2) The Probability of False Alarm

In a similar manner of calculation of the probability of detection for the equal gain combining p_{Dfe}, we can obtain the probability of false alarm for the same combining scheme as [10]:

$$p_{Fafe} = \sum_{m=0}^{M-1} \left(\gamma^*\right)^m \frac{\exp\left(\gamma^*\right)}{Fact(m)} \tag{11}$$

$$with \; v'_{m,0} = v_{m,0} + \frac{m_{m,0}{}^2}{2} \; since \; m_{m,0} = 0$$

B. The Probabilities of Detection and False Alarm for the Selecting Combining

1) The Probability of Detection

We obtain the probability of detection (under hypothesis H_1) for the selecting combining as:

$$p_{Dfs} = \int_{\gamma*}^{+\infty} \left(\frac{M}{(B+a)}\right) \exp\left(-\frac{z*}{(B+a)}\right) \left[1 - \exp\left(\frac{z*}{(B+a)}\right)\right]^{M-1} dz* \tag{12}$$

With $z* = \frac{z}{N}$, have been defined in (10)

2) The Probability of False Alarm

Using the same calculi of the probability of detection for the selecting combining p_{Dfs}, the probability of false alarm for the same combining scheme p_{Fafs} is equal to:

$$p_{Fafs} = \int_{\gamma^*}^{+\infty} M \exp\left(-z^*\right)\left[1 - \exp\left(-z^*\right)\right]^{M-1} dz \tag{13}$$

5. Numerical Results

In this section we make a comparison between the probability of detection and false alarm in Rayleigh Fading channel with no- partial band interference for both multi-carrier and single-carrier systems in presence of the multiple access interference. The period of correlation of the single carrier has been chosen equal to $N_1=521$, but for the multi-carrier $N = {N_1}/{M}$, the roll-off factor α of the raised-cosine filter is chosen equal to 0.5 [5]-[8].

The probability of detection and false alarm for multi-carrier and single-carrier systems for both the selecting combining see Figure 3 and equal gain combining see Figure 5 respectively.

In Figure 3, the performance of single carrier system is better than the performance of multi- carrier, because the probability of detection in single-carrier system is higher than the probability of detection for multi-carrier system. However, the probability of false alarm increase when M increase.

Since in figure 4, when the number of sub bands increases, the probability of detection increases and the probability of false alarm degrees. So the multi-carrier system is better than the single-carrier for the equal gain combining.

These results are the same results obtained in reference [5] by calculi the probability of error for the equal gain combining and the selecting combining.

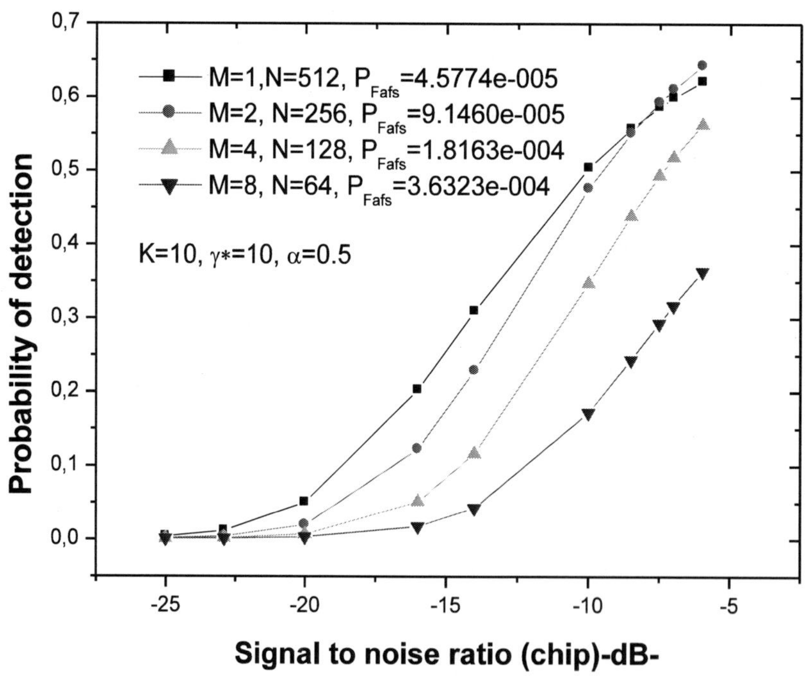

Figure 3. Probability of detection for Rayleigh Fading channel for the multi-carrier system ($M \geq 2$) and the single carrier ($M=1$) with the selecting combining.

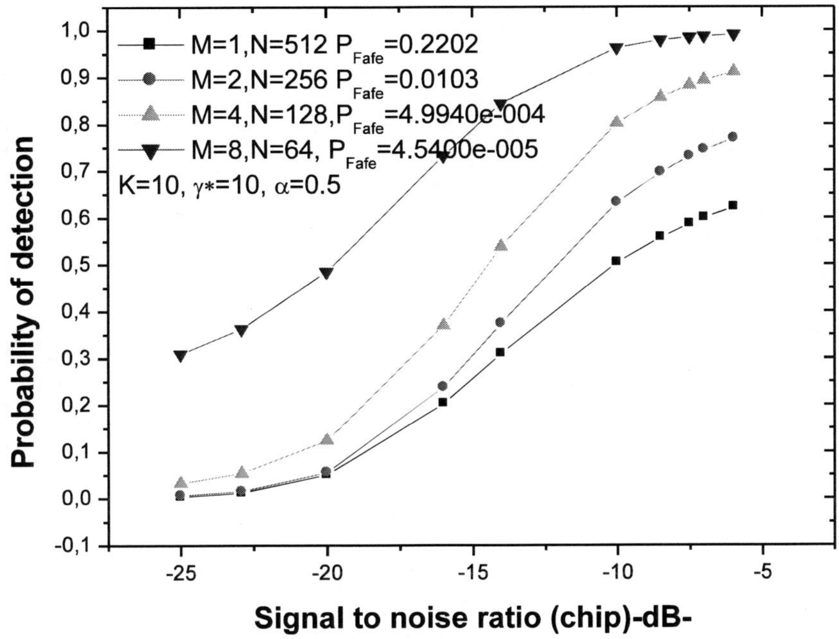

Figure 4. Probability of detection for Rayleigh Fading channel for the multi-carrier system ($M \geq 2$) and the single carrier ($M = 1$) with the equal gain combining.

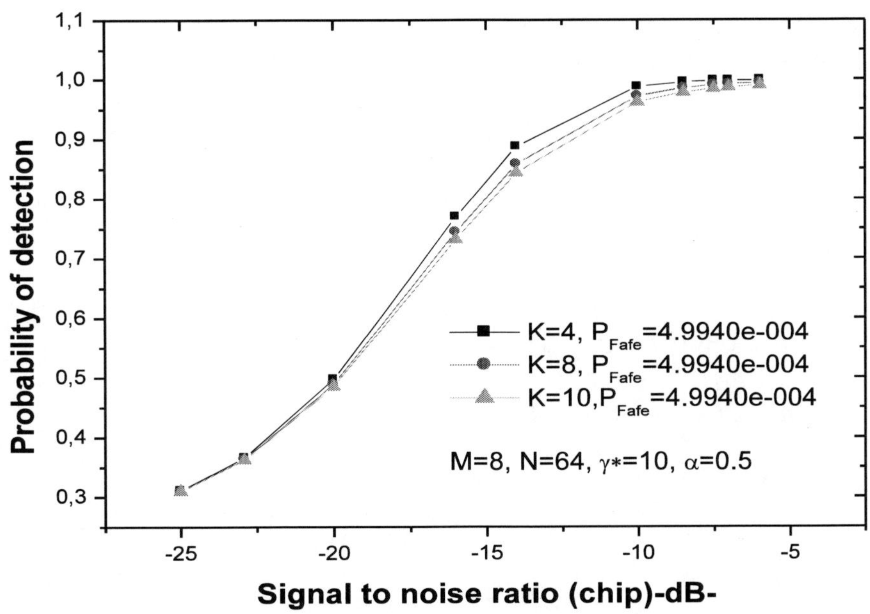

Figure 5. Probability of detection for Rayleigh Fading channel for the multi-carrier system ($M \geq 2$), with selecting combining using different values of *K*.

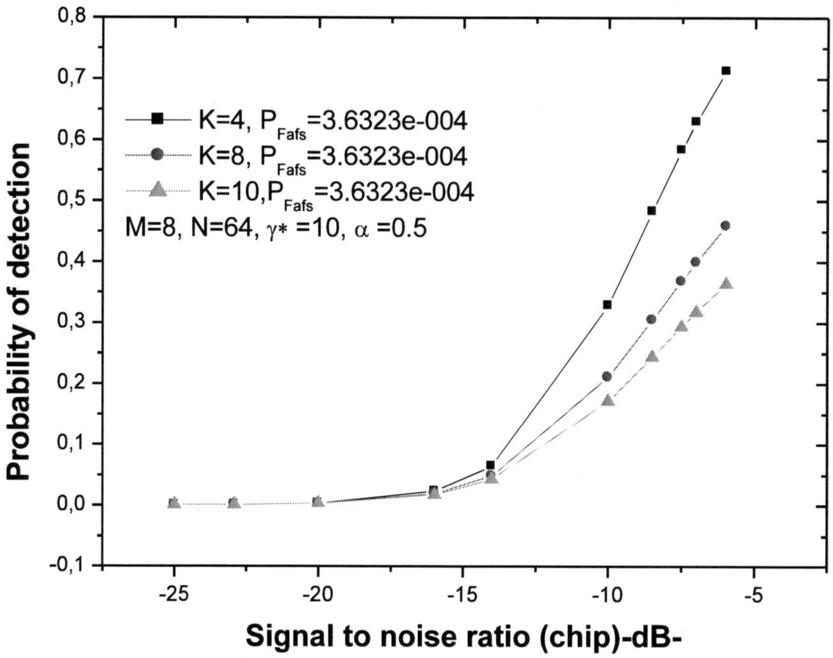

Figure 6. Probability of detection for Rayleigh Fading channel for the multi-carrier system (M≥2), with equal gain combining using different values of *K*.

The significant inconvenient of the CDMA scheme is the multiple access interference (MAI). The effect of the MAI on both probabilities; of detection and false alarm using equal gain combining is shown in Figure 5.

It can be observed that when the number of interferes (undesirable users) increases, the probability of detection decreases slowly. However the probability of false alarm remains constant, this could be explained by the fact that this probability doesn't depend on the value of *K* (active users). However, the effect of the MAI on both probabilities; of detection and false alarm using selecting combining is shown in Figure 6. It can be observed that the influence of the interferes is the same of that of the equal gain combining but, the decrease of the probability of detection is more significant for the first kind of combining scheme. We conclude that MC-DS-CDMA system using the EGC is robust against the MAI interference in comparison with that using the SC.

CONCLUSION

In this paper, we have studied two types of combinations: the first is the equal gain combining and the second is the selecting combining. Both are used for multi-carrier and single carrier systems. The results are compared with the results obtained in [5].

In the Rayleigh Fading channel, the performance of the multi-carrier system is better than the single carrier system for the equal gain combining.

However, the performance of the single carrier system is better than the multi – carrier system for the selecting combining.

We conclude that MC-DS-CDMA system using the EGC is robust against the MAI interference in comparison with that using the SC.

Finally, we propose the equal gain combining for the multi-carrier system and the selecting combining for the single carrier in the Rayleigh Fading channel.

APPENDIX

For a Rayleigh Fading channel, is a Rayleigh random variable with a unit second moment, the conditional (p.d.f) of z_m conditioned on hypothesis H_i is:

$$p\left(z_m/H_i\right)=\int_0^{+\infty} p(\alpha_{1,m})\,p\left(z_m/H_i,\alpha_{1,m}\right)d\alpha_{1,m} \tag{A1}$$

with : $p(\alpha_{1,m})=\alpha_{1,m}\exp\left(-\alpha_{1,m}^2\right)$ $for\ \alpha_{1,m}\succ 0$

$$p\left(z_m/H_i\right)=\int_0^{+\infty} 2\alpha_{1,m}\exp\left(-\alpha_{1,m}^2\right)\frac{1}{2v_{m,i}}I_0\left(\frac{\sqrt{z_m}\,m_{m,i}\alpha_{1,m}}{v_{m,i}}\right)\exp\left(\frac{-z_m-m_{m,i}^2\alpha_{11,m}^2}{2v_{m,i}}\right)d\alpha_{1,m}$$

$$=\int_0^{+\infty} 2\alpha_{1,m}\exp\left(-\alpha_{1,m}^2\left(1+\frac{m_{m,i}^2}{2v_{m,i}}\right)\right)\exp\left(-\frac{z_m}{2v_{m,i}}\right)I_0\left(\frac{\sqrt{z_m}\,m_{m,i}\alpha_{1,m}}{v_{m,i}}\right)d\alpha_{1,m} \tag{A2}$$

We divide and multiply the equation by $\left(1+\frac{m_{m,i}^2}{2v_{m,i}}\right)$, we choose a new variable and new variance and respectively as:

$$\alpha'_{1,m}=\alpha_{1,m}\left(1+\frac{m_{1,m}^2}{2v_{1,m}}\right)$$

$$v'_{m,i}=v_{m,i}\left(1+\frac{m_{m,i}^2}{2v_{m,i}}\right) \tag{A3}$$

(B.2) became:

$$p\left(z_m/H_i\right)=\frac{1}{\left(1+\frac{m_{m,i}^2}{2v_{m,i}}\right)}\int_0^{+\infty}\frac{2\alpha'_{1,m}}{2v'_{m,i}}\exp\left(\frac{-\alpha'^2_{1,m}}{\left(\left(1+\frac{m_{m,i}^2}{2v_{m,i}}\right)\right)}\right)\exp\left(-\frac{z_m}{2v'_{m,i}}\left(1+\frac{m_{m,i}^2}{2v_{m,i}}\right)\right)I_0\left(\frac{\sqrt{z_m}\,m_{m,i}\alpha'_{1,m}}{v'_{m,i}}\right)d\alpha'_{1,m} \tag{A4}$$

We define tow new variable and respectively,

$$\alpha''_{1,m} = \frac{\alpha'_{1,m}}{\sqrt{1+\frac{m_{m,i}^{2}}{2v_{m,i}}}} \text{ and } \alpha'''_{1,m} = \sqrt{2}\,\alpha''_{1,m}$$

(B.4) is given by:

$$p\left(z_m / H_i\right) = \frac{1}{2v'_{m,i}} \exp\left(\frac{-z_m}{2v'_{m,i}}\right) \int_0^{+\infty} \alpha'''_{1,m} \exp\left(-\frac{\alpha'''^{2}_{1,m} + a^2}{2}\right) I_0\left(a\alpha'''_{1,m}\right) d\alpha'''_{1,m} \tag{A5}$$

with $a = \dfrac{\sqrt{z_m}\, m_{m,i}^{2}}{\sqrt{2}\, v_{m,i} \sqrt{1+\frac{m_{m,i}^{2}}{2v_{m,i}}}}$

Marcum's function is defined as:

$$Q(a,b) = \int_b^{+\infty} x \exp\left(-\frac{x^2 + a^2}{2}\right) I_0(ax)\, dx \tag{A6}$$

For the special case when $b=0$, we have [9]:

$$Q(a,0) = 1$$

Finally, we obtain the equation (5).

REFERENCES

Periodicals

[1] Charles E. Cook, Fred W. Ellersick, Laurence B. Milstein, Donald L. Schilling, "Spread-spectrum communications," *IEEE press*, Inc., New york, copyright 1983.

[2] Douglas N. Rowitch, Laurence B. Milstein, "Convolutional coded multicarrier DS-CDMA in-a multipath channel-part I," *IEEE Trans. Commun.*, vol.47, no.10, pp.1570-1582, Oct. 1999.

[3] R.L Pickholtz, D.L. Schilling and L.B. Milstein, "Theory of spread spectrum communications-A tutorial," *IEEE. Trans. Commun.*, vol. COM-30, no.5, May 1982.

[4] Lie-Liang Yang, Lajos Hanzo, "Serial acquisition performance of single-carrier and multicarrier DS-CDMA over Nakagami-m Fading channels," *IEEE. Trans. Commun.*vol.1, no.4, pp. 692-702, Oct 2002.

[5] Dongwook. Lee, Laurence B. Milstein, Hun. Lee," Analysis of multicarrier DS-CDMA code –acquisition system,"*IEEE Trans. Commun.*, vol.47, no.8, pp.1233-1243, Feb.1999.

[6] S. Kondo and L. B. Milstein, "On the performance of multicarrier DS-CDMA systems," *IEEE Trans. Commun.*, vol.44, pp.238 - 246, Feb.1996.

[7] J. Bingham," Multicarrier modulation for data transmission: An idea whorse time has come," *IEEE. Commun.* Mag., May 1990.

[8] Tan F. Wong, Tat M. Lok, James S. Lehnert,"Asynchronous multiple-access interference suppression and chip waveform selection with aperiodic random sequences," *IEEE, Trans. Commun*, vol. 47, no.1, pp.103-114, Jan 1999.

Books

[1] Frederic Coulon,"*Theory and treatment of signals*," Edition Georgi. Paris: Dunod, 1984.

[2] J. Viterbi,"*CDMA principles of spread spectrum communication*," Copyright 1995 by Addison Wesley Longman, Inc.

In: Knowledge Management and Communication … ISBN: 978-1-62808-250-0
Editor: Rafiqul Islam

Chapter 17

LEAN PERFORMANCE IMPROVEMENT FOR THE PRODUCTION OF 5-BURNER GAS OVENS

Mohammad D. Al-Tahat*[*,1], Al-Refaie Abbas[1] and R. H. Fouad[2]

[1]University of Jordan, Faculty of Engineering and Technology
Industrial Engineering Department, Amman, Jordan
[2]Hashimite University, Faculty of Engineering,
Industrial Engineering Department, Zarqa, Jordan

ABSTRACT

A lean analysis for the production method of producing 5-burners gas oven at one of the biggest Jordanian companies hereinafter referred to as SABA is carried out. A new approach for process flow and activity relationship analysis is followed. Flow process chart, and Gantt chart are used as a lean tools for documenting the production process. The current method is questioned and brainstormed with the senior management of SABA, consequentlty lean performance measures like Tact time, utilization rate, queuing time etc. were developed, discussed, then used. Based on that, a new design for the production process that can meet the gradual increase of customer demand on 5-burners gas oven is introduced. The main advantages of the new designed production process is that it reduces the average production time by 46%, increases the average utilization of production resources by 36%, and increase production capacity of the production line by 100%. Another important advantage of the new design is that Work in Process (WIP) is decreases, bottleneck is diminished and waiting time of WIP between production stages is totally eliminated.

Keywords: Lean assembly/ Design for assembly/ Productivity improvement/ Gas oven

* E-Mail: altahat@ju.edu.jo.

1. INTRODUCTION

Lean production is stemmed from mass production developed by Henry Ford and other pioneers (Groover M. P. 2001). Womack et al. (1990) was described the term "lean production". Broadly considered practices of lean production have been characterized recently by Browning T. R. and Heath R. D. (2009). Other literature sources (Al-Tahat M. D., and Eteir M. F, 2010, Al-Tahat M. D., 2010; Tahat M. D., 2009; Chase et al., 2006) described different lean method followed for the elimination of non-value-adding tasks.

SABA is a Jordanian public share holding company and one of the pioneers in the sheet metal industry in Jordan, SABA was established in 2000 with a registered capital of $8 million, and is currently staffed with more than 250 employees in its plants. SABA owns a well-equipped four metal processing workshops. High quality products are produced by SABA to the latest technology on an area of 15000 m. sq. 5-burners gas cooking oven, Tabletop cooking oven, Oil radiators, Water heaters and many others are manufactured according to international standards (ASTM, BS, DIN, JIS, GOST, etc.). SABA produces a high volume of a small variety of outputs, thus it can organize its operation as a flow shop (Meredith and Shafer, 2011). A quick response to customers is very important for the company as well as meeting all customers' requirements to achieve their satisfaction and to exceed their expectations. SABA is always searching for lean and agile principles such as manufacturing strategies, production methods, materials flow control techniques, etc. that help in improving out the company's processes, and reduce costs. One goal SABA is looking to achieve is to increase the overall utilization of ovens production line with the same consumed production resources. Operations, procedures, and standard methods that added no value to the final ovens should be avoided.

2. PROBLEM FORMULATION AND OBJECTIVES

The general scope of this paper is to search for more efficient lean manufacturing approach that concerns with: the analysis of the methods and equipments used in performing production processes; the design of a new improved production method; and the standardization of the new proposed method. The new proposed production method is desirable to ensure that the work is being done in the easiest, safest, and most productive way with least possible waste. Improvements of the existing production process, consequently design of a new production plant, are considered as the main mission of this paper.

Production of gas cooking oven at SABA is currently carried out by an production line operating with a total capacity of 100 ovens cocker per 8 hours shift per day. A new production line with capacity of 300 ovens per 8 hours shift per day is desired. Assistance from the technical experience available is provided in the field of production system design and analysis design. Therefore an approach for process flow and activity relationship analysis is followed to build and to introduce a new production process that can meet the gradual increase of customer demand of gas cooking ovens.

This paper is concluded with a new recommended production process operates with larger production capacity, shorter production lead-time, responds rapidly to customer concerns, and includes the following main characteristics:

(1) Produces uncomplicated ovens.
(2) Uses minimum number of components in the product
(3) Assembles 300 ovens per 8 hours shift per day with the lowest possible waste.
(4) Maintains and repairs ovens with problems during the production process without interruption the production rate.

3. Proposed Methodology

The following work phases was carefully investigated and explored as a systematic approach for achieving and a accomplishing the objective of this paper:

1. As stated in the previous section goals, objectives, and mission of paper was clearly stated and identify and a full analysis of problem and of the existing situation was carried out. This phase includes:
   a) Mission: Understanding problem scope and objectives
   b) Output: Identifying output of the production process. Types of ovens and list of components (usage quantities) included in each oven type (Bill of Material). Analyzing the production drawing and specifications for each oven type
   c) Input: Considering and recognizing input resources and materials for the production process and studying the functional description of each detailed component
   d) Sequence: Detailing production plans for each oven type.
   e) Environment: Recognizing the current layout type including all elements of the production line, and specifically with the following information; Name of each element in the production process; Production capacity of each element. Size and dimension of each element; Resources needed for each element (Workers); Operating method of each element (manually or mechanically); and Handling equipment involved at each element. (Condition of work, physical, attitudinal organizational, contractual, cultural, political, and legal)
   f) Workers: Human agent and labor employed for operation the line
   g) Physical catalysts: Machines, equipments, instruments, tools, and other physical entities without becoming a part of the output product.
   h) Information aids: Knowledge and know- how of operating the line.
2. Based on the above, process flow and activity relationship analysis was carried out. Some lean tools and techniques for methods analysis were reviewed and the proper tool and technique was selected. Lean tools and techniques may includes; Charting techniques (Flow process chart, Right and left hand operation chart, SIMO-simultaneous motion chart, and multiple activity chart); Human factors engineering; Work simplification programs; Job enlargement and enrichment; Value analysis; and suggestion systems (Krajewski et. al, 2010; Montgomery et. al, 2011; Groover, 2011)
3. Questioning with senior management of SABA the present method, and discussing productivity measures, key performance indicators and key output indicators.
4. Evaluation of the current performance of the production line, and synthesize a proposed alternative redesign models for enhancing the existing ovens production

processes, procedures, methods, standards etc. (Korsakov and Zamyatin 1987). Presenting the redesign proposals to SABA' management and agreeing with that on the best course(s) of action.

5. Applying the proposed method (to be carried out by SABA)

4. Current Production Process

The current production system is working under total capacity of 100 products per 8 hours daily shift, to produce modern competitive high quality cooking ovens, supported by a high caliper efficient after sales service to cater for the customer needs. The production yard area is about 50 M X 12 M, as in the production plant layout drawing shown in Figure 1. The production yard includes inventory buffers of: needed materials; sub assembly products, tools and equipments to operate the system, tools and instruments for in process quality control; sufficient area as a buffer storage for final produced oven; labor power; and conveyor handling system transfers the work in process from production station to the next. Raw material flow within this production line from left to right, conveyor of a constant traveling speed of 0.66 M/Minute (1.53 Minute/1 M) is used. More than 20 workers are employed in the line distributed between more than 20 production stages and control operations. Three similar types of gas cooking ovens are produced with the same operations and the same flow of material. Only, 5-burners gas cooking oven is considered in this paper.

The production items that constitute the oven are 104 items, some of these used as a single item, the other are consumed as a group or sub assembly. For instance, the item named "Front Panel profile" is consumed as single item with a usage quantity of one, while the ceiling and the upper grill of the oven consumed together as a subassembly represent. Joining screws, electrical connections and components, paints, rock wool, and many others are also needed as input resources to operate the assembly process.

The flow of the production process is routed as indicated on Figure 1 by the solid black arrow starting from oven frame building stage ending at packaging area. Oven during his progressive journey up to be a complete assembled oven at the end of the line passes through 18 stages. The sequence of these stages is detailed and analyzed. The needed resources of children components and sub assembly are identified in all stages of producing the oven. For example, at the first stage item 4, sub assembly 1.1, sub assembly 1.2, sub assembly 1.3, sub assembly 1.4, and sub assembly 1.5, are used as inputs, the output of the first stage is a full oven box and is designated by, WIP (1).

Flow process chart (Tompkins et. al, 2003) will be used as a technique for analyzing and for searching for opportunity to improve, as well as to design a new production plant that achieve the objectives of this paper.

5. Process Flow Chart

Flow Process chart (Tompkins et. al, 2003) is one of the most useful tools that being used for studying of macro motion, it provides a big picture of the movements of materials between stages of the considered production process.

Table 1. Partial representation of the process chart

Stage No.	Stage Describtion	Stage Time (Min)	Step No.	StepTime (Min)	Input	●	➡	■	D	▼	Step Description	Output
2	Clemens, exhausts, and rock wool	3.50	21		WIP(1)		X	.	.	.	Transfer of WIP(1) into stage 2	WIP (2)
			22		?	X					Adding the electrical cord ()	
			23		44	X					Fixing the Oven lamp into lamp group (44)	
			24		19	X					Fixing rear chimney exhaust(19)	
			25		45	X					Fixing front chimney exhaust(45)	
			26		18	X					Covering the whole with Rock Wool (18)	

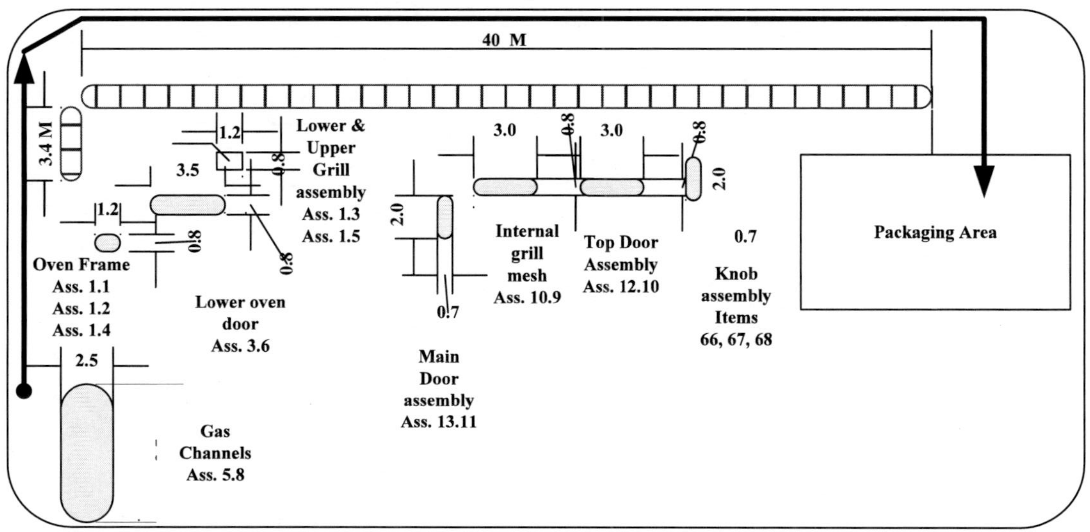

Figure 1. Plant layout of the current production process.

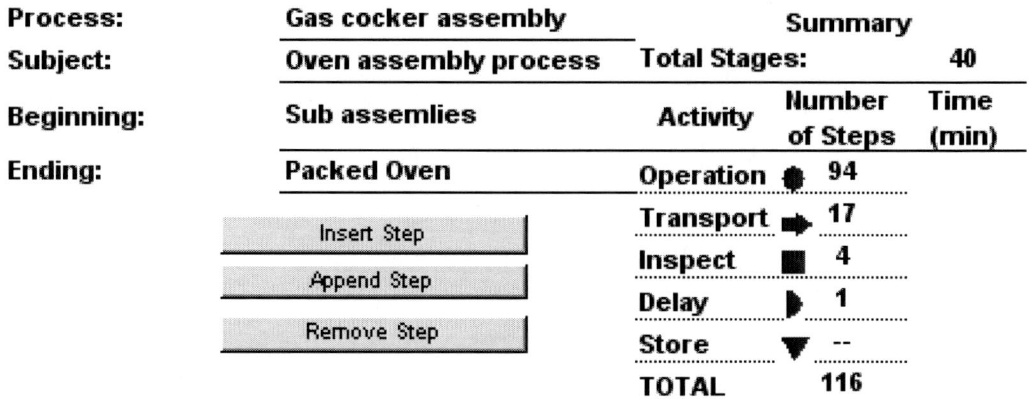

Figure 2. Summary of the process chart.

Operation by operation records is obtained, that may be performed by visual observation. Excel sheet is used and process chart is developed for all stages of production that imply a number of production steps, a brief summary of the process chart is presented in Figure 2.

Total time to accomplish the whole production process is estimated in last column, 94 operations, 17 transportation activities, and 4 inspections were incorporated. For more explanation of the process chart, stage 2 is considered as a sample, as shown in Table 1.

The fixation process of clement, exhaust, and rock wool is carried out in this stage 2. There are 6 steps in this stage; these are – 21, 22, 23, 24, 25, 26--, five are operational step, and one is transportation. Also stage 2 consumed WIP (1), item 44, item 19, item 45, and item 18. The output of this stage is WIP (2) which will continue its bath downstream.

6. Performance Evaluation of Current Production Process

It is shown clearly at top of Figure 2 that a total of 40 minutes are required to assemble a whole oven starting from stage one ending up with a complete oven. Figure 3 shows the distribution of these 40 minutes between the different production stages. Assume ti, is the unit processing time at stage I, then:

$$Oven\,assembly time = \sum_{i=1}^{17} t_i = 40 \tag{1}$$

For a complete production cycle under steady state working conditions, the production line can be represented by Gantt chart as in Figure 4. The chart shows the utilization and idle time for each stage. Also we can conclude from the chart that there is a bottleneck in stage 3, therefore parts queued up in front of stage 4. The average unit waiting time in front of stage 4 for production quantity of Q ovens is estimated by equation 2,

$$Average oven waiting time = \frac{\sum_{q=1}^{Q}(q-1)(t_3)}{Q} \quad (2)$$

As a result production cycle time is not constant and it depends on the quantity of production (Groover M. P., 2001) this result is represented by equation 3.

$$\begin{matrix} Average\ production \\ cycle\ time \end{matrix} = \sum_{i=1}^{17} t_i + \frac{\sum_{q=1}^{Q}(q-1)(t_3)}{Q} \quad (3)$$

For production quantity of 100 ovens, average production cycle time will be:

$$\begin{matrix} Average\ production \\ cycle\ time \end{matrix} = 40 + \frac{\sum_{q=1}^{100}(q-1)(0.8)}{100} = 80 Minute \quad (4)$$

Each oven needs 40 minutes for production and each oven spends an average of 40 minutes waiting in the line. The first part spends only 40 minutes, part number 100 spends 80 minutes. According to the previous analysis, and reference to Figure 4, tact time, which is the time needed for an output of 1 oven, or the interval of time between two successive products is estimated as in equation 5,

$$Tact\ time = \frac{4.0+4.0+0.4}{2} = 4.2\ Minute \quad (5)$$

For 8 hours shift of production the total output can be estimated as in (6)

$$\begin{matrix} Output/ \\ 8\ hours/shif \end{matrix} = \frac{(8\ hours)60}{(Tact\ Time)} = \frac{480}{4.2} = 114\ oven \quad (6)$$

In same manner utilization of each stage of the production process can be estimated, the result are presented in table 2. The over all estimated utilization of the production line is 58%.

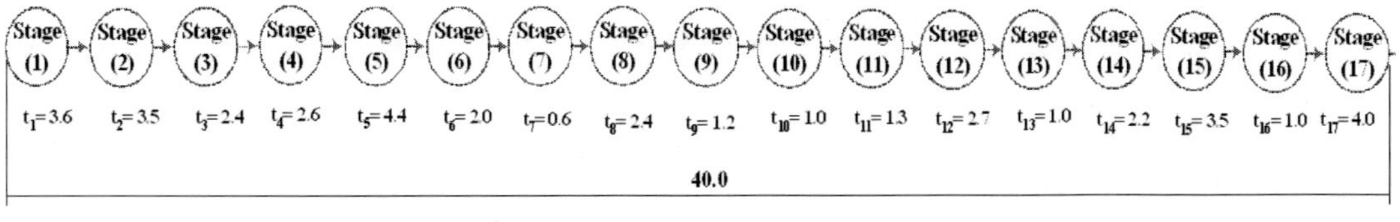

Figure 3. Processing time at each production stage.

Table 2. Estimated utilization rates

Stage		Stage	
Index	Utilization (%)	Index	Utilization (%)
1	100	9	43
2	100	10	23
3	67	11	30
4	72	12	61
5	100	13	23
6	46	14	50
7	14	15	80
8	55	16	23
		17	91
Average Utilization (%)			58

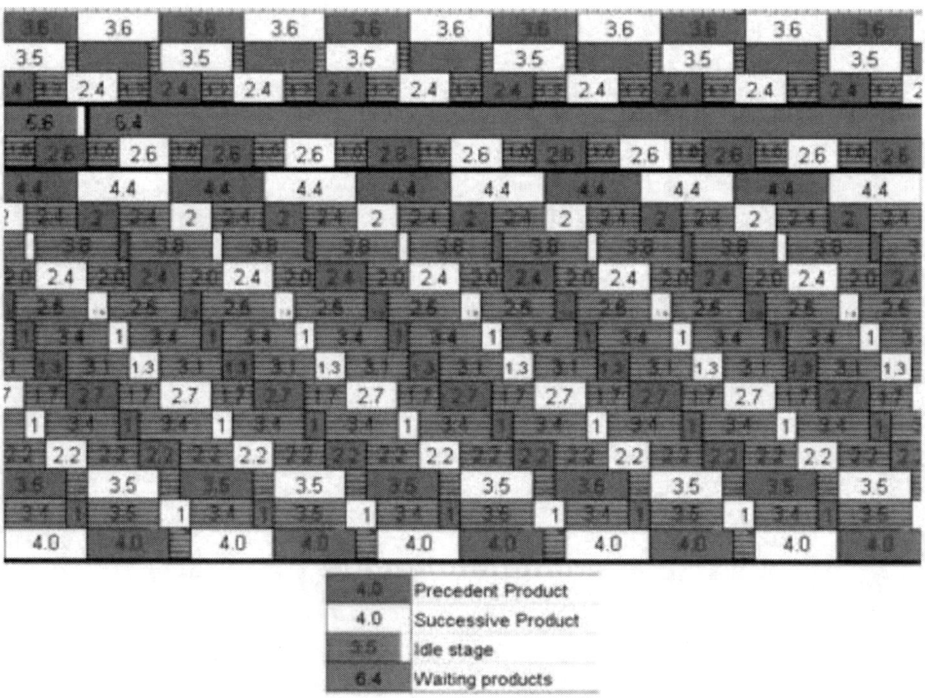

Figure 4. Gantt chart of production during the complete production cycle time.

7. Design a New Production Process

A new modified assembly process resulted from questioning each stage involved in the production line. The following questions are raised about each step (Meyers, and Stewart, 2001).

1. Can a stage be eliminated?
2. Can a stage be combined with another operation?
3. Can the production sequence be changed?

4. Can a stage be simplified?
5. Can a movement be eliminated or shortened?
6. Can a delay be eliminated?
7. Can storage be eliminated?

The most constructive question was question 1, question 2 and question 4. The main objective of this phase of the work is to design a new production process with a capacity of 300 ovens per 8 hours shift. This goal is achieved through the following steps:

1. Producing 300 ovens in 840 minutes, means that the tact time should be as in equation 7. Every 1.6 minute a complete oven must exit the line.

$$(Tact\ time) \le \frac{480\ Minute}{300\ Oven} \le 1.6\ Minute/Oven \qquad (7)$$

2. To achieve equation 7, stage 17 should be simplified and broken into more than one unit such that the output of the stage must be more than one oven per 1.6 Minute.
3. Combine stage 16 and stage 15 then decomposed the result into three stages of processing rate of 1.5 for each one.
4. Combine stage 14 with stage 13 and split the result into two stages of 1.6 processing rate.
5. Combine 12 and 11 and decompose the result into two stages of processing rate of 2.
6. Combine 7, 8, 9 and 10 and decompose the result into four stages of processing rate of 2.
7. Keep stage 6 as is.
8. Combine 3, 4 and 5 and decompose the result into 3 stages of processing rate of 3.2.
9. Don't change any condition in stages 1, 2, and 6.

The above modifications led to Figure 5 which represents a Gantt chart for the new design production system (Babcock and Morse 2002).

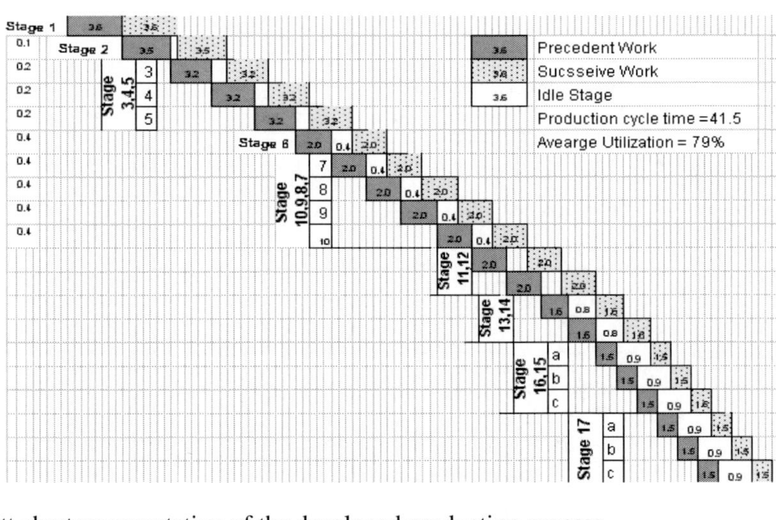

Figure 5. Gantt chart representation of the developed production process.

8. PERFORMANCE EVALUATION OF THE NEW PRODUCTION PROCESS

As shown in Fig. 5, 41.5 minutes are required to process the total oven starting from stage one ending up with a complete oven. Number of stages is increased to 20 stages; therefore the total unit processing time is;

$$\textit{Unit processingtime} = \sum_{i=1}^{20} t_i = 41.5 \qquad (8)$$

There is no waiting time, as well as there is no bottleneck inside the new production line, therefore equation (1) equal to zero all the time, and hence,

$$\textit{Averageunit waiting time} = 0.0 \qquad (9)$$

When Q =100, there is a reduction in average production cycle time of 46%, as seen in equation 10:

$$\begin{matrix}\textit{Average production} \\ \textit{cycle time}\end{matrix} = \sum_{i=1}^{20} t_i + 0.9 = 42.4 \qquad (10)$$

Tact time is;

$$\textit{Tact time} = 1.5 + 0.9 = 2.4 \textit{Minute} \qquad (11)$$

Table 3. Estimated utilization rates for the new design

Stage		Stage	
Index	Utilization (%)	Index	Utilization (%)
1	100	11	83
2	100	12	83
3	94	13	66
4	94	14	66
5	94	15	63
6	83	16	63
7	83	17	63
Stage		Stage	
Index	Utilization (%)	Index	Utilization (%)
8	83	18	63
9	83	19	63
10	83	20	63
Average Utilization (%)			79

For 8 hours shift of production the total output can be estimated as follows;

$$\frac{Output}{8hours/shft} = \frac{(8\ hours)60}{(Tact\ Time)} = \frac{480}{2.4} = 200\,oven \tag{12}$$

Utilization percentages are presented in table 3. The over all estimated utilization for the new production line is 79% with an increase of 36%.

Finally, a simple comparison between the current way and the proposed way of production is presented in table 4.

Table 4. Comparisons between current and new method

Performance Measure	Current Method	New Method
Average Utilization %	58.0	79.0
Production Capacity	100.0	200.0
Average Production Cycle Time	80.0	41.5
Tact Time	4.2	2.4
WIP Waiting Time	40.0	0.0
Processing Time	40.0	42.4

CONCLUSION

An analysis of the production methods for producing 5-burners gas cooking oven at SABA Company is carried out. A new approach for process flow and activity relationship analysis is followed. Flow process chart, and Gantt chart are used as a lean tools for process documentation. The current method is questioned and brainstormed with senior management of SABA, consequentlty lean performance measures like Tact time, utilization rate, queuing time etc. were discussed. Modification in the current method of production is carried out. These modifications were focused on the increase of customer demand, and on the elimination of some types of wastes. Lean manufacturing principles were considered and finally new methods for operating and scheduling the production line is introduced in such away that that can meet the gradual increase of customer demand of ovens. The main advantages of the new designed production line is that it reduces the average production time by 46%, increases the average utilization of production resources by 36%, and increase production capacity of the production line by 100%. Another important advantage of the new design is that Work in Process (WIP) is decreases, bottleneck is diminished and waiting time of WIP between production stages is totally eliminated.

REFERENCES

Al-Tahat M. D., and Eteir M. F., 2010. An Investigation of the Potential of Implementing Kaizen Principles in Jordanian Industries. *Int. J. Product Development*, Vol. 10, Nos. 1/2/3, pp.87–100.

Al-Tahat M. D., 2010. Effective Design and Analysis of Pattern Making Process Using Value Stream Mapping. *Journal of Applied Sciences*. 2010. Volume: 10. Issue: 11.pp: 878-886.

Al-Tahat M. D., 2009. Re-engineering of production processes using Value Stream Mapping tool, case study: "ABC" Plant for producing tissue and hygienic paper. *Journal of Information, Intelligence and Knowledge (JIIK),* Volume 1, Issue 3. ISSN 1937-7983.

Babcock D. L., and Morse L. C., 2002. *Managing Engineering and technology*, 3rd ed., Prentice Hall, pp.338– 339. ISBN 0-13-061978-7.

Chase, R.B., Jacobs, F.R., Aquilano, N.J., 2006. Operations Management for Competitive Advantage, 11th ed. McGraw-Hill/Irwin, New York.

Groover M. P., 2011. *Principles of Modern Manufacturing*, 4th ed., John Wiley and Sons, Inc., pp.913– 917. ISBN: 978-0-470-50592-2.

Groover M. P., 2001. *Automation, Production Systems and Computer Integrated Manufacturing,* 2nd ed., Prentice Hall, pp.32– 38. ISBN 0-13-089546-56.

Korsakov V. S., and Zamyatin V. K., 1987. Assembly Practice in Machine Building, 1st ed., *Mir Publisher Moscow,* pp.256– 278.

Krajewski L. J., Ritzman L. P., and Malhotra M. K., 2010. *Operation Management: Processes and value chains,* 9th ed., Pearson, pp.142–175. ISBN-10:0-13-245891-8.

Meredith J. R., and Shafer S. M., 2011. Operation Management, 4th ed., John Wiley and Sons, Inc., pp.56–57. ISBN: 978-0-470-52457-2.

Meyers F. E., and Stewart J. R., 2001. *Motion and Time Study for Lean Manufacturing*, 3rd ed., Prentice Hall, pp.32– 38. ISBN: 0-13-031670-9.

Montgomery D. C., Jennings C. L., and Pfund M. E., 2011. *Managing, Controlling, and Improving Quality,* 1st ed., John Wiley and Sons, Inc., pp.74–94. ISBN: 978-0-471-69791-6.

Tompkins J. A., White J. A., and Tanchoco J. M. A., 2003. *Facilities Planning*, 3rd ed., John Wiley and Sons, Inc., pp.36– 47. ISBN 0-471-41389-5.

Womack, J.P., Jones, D.T., Roos, D., 1990. *The Machine that Changed the World.* Rawson Associates, Harper Collins, NY.

In: Knowledge Management and Communication … ISBN: 978-1-62808-250-0
Editor: Rafiqul Islam

Chapter 18

IMPROVING CONVERGENCE IN A COMPLEX PRODUCTION PLANNING PROBLEM WITH TWO HYBRID METHODS

Mikael Nyberg*[*,1] and Kaj-Mikael Björk[2]
Process Design and Systems Engineering / Åbo Akademi University, Finland
[1]IAMSR / Åbo Akademi University, Finland
[2]Arcada University of Applied Sciences, Finland

ABSTRACT

This paper presents two hybrid methods to solve a production planning problem. The problem is found in a producer of food-sweeteners, with several different kinds of products. The case study plant is found in the USA. The aim is to improve convergence for a big optimization problem. The first method uses a Genetic Algorithm in combination with an existing rigorous MILP model. The fitness value of each chromosome is evaluated by solving the MILP model with a highly reduced number of binary variables. The second method is a Tabu search that also uses the rigorous MILP to evaluate each node. The meta-heuristic models are presented and used to solve a few case problems and the results are discussed. Furthermore the two methods are compared in terms of solution times and convergence rate.

Keywords: Genetic Algorithm, Tabu search, hybrid method, meta-heuristic, soft computing, production planning

1. INTRODUCTION

Good and effective production planning has become increasingly important in the production industry. The capital intensive and complex production processes within the chemical industry is no exception. Discussion with representatives from companies in this

[*] Email: mikael.nyberg@abo.fi.

industry segment has revealed that efficient production plans are a key issue when trying to improve overall performance. The SmartOpt (Smart Optimization of Production-Distribution Networks) project (http://iamsr.abo.fi/smartopt/) aims to optimize the entire production-distribution network (PDN) of a company. In order to achieve the goals we have chosen to start by modeling the production and then expand the scope to involve the whole PDN. We believe this approach will lead to better results as the production process in the fine chemical plant is very complex, and needs to be modeled in detail to be able to generate accurate and effective production plans (a major component in finding good overall solutions).

The case company, a food-sweetener producer, in this paper has a complex, multi-stage production facility that we needed to model. The initial, rigorous MILP model presented in Nyberg and Björk (2010) produced good results but solution convergence was utterly slow. To mend this problem, two hybrid methods using a heuristic approach combined with the rigorous MILP were created. The first method (M1) uses a Genetic Algorithm (GA) approach which uses the MILP-formulation mentioned above to evaluate the fitness function and the second method (M2) uses a Tabu search (TS) in combination with the same MILP model as M1. In this paper these two methods are presented and discussed. The paper is structured as follows: first a brief literature review is conducted. Then the SmartOpt project as well as the case company is presented, after which the two hybrid methods are presented. These models are used to solve four real production-inventory problems at the producer of food-sweeteners and the results and models are compared. Finally, some conclusions are drawn.

2. Literature Review

Production planning through optimization has an important role in today's competitive manufacturing industry world. An increasing number of companies are relying on sophisticated optimizations models to create good production plans. The process industry segment has, according to us, not obtained enough attention in this field of research.

The scale of the problems to optimize has grown as computers have become faster. Two main categories of optimization exist; rigorous techniques that guarantee global optimum and heuristic methods (soft computing, Zadeh, 1998) that provide good solutions but cannot guarantee optimality. The desire to solve ever larger problems has resulted in a wide use of heuristic methods because of their immense advantage in speed compared with rigorous methods. This makes it possible to solve problems that are too time-consuming to solve using rigorous methods. When a heuristic approach is chosen a rigorous method is often used to benchmark the heuristic to give indications about the quality of the solution. Rigorous methods have, however, the advantage of being able to guarantee that the solution found is globally optimal. Thus, a combination of these two techniques can provide very effective new methods capable of solving bigger problems with better results than by using the techniques separately. As stated before real life problems are often too big to be solvable using only rigorous methods. One example is found in Björk and Nordman (2005) who present a hybrid method using a GA combined with a deterministic MINLP model to solve large-scale heat exchange network problems.

The model presented in this paper is a hybrid soft computing model that combines a heuristic method and a rigorous MILP method. The term hybrid method is used for a variety of models that combine the benefits of two or more separate methods. There are a number of similar contributions found in the literature.

In Björk and Carlsson (2007) a Genetic Algorithm is used to solve a PDN problem in the process industry. The main focus is similar to the one in this paper; to reduce the production and inventory costs of the producer. The methods are also somewhat similar to those of M1 in this paper. The contribution of Russel, Chiang and Zepeda (2007) is very interesting as it uses the same approach as we have used in M2 but on a vehicle routing problem. The solutions found by a rigorous model are improved using a Tabu search.

Józefowska and Zimniak (2008) used a slightly different approach. In their model they decided to not use a rigorous method but two soft computing techniques to form a hybrid method. They used an expert system module to reduce the search space of a pure GA. Lukac et al (2008) used a two stage Tabu search model to solve a production planning problem, similar to our case, minimizing setup, production and storage costs.

Other examples found in the literature are for instance Gen and Syarif (2005) who use a hybrid Genetic Algorithm to solve a production-distribution problem. The initial GA model is improved by using a fuzzy logic controller to auto-tune the parameters. Another example where fuzzy computing is combined with a GA is found in Aliev et al. (2007). In this model inaccuracies in market demand and its impact on the overall solution are greatly reduced by using fuzzy programming. Other work in the field of hybrid methods and meta-heuristic that can be of interest can be found i.e. in the following research papers: Defersha and Chen (2008) (embedded GA), Sachdeva et al (2008) (GA with Stochastic Petri nets), Akrami et al (2006) (two heuristics; GA and Tabu search), Pradenas et al (2004) (Tabu search) and Hung et al (2003) (hybrid Tabu search).

3. CASE

This case is a part of the larger SmartOpt project carried out by Åbo Akademi University in cooperation with several process industry companies. SmartOpt consists of three cases; the plastic raw-material used for wrapping tissue paper (SC1), the supply and production of wood to pulp-mills (SC2) and the production-distribution of food-sweeteners (SC3). In all three cases the goal is to investigate how the complex supply chains can be made more effective. In this paper we will discuss the research done in SC3.

The company for this case study produces bulking agents for the food industry. The company is one of the market leaders in bulking agents with their biggest market areas in USA, Europe and Asia. The awareness of customers and the demand for non-genetically modified products, referred to as non-GMO or identity preserved, has increased rapidly during the past years.

The growth of this non-GMO market, especially in Europe, has forced the company to develop two almost identical product lines; one using genetically modified organisms (GMO) raw materials and another using non-GMO raw material. The two product lines have introduced new challenges to the production planning. Partly because of the increase in products and partly because of the new restrictions introduced to insure the integrity of the identity preserved products. The case company's assortment consists of a fairly small number of base products. However, these base products can be grinded, blended into solutions and packaged in a multitude of ways resulting in a large number of finished products.

The production is a multi-stage process starting with raw materials entering the chemical reactors. The intermediate products pass through several processes on their way to a finished base product. Finally the refined products are grinded, packaged and stored for shipping.

In figure 1 we see the different stages in the production of food-sweeteners in the case plant. The whole process was initially modeled as a discrete-time MILP-problem.

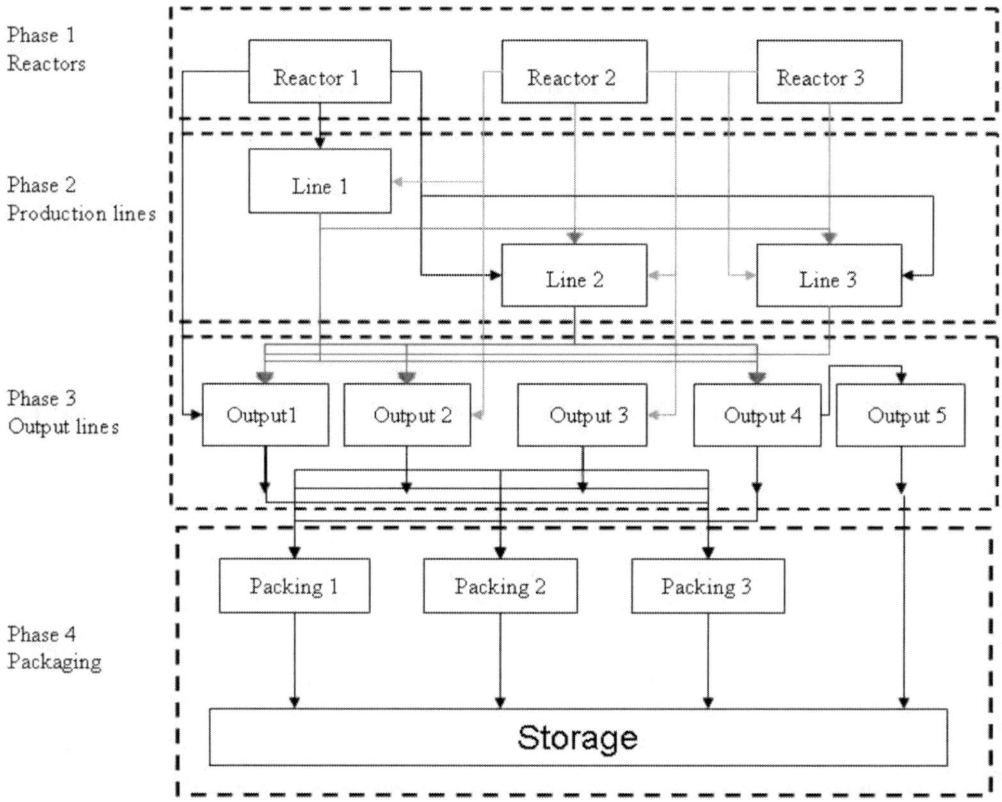

Figure 1. A schematic picture of the production process.

In order to achieve the level of detail required to produce accurate production plans for the facility the MILP-formulation became complex and computationally heavy. It consists of 20 sets of constraints describing the production process and includes over 4000 variables of which more than 1500 are binary. A detailed description of the MILP can be found in Nyberg et al. (2010). Subsequently the models presented in this paper span the whole production process as does the initial discrete-time MILP.

4. The Hybrid Methods

The problem at hand is difficult to solve using a rigorous method because of the size of the problem. On the other hand the complicated structure of the production process made it difficult to apply a pure heuristic method. It proved to be nearly impossible to find ways of including all restrictions using only a heuristic approach. The solution was to combine the existing MILP model and reduce the computational complexity of it by harnessing it with a heuristic method. We chose to use the concept of chromosomes used in Genetic Algorithms as our starting point. The biggest challenge was to find a simple chromosome structure that would allow us to use the existing MILP with a reduced solution space to speed up the convergence. After examining the properties of the production facility and the MILP model the following chromosome was chosen.

4.1. Improving the Initial MILP-Formulation

The chromosome determines in which order products are produced on the reactors. In the example chromosome in figure 2 each reactor has four alleles ($i_{max} = 4$) and therefore each reactor can produce a maximum of four products during the optimization window in the order given by the chromosome. This structure can easily be expanded to allow more products per reactor. The number of reactors can also be changed if need be.

Let $p_{i,j}$ be the i:th product on reactor j given by the chromosome. To reduce the number of variables in the MILP problem we only allow the products of two consecutive alleles ($p_{i,j}$ and $p_{i+1,j}$) to be produced for a number of time steps, called a slot $s_{i,j}$. In our case the slot size is determined by the minimum number of days ($\pi_{p_{i,j}}$) a product has to be produced once production has started. To make sure we do not violate these minimum production constraints we always choose the longer minimum time for the two allowed products as the slot size. The MILP model has the freedom to determine for how long a production is produced as a product is available for production in two consecutive slots. After the minimum production can be terminated at any time within a slot and production of the other allowed product for that slot can be started. E.g. product 1 in allele $p_{1,1}$ (in figure 2) is available for production in slots $s_{1,1}$ and $s_{2,1}$. The MILP can choose to produce $p_{1,1}$ for the whole duration of slot $s_{1,1}$ and for a part or the entire duration of slot $s_{2,1}$. If $p_{1,1}$ is only produced for a part of slot $s_{2,1}$ the model can chose to start production of $p_{2,1}$ (and continue production into slots $s_{3,1}$ if necessary to meet the minimum production time requirement) or skip $p_{2,1}$ all together and commence with $p_{3,1}$ in slot $s_{3,1}$. For even longer production cycles two consecutive alleles can be given the same value thus allowing a product to be produced in three slots. We also know what product has been produced on reactor j the time step prior to the optimization, giving us a starting product $p_{0,j}$ giving us { $p_{0,j}$, $p_{1,j}$} as the allowed products for slot $s_{1,j}$. We can now reduce the number of possible variables by the following rules:

1. Set i=1 and the starting time for slot $s_{i,j}$ to $t_j^{start} = 1$
2. Set the ending for slot $s_{i,j}$ to $R(P_S) = \frac{\Gamma_A}{\Gamma_{A0}}$
3. Set all decision variables to zero except for products $p_{i,j}$ and $p_{i-1,j}$ for the duration of slot $s_{i,j}$
4. Set the starting time of slot $s_{i+1,j}$ to $t_j^{start} = t_j^{start} + t_j^{end}$
5. Increment i with 1 until $i = i_{max}$ when $i = i_{max}$ go to 7

6. Go to step 2
7. Increment j with 1 until $j = j_{\max}$ if $j = j_{\max}$ go to 9
8. Go to step 1
9. Stop

Reactor 1				Reactor 2				Reactor 3			
1	2	3	...	1	2	3	...	1	2	3	...

Figure 2. Chromosome structure used for the hybrid methods.

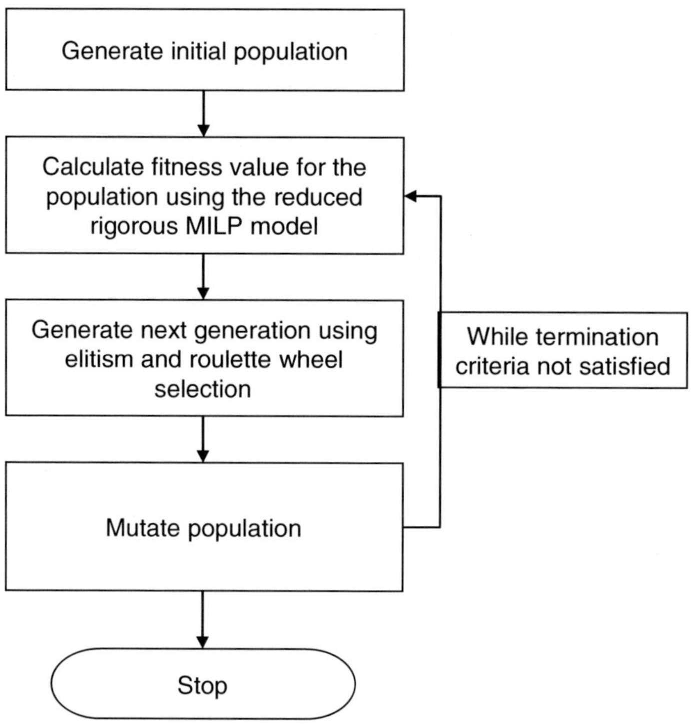

Figure 3. Flow chart of model M1 using a Genetic Algorithm in combination with a MILP model.

After the above procedure is finished the solution space of the original MILP problem is greatly reduced and the resulting MILP can now be solved within seconds compared to several days for the original unreduced problem. With this method the rigorous MILP can choose to skip a product given by the chromosome simply by not producing anything or by first producing and then proceeding with $p_{i+1,j}$.

In our case this approach reduces the number of non-zero main decision variables by 85% on average. Initially the unreduced MILP has 1570 main variables for a 30 time step optimization. After variable reduction according to the above algorithm the number of non-zero variables is on average 240. As solution times grow exponentially to the number of binary decision variables this reduction has a great impact on solution times and convergence.

In order for this reduction of the search space to yield good solutions to the production planning problem the reduction scheme described has to be applied iteratively. Furthermore, the evolution of the chromosomes has to be guided towards a more optimal product sequence. This was done by creating two hybrid methods combining the improved MILP-formulation with two different heuristic approaches.

4.2. Genetic Algorithm (M1)

The first heuristic model, M1, is a hybrid Genetic Algorithm. The general concept for the model is a normal GA but the fitness of each chromosome is evaluated with the rigorous MILP model presented in Nyberg et al (2010).

As explained earlier the solution space for the MILP is greatly reduced to speed up the solution times. With the chosen chromosome structure it is possible to have different sets of products for different reactors. In our test cases reactor 3 has a reduced set of products compared to reactors 1 and 2. As long as the mutation is controlled so that alleles are mutated within the set of allowed products all offspring from suitable parents will also be suitable.

4.2.1. Fitness Function

The fitness function for the GA is the objective function (1) of the MILP-problem which minimizes the total production costs consisting of; setup costs, inventory costs and penalty costs caused by late deliveries.

$$\min \sum_{k}\sum_{j} a_{j,k} + \sum_{k}\sum_{i} \delta_i * u_{i,k} + \sum_{k}\sum_{i} \beta_i * t_{i,k} \tag{1}$$

where $a_{j,k}$ describes setup costs, $u_{i,k}$ is the negative inventory, i.e. late deliveries, multiplied by a penalty cost factor δ_i and $t_{i,k}$ is the inventory level multiplied by a inventory cost factor β_i. In order to calculate the fitness of each individual the optimal production plan for the product sequence given by the chromosome has to be determined. This is done by solving the improved MILP-problem using the scheme described in section 4.1. In our case it is obvious that the lower the fitness value the better the individual.

4.2.2. Initial Population

The first step in the algorithm is to generate an initial, good population. Testing clearly showed that a completely random starting population leads to slow convergence. Therefore a rather extensive selection process is made during the initialization stage of the GA. Let the population size be *k*. The initial population is composed of chromosomes generated in three different ways; totally random chromosomes, good randomly generated chromosomes and manually entered chromosomes.

To guarantee a good and diverse starting population the majority of the population consists of good random chromosomes. These are created by generating random chromosomes that are evaluated with the rigorous MILP to determine their fitness. The best $1/4^{th}$ of these chromosomes are then chosen to the initial population and the rest is discarded. In order to introduce enough genetic diversity into the population a number of completely random chromosomes are also included in the first population. It is also possible to enter chromosomes manually. This makes it possible to test a certain solution or include a desired solution into the gene pool. Even though the pre-selection process is somewhat time consuming it ultimately leads to faster convergence.

4.2.3. Generating the Next Generation

The next generation is created by using three methods; elitism, roulette wheel selection and mutation. To ensure that the best specimens are carried on to the next generation the e best chromosomes are transferred directly to the next generation. The rest of the population is created by selecting two parents through roulette wheel selection. All chromosomes in the population are given a ranking number f according to their fitness value with the fittest chromosome getting the value $f = 1$. Two distinct parents are chosen using a roulette wheel selection. Every chromosome has a possibility p_f, depending on their fitness, to be chosen as a parent. The following formula is used:

$$p_f = \frac{1}{\sum_{i=1}^{k}\left(\frac{1}{2+f}\right)*(2+f)} \qquad (2)$$

where k is the population size and f is the fitness rank of the chromosome. Equation (2) assigns each individual a mating probability based on the individual's fitness rank. The following is true for equation (2),

$$\lim_{k\to\infty}\sum_{f=1}^{k} p_f = 1$$

which means that the sum of all mating probabilities for the population is equal to one. Therefore, each individual has a p_f * 100 percent chance of being selected as a parent. The probability distribution created by using (2) is relatively flat which prevents premature convergence as the best individuals in the population are not too heavily favored in the selection process. The choice of the probability function is backed up by test showing steeper distributions easily caused premature convergence to local optima.

When two parents are found a one point crossover is conducted to produce two children. A randomly chosen crossover point is elected to determine what part of each parent the children will inherit. As long as the crossover is done so that allele positions are preserved two acceptable parents will always produce acceptable offspring even when reactors use different sets of feasible products as alleles will not change reactor in the crossover process.

The roulette wheel selection is repeated until the next generation reaches the predefined population size k. When we have a complete generation every allele is mutated with a given possibility. Mutations prevent the population from becoming too homogenous and introduce new chromosomes into the gene pool which reduces the risk for the Genetic Algorithm to get stuck in a local optimum.

The Genetic Algorithm is repeated a predefined number of iterations. Once the termination criteria are reached the GA terminates.

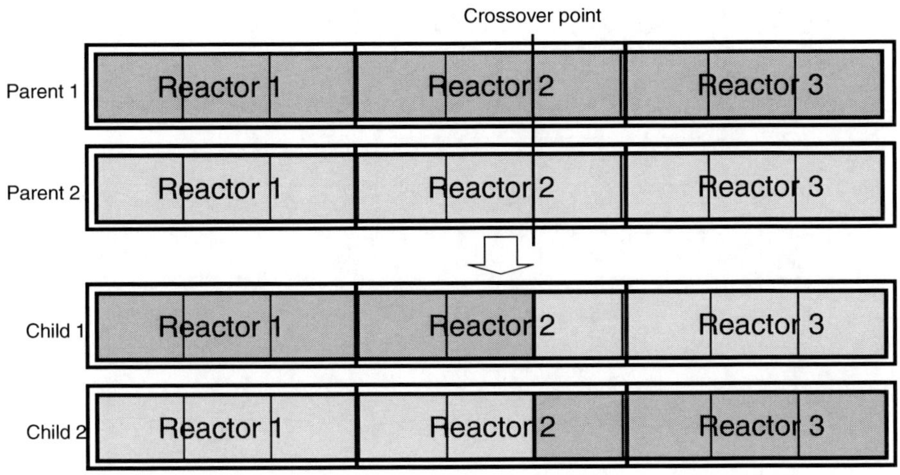

Figure 4. One point crossover two produce two children from two parents chosen by roulette wheel selection.

4.3. Tabu Search (M2)

The second heuristic, M2, uses a neighborhood Tabu search in combination with the rigorous MILP. In M2 we use the exact same chromosome structure as for M1 but this time the "population" consists of only one chromosome. This chromosome is mutated to find a neighbor and the fitness of the neighbor is evaluated. If the neighboring chromosome's fitness is poorer we make a new mutation on the initial chromosome. If, on the other hand, the fitness of the neighbor is better than (or equal to) its predecessor's we move to the neighbor and the predecessor is tabued. It is imperative to ban the previous chromosome to prevent the algorithm from going back to old solutions causing cycling. In our case this is even more crucial as numerous chromosomes can have the same fitness value. By tabuing old solutions we guarantee that the chromosome keeps evolving and does not get stuck in a loop of chromosomes all having the same fitness value.

This mutation process is iterated until the termination criteria are met. In our case we define the number of mutations to make. Another possibility would be to stop the algorithm after a given time.

The biggest advantage with M2 is that new solutions are tested on every fitness evaluation where as the GA in M1 has to evaluate the whole population to find a new solution. On the other hand, the drawback with a TS is that the solution may evolve more slowly as only a few alleles are mutated between fitness evaluations opposed to an entire new generation in a GA. In order to accelerate the convergence of the Tabu search in M2 we randomly mutate 1 to 3 alleles with a probability of 3:4:3. If more than one allele is to be mutated we have implemented a controller making sure the mutations occur on different reactors. With this procedure we ensure that the influence of the mutations are evenly distributed over the whole production process.

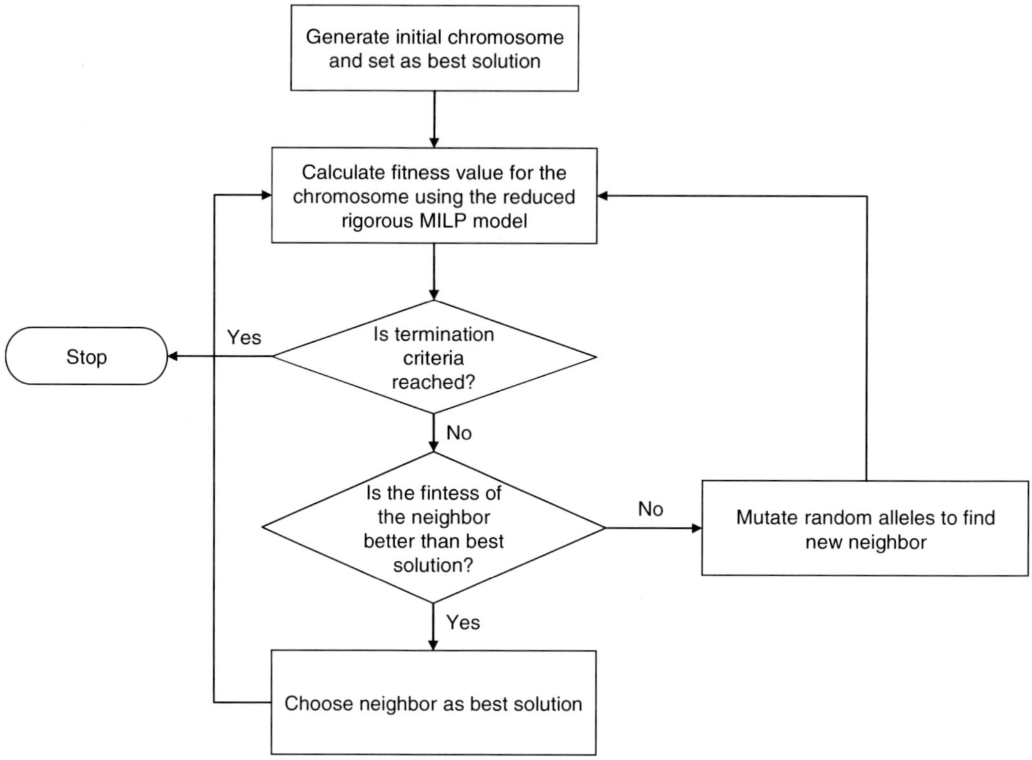

Figure 5. Flow chart of the Tabu search model (M2) combined with a rigorous MILP.

5. Test Cases

In order to test the hybrid methods presented we devised four test cases. We chose three key parameters that were varied to derive the test cases. These parameters were; start inventory, demand and minimum production time. Each of these parameters increase or decrease the difficulty of finding near-optimal solutions depending on how they are set. The impact of existing start inventory is quite clear; a high start inventory makes it easier to find production plans that meet all customer orders while a low starting inventory requires a more efficient production schedule to meet the demand. The amount of demand also impacts the optimization, making it harder to solve if demand is high and easier if it is low. The third parameter might not be as self evident. The biggest challenge caused by varying the minimum production plan is the number of alleles needed to represent an individual as this has a direct effect on how much the search space is reduced in the MILP problem. Short minimum production times require more alleles than longer minimum times thus making the problem harder with short minimum production times. Table 1 shows the setup for each test case and how the parameter settings affect the optimization problem. All cases were first solved using the original, unimproved MILP to obtain a lower bound (LB) for the problems. Thereafter we solved the case problems with both hybrid methods. We will present the results and compare methods M1 and M2 both in terms of speed and solution quality.

Table 1. The four example problems

Case name	Start inventory	Demand	Min production time
C1	High – (Easy)	Low – (Easy)	Long – (Easy)
C2	High – (Easy)	Low – (Easy)	Short – (Hard)
C3	Low – (Hard)	High – (Hard)	Long – (Easy)
C4	Low – (Hard)	High – (Hard)	Short – (Hard)

Cases C1 and C2 are derived from an actual production planning situation provided by the case company. The relatively high starting inventory and low demand for the period at hand provided a problem where the optimization model disables setup costs to prevent unnecessary production. We also wanted to test how different minimum production times affected the solutions. The short minimum times is the absolute minimum time for production as dictated by the physical properties of the case plant. The longer minimum times are desired minimum times given by the companies production planner. The longer minimum times are twice, or more than twice, as long as the short ones. In order to test the models on a more difficult problem including setup costs we manipulated the data from the real case used for C1 and C2. We lowered the starting inventories considerably and increased the demand to create an example where meeting all orders would be hard. Again we chose to test how longer minimum production times would affect the results. The reason for experimenting with minimum production times is that every production change poses a risk of mechanical failure and the case company therefore wants to avoid unnecessarily short production cycles that lead to an increased number of product change-over.

The gap shown in table 2 is calculated as the relative gap according to the definition given by GAMS. The relative gap is calculated by the following formula:

$$gap = \frac{Best^{estimate} - Best^{integer}}{Best^{estimate}} \tag{3}$$

where $Best^{estimate}$ is the LB acquired by solving the unimproved MILP and $Best^{integer}$ is the best integer feasible solution found.

Table 2. Solution times, objective values and gap for the example cases solved with the MILP model

	Objective value	Gap	Solution time (sec)
C1	38 672 064.00	2 %	69 120
C2	23 468 891.92	0.41 %	167 400
C3	1 987 445 591.60	0.94 %	195 674
C4	1 751 743 523.60	2.41 %	432 060

Table 3. Parameter for the Genetic Algorithm in M1

GA parameters	
Population size	14
Good random chromosomes	12
Completely random chromosomes	2
Number of iterations	20
Mutation probability	5%
Elitistically chosen chromosomes	2

As we can see from Table 2 problems C3 and C4 are more difficult to optimize (longer solution times) as it is challenging to find a production plan able to meet all orders. The big leap in objective values is caused partly by the introduction of setup costs and because the MILP model tries to enforce a certain minimum inventory at the last time period. When these minimums cannot be reached, due to the high demand and low starting inventory, the model is punished with a penalty cost for not being able to reach the desired inventory level. Furthermore we can see that there is a noteworthy difference in the solution values depending if we use long or short minimum production times in cases C1 and C2. The reason is that the demand is so low that no product needs to run more than its minimum production time and when those times are twice as long in C2 opposed to C1 the difference is big. In cases C3 and C4 the difference is smaller as we have a big order stock and all reactors need to run on full capacity.

After experimenting with different parameters for M1 all test cases were conducted with the same set of parameters. The improved MILP model's solution times were limited to 60 seconds for each fitness evaluation to ensure fast results. This maximum time was only reached in approximately 3% of the optimizations.

All runs that reached maximum time were early in the optimization when poor chromosomes were evaluated. These were quickly sifted out when new generations were created.

We decided to use a fairly small population and generate most of the initial population by picking the good random chromosomes. To ensure diversity two completely random chromosomes were also included. To avoid stagnation due to homogenous chromosomes we used a 5% chance that an allele would be mutated after crossover. To guarantee that the best known chromosome would survive to the next generation the two best fitted chromosomes were transferred unaltered to the new generation.

For M2 the same 60 second maximum time limit was used for the MILP with less than 1% of all optimizations reaching the limit. The tabu list was 10 chromosomes long, meaning that the last 10 best solutions were tabued. This length proved to be sufficient to prevent the model from returning to old solutions.

5.1. Case Results

We start by presenting the solution progress for all four cases and the cases are analyzed and compared shortly. After this the solution quality and the solution times are presented and discussed.

It is clearly visible that the extensive pre-selection done in M1 generates a very good starting population. Generally it only takes only took a few iterations (one iteration being 14 optimizations) to improve this solution. In both cases C3 and C4 a mutation of the best known chromosome was needed to find a better solution. This indicates that the high mutation probability is useful to avoid getting stuck in a locally optimal solution. For the Tabu search in M2 the initial starting point is significantly higher (worse) because of the use of only once chromosome. M2 also uses the pre-selection but in a smaller scale than M1. In all cases except C4 the TS very quickly improved the solution and in less than 50 optimizations found a better solution than the GA with the same amount of optimizations. This shows that it is probably not worth consuming more time on a more sophisticated pre-selection to find better starting points.

When we compare the final solutions found by M1 and M2 in three cases out of four the Tabu search (M2) found a better solution than the GA (M1). For some reason case C4 proved hard for the Tabu search; in this case the convergence was slower than in the three other cases and the best found solution was poorer than the solution of the GA.

As we can see from figures 6-9 the convergence of both method M1 and M2 is good and the final solution is close to the LB found by the rigorous MILP alone. In all four cases both hybrid methods were able to found solutions within 4% (for cases C1 and C2 the solutions were within 2%) of the lower bound. If the solutions are compared to the best integer feasible solutions found by the MILP alone (after 5-8 days of optimization) the results are even better. The gap in table 4 is calculated as the relative gap as explained earlier. Table 4 shows that method M2 in most cases is faster than M1. When we combine this with the fact that M2 also produced better results on three of the cases we can conclude that the hybrid method using Tabu search is more effective in our case than the hybrid GA. The solution times for the rigorous MILP model are extremely long as the last few percents of closing the gap takes long to solve.

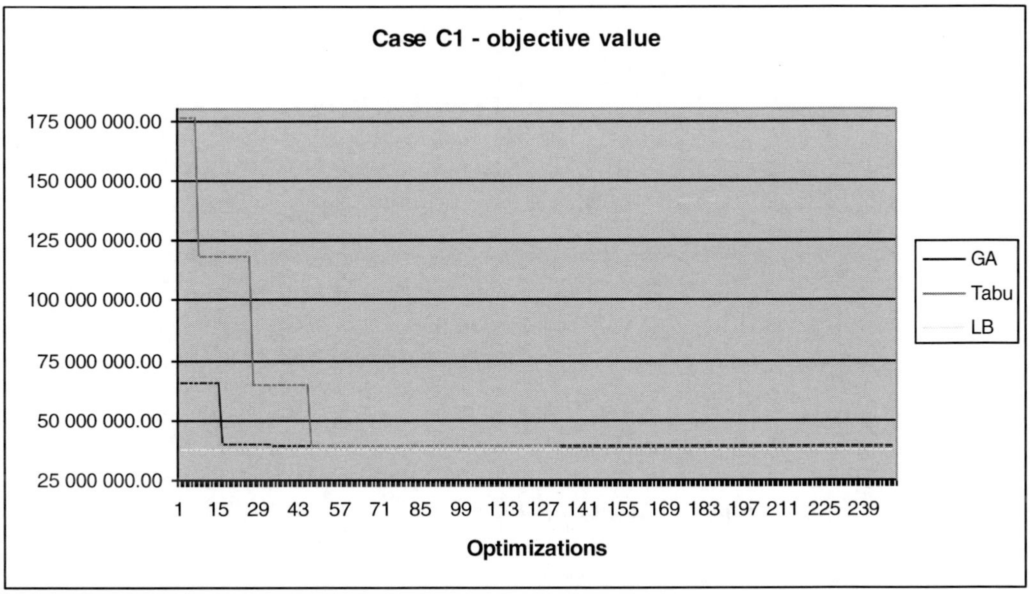

Figure 6. Solution progress for case C1.

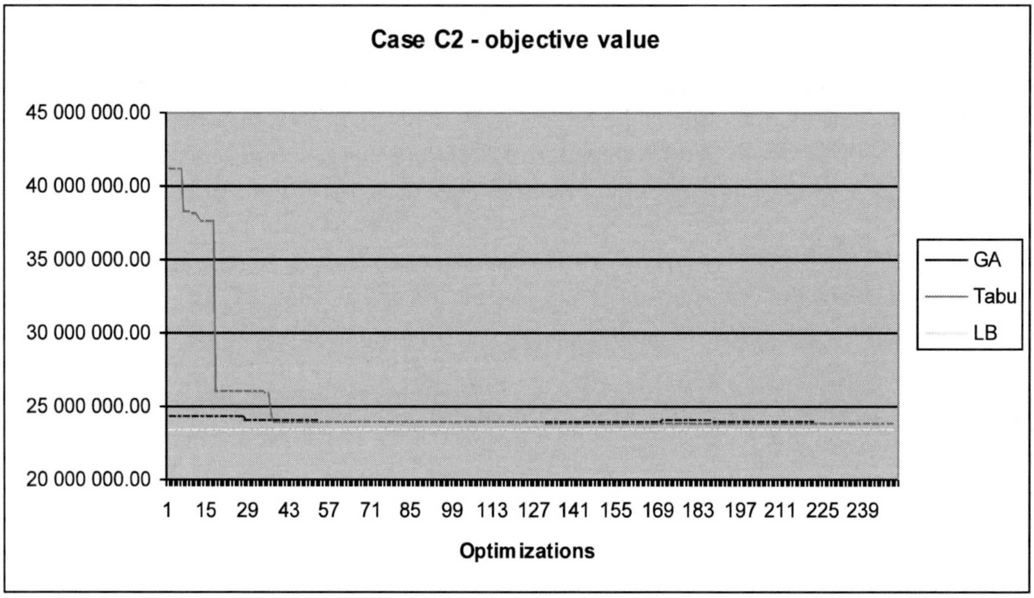

Figure 7. Solution progress for case C2.

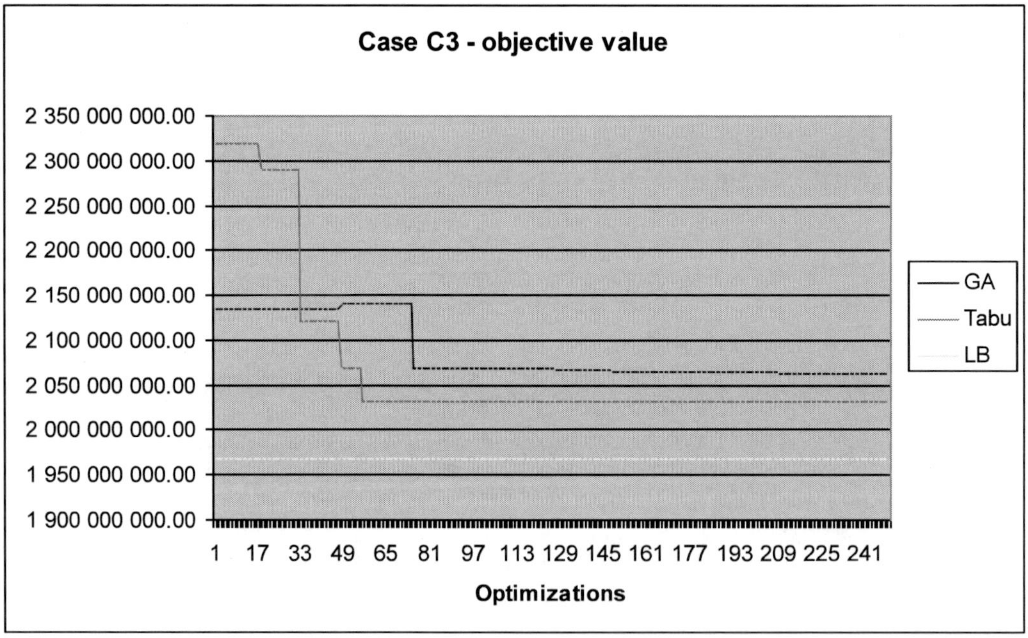

Figure 8. Solution progress for case C3.

It is worth to note that in 1/10 of the total solution time the MILP model was able to find solution within 8% of optimum. This means that both hybrid methods are faster than the MILP model but not as fast as we had hoped.

It is imperative to point out that neither of the two hybrid models was optimized for speed. We approximate that code optimization could lower the solution times for both M1 and M2 with at least 30% and up to 50%. This can be achieved by more effective code and better reuse of earlier solutions. The coding effort for doing these improvements was deemed to big to be carried out at this stage. Later on when a production planning tool is developed the coding will be optimized to ensure faster results.

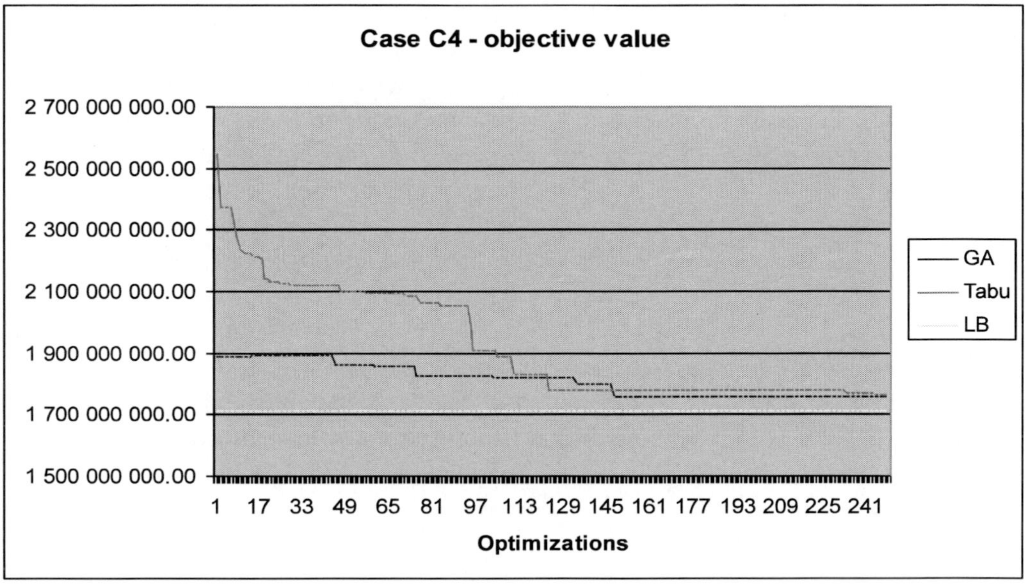

Figure 9. Solution progress for case C4.

Table 4. Solution time and gap for all cases and methods

		Gap	Solution time (sec)
C1	MILP	2.00%	69 120
	M1 (GA)	2.60 %	7 200
	M2 (TS)	2.00%	3 480
C2	MILP	0.41 %	167 400
	M1 (GA)	2.48 %	3 400
	M2 (TS)	1.63 %	1 280
C3	MILP	0.94 %	195 674
	M1 (GA)	4.53 %	3 945
	M2 (TS)	3.12 %	5 121
C4	MILP	2.41 %	432 000
	M1 (GA)	2.74 %	3 640
	M2 (TS)	3.14 %	1 650

CONCLUSION

The SmartOpt case SC3 proved to be so large that conventional rigorous methods were not able to find optimal or near optimal solutions fast enough for our purposes. This is especially true when we look forward when the production planning models now created will be combined with other models to span the whole PDN of the food-sweetener company (overall goal of the SmartOpt project). Because of the complex nature of the fine chemical plant we are modeling a pure heuristic approach would have been both difficult and time consuming, therefore we settled for formulating two new hybrid methods, M1 and M2 using GA and TS respectively, combined with the existing rigorous MILP to improve convergence.

After testing both models with four test cases, all having different characteristics we can conclude that both methods (M1 and M2) work and are able to improve the convergence for the problem. Method M2 came ahead both in terms of speed and solution qualities for 3 out of 4 of the test cases. The main problem with the Genetic Algorithm is that many optimizations, the most time consuming part in the model, are needed to find a new solution whereas the Tabu search can find an improved solution on every optimization. Initially we feared that convergence for the TS might be slower than for the GA, especially if the starting point was far from optimal, but these fears proved to be needless. Even though convergence was not as fast as we hoped for in the beginning we are confident that by code optimization and further research we can speed the methods up. The next step will be to continue the promising research with method M2 in order to improve it further.

The need to solve ever bigger problems first gave birth to heuristic methods. A natural step forward has been to start combining the benefits from different methods leading to a large variety of hybrid methods. This case study presents two such hybrid methods and the test results show significant improvements in convergence compared to the deterministic approach.

REFERENCES

Akrami B., Karimi B., Mottar Hosseini S.M., 2006. Two metaheuristic methods for the common cycle economic lot sizing and scheduling in flexible flow shops with limited intermediate buffers: The finite horizon case, *Applied Mathematics and Computation*,183, pp.634–645.

Aliev R., Fazlollahi B., Guirimov B. and Aliev R., 2007. Fuzzy-genetic approach to aggregate production-distribution planning in supply chain management. *Information Sciences*, 177, pp. 4241-4255.

Björk K-M. and Carlsson C., 2007. The Effect of Flexible Lead Times on a Paper Producer. *International Journal of Production Economics*, 107, No. 1, pp. 139-150.

Björk K-M. and Nordman R., 2005. Solving large-scale retrofit heat exchanger network synthesis problems with mathematical optimization methods. *Chemical Engineering and Processing* vol. 44, Issue 8, pp. 869-876.

Defersha F,M., Chen M, 2008. A linear programming embedded Genetic Algorithm for an integrated cell formation and lot sizing considering product quality, *European Journal of Operational Research,* 187, pp. 46–69.

Gen M. and Syarif A., 2005. Hybrid Genetic Algorithm for multi-time period production/distribution planning. *Computers and Industrial Engineering*, 48, pp. 799-809.

Hung Y-F., Chen C-P., Shih C-C., Hung M-H., 2003. Using Tabu search with ranking candidate list to solve production planning problems with setups, *Computers and Industrial Engineering*, 45, pp.615–634.

Józefowska J., Zimniak A., 2008. Optimization tool for short-term production planning and scheduling, *Int. J. Production Economics*, 112, pp.109–120.

Lukac Z., Soric K., Rosenzweig V.V., 2008. Production planning problem with sequence dependent setups as a bilevel programming problem, *European Journal of Operational Research*, 187, pp.1504–1512.

Nyberg M., Björk K-M, 2010. Solving a complex production planning problem with mathematical programming techniques, *Proceedings of the 2nd International Conference on Applied Operational Research,* pp. 123-144.

Pradenas L., Penailillo F., Ferland J., 2004. Aggregate production planning problem. A new algorithm, *Electronic Notes in Discrete Mathematics*, 18, pp.193–199.

Russel R., Chiang W. and Zepeda D., 2007. Integrating multi-product production and distribution in newspaper logistics. *Computers and Operations Research*, AIP.

Sachdeva A., Kumar D., Kumar P., 2008. Planning and optimizing the maintenance of paper production systems in a paper plant, *Computers and Industrial Engineering.*

Zadeh L., 1998. Some Reflections on Soft Computing, Granular Computing, and their Roles in the Conception, Design and Utilization of Information/Intelligent Systems. *Soft Computing*. 2, pp. 23-25.

In: Knowledge Management and Communication … ISBN: 978-1-62808-250-0
Editor: Rafiqul Islam

Chapter 19

BUILDING OF TELEPROTECTION SYSTEM MODELS

***Bander Mahfouz Saman, Mosleh M. Al-Harthi, Nadjim Merabtine and A. Said*[*1]**
Electrical Engineering Department, Faculty of Engineering, Taif University, Taif, Saudi Arabia
[1]UAE University, United Arab Emirates

ABSTRACT

The sizes, complexity and power ratings of high voltage networks emphasize the need for highly reliable protection systems. Protection is based on information derived from the power system at one or more points, and fast selective protection applied to circuits with geographically separated terminals, such as cables and overhead lines, requires information interchange between these terminals. A secure and uninterrupted supply of electricity is only possible with the help of comprehensive protection and control functions, which ensure the reliable operation of the power system. As the complexity and ratings of electrical power systems increase, so do also the demands on the protective devices and systems which have to protect them from damage and to preserve the power system stability. Protection relays in conjunction with communication links [1] ,[2] ,[3] provide the best possible means of selectively isolating faults on high voltage (HV) lines, transformers, reactors and other important items of electrical plants [4],[5] ,[6]. To prevent the system from failure and damage, the tele protection system (TPS) has to selectively disconnect the faulty part within the shortest possible time.

Keywords: Tele protection system (TPS), audio telecommunications links, Tone Commands, Direct Transfer Tripping

[*] E-mail: Abdellatifsd@yahoo.com.

1. INTRODUCTION

Audio Tone tele protection is a method of communicating binary states from one substation to another over audio telecommunications links [7], [8], and [9]. These binary states are used, for example, to control or monitor remote power line circuit breakers, pipe valve, machine and any element in important area. This practice has existed for over 30 years.

In the guard state, i.e. when there is no command actuated at the inputs, the TPS transmits a guard signal. The guard signal is continuously processed by the receiver in the opposite station, which gives alarm should the signal quality become inadequate. In the command state, the TPS interrupts the guard signal to transmit the command signal within the chosen communication channel (analog, digital or optical). If the continuous command supervision is enabled and the duration of a command transmission exceeds a predefined value, an alarm will be given at the transmit end and the guard is switched on again. As soon as the receiver recognizes the missing guard signal and simultaneously detects a valid command signal of sufficient quality, the command is relayed to the designated output. Simultaneous reception or simultaneous loss of a command signal and the guard signal leads to an alarm. In the latter case, the output(s) programmed for unblocking function are activated. An unblocking (de-blocking) function is standard on all units, which closes a contact for 200 ms (default setting; configurable) in the event of loss of signal. De-blocking is sometimes used in the case of permissive overreaching schemes. All interfaces are potential free and DC isolated. A command is transmitted by applying a DC voltage (contact + battery). At the receiving end, each command may be mapped to a main (solid state) and an auxiliary (relay) contact on the relay interfaces. Outputs may also be used to signal various alarm sources or to acknowledge a Tx or Rx command signal.

1.1. Analog Operating Mode

The TPS Analog operates according to the frequency shift principle. Guard and command signals are transmitted within the selected frequency band. The guard signal is always represented by a single tone. Command signals can either be a single-tone or a dual-tone signal, depending on the operating mode. Command signals can be transmitted at a higher power than the guard signal (power boosting). TPS equipped for several commands transmits each signal or signal combination with the same full power so that the maximum possible signal-to-noise ratio is achieved at the receiver. Alarm is given at the receiver if the signal level is out of a pre-defined level range. The TPS can be set programmed for the following operating modes.

1.1.1. Single Tone Command

The main advantage of this mode less complexity, but no more than one command can be sent at the same time.

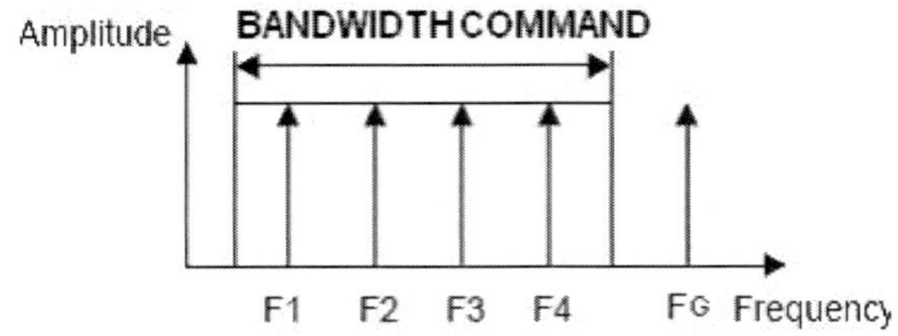

Figure 1.1. Single Tone Command.

1.1.2. Independent Dual Tone Commands

The main advantage of this mode simulations transmission can be used, but it is more complex.

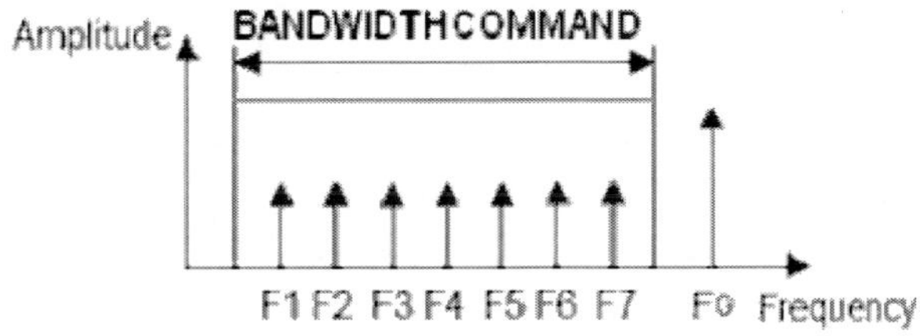

Figure 1.2. Independent Dual Tone Commands

Table 1.1. Single Tone Command Frequency Locations

Injected	Transmitted	b1	b2	b3	b4	F1	F2	F3	F4	F5	F6	FG
-	-	0	0	0	0							X
A	A	0	0	0	1	X						
B	B	0	0	1	0		X					
A B	A	0	0	1	1	X						
C	C	0	1	0	0			X				
A C	A	0	1	0	1	X						
B C	B	0	1	1	0		X					
A B C	A	0	1	1	1	X						
D	D	1	0	0	0				X			
A D	A	1	0	0	1	X						
B D	B	1	0	1	0		X					
A B D	A	1	0	1	1	X						
C D	C	1	1	0	0			X				
A C D	A	1	1	0	1	X						
B C D	B	1	1	1	0		X					
A B C D	A	1	1	1	1	X						

Results Interpretation

The above Figures and tables show examples of different command signal schemes b1, b2, b3, and b4. The arrows A, B, C, D represent symbolically the command signals (single tones or dual tones) for commands A to D of the TPS. For ease of reference, only single-tone signals (one frequency per command signal) are depicted in figure 1.1 and table 1.1; the frequency values F1, F2, F3 and F4 are depending on channel bandwidth.

Other signal types "dual tone", two single-tone signals (two frequencies per one command signal) are depicted in figure 1.2 and table 1.2. For example command A will represented by F1 and F2. Also the frequency values F1, F2, F3 and F4 are depending on channel bandwidth.

As table 1-1 and table 1-2 show in single tone, if two commands are injected the only one command "highest priority" will be sent , but in dual tone all command possibility are available.

Table 1.2. Independent Dual Tone Commands Frequency Location

Injected	Transmitted	b1	b2	b3	b4	F1	F2	F3	F4	F5	F6	FG
-	-	0	0	0	0							X
A	A	0	0	0	1	X	X					
B	B	0	0	1	0		X	X				
A B	A B	0	0	1	1	X		X				
C	C	0	1	0	0			X	X			
A C	A C	0	1	0	1			X		X		
B C	B C	0	1	1	0		X		X			
A B C	A B C	0	1	1	1	X			X			
D	D	1	0	0	0					X	X	
A D	A D	1	0	0	1	X				X		
B D	B D	1	0	1	0		X				X	
A B D	A B D	1	0	1	1			X			X	
C D	C D	1	1	0	0				X		X	
A C D	A C D	1	1	0	1		X			X		
B C D	B C D	1	1	1	0				X	X		
A B C D	A B C D	1	1	1	1	X					X	

1.2. The Time Calculation Principle

1.2.1. The Transmission Time

The transmission time is the duration between change of state at the input terminals of the tele protection interface and the corresponding change of state at the output terminals.

1.2.2. Command Prolongation Time

A command prolongation, which can be ensures a steady output command even in the event of discontinuations of the signal being received. The receiver of the TPS compensates the command duration without prolongation for equal pulse width as the transmitted command at the remote station.

1.3. TPS Application

1.3.1. Accelerate DIST Zone

It used to extend the zone 1 relay reach to typically 130 percent of the protected circuit length, or to bypass the time lag of an independently operating zone 2 relay [10], [11], [12], and [13].

1.3.2. Direct Transfer Tripping (DTT)

It is necessary to employ telecommunication techniques to reduce the operating time of back-up protections to a few power frequency cycles.

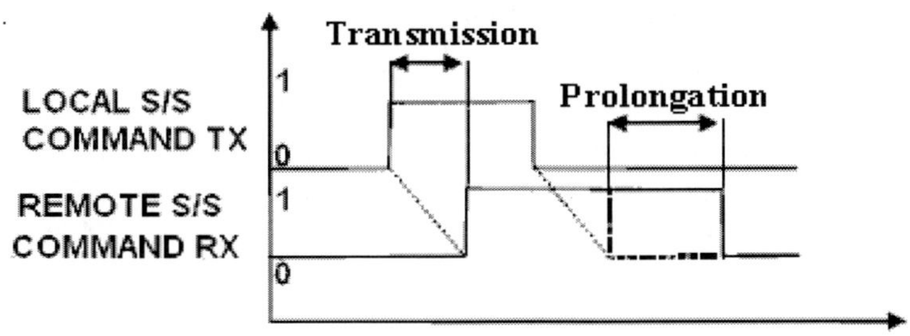

Figure 1.3. Transmission Time and Command Prolongation Time.

2. System Model

With MATLAB SIMULINK version 6.0 (R14) dated 05-May-2004, we could able to design TPS, as dynamic systems model in a block diagram format. Our model is composed of three major equipments, which are: TPS TX, TPS RX and with command tester.

This part will illustrate each circuit with further information needed. Also we added additional blocks such as, Power Line Carrier "PLC" Attenuator, white noise and, signal Amplifiers, The all above blocks designed in one file that called system tool. This file used in each model according to the application requirements.

2.1. TPS Equipment

The TPS equipment is divided in two parts, transmitter part (TPS TX) and receiver part (TPS RX), this division will let us now signal follow between protection relays and all related interfaces[14],[15] ,[16],[17]. In the real system the two parts compact in on one card or device. As showing in previous chapter the protection relay (PR) is monitoring the electrical network, in the normal condition hold the signals follow between PR and local circuit breaker (CB) and also between PR and local TPS TX.

In case of any fault PR send the 110 or 60 dc voltage to local CB for closeting the fault locally, but for speeding remote closing, the PR send the 60 dc voltage (command input) to local TPS-TX as command input. Then TPS-TX sends the assigning AF signals to remote TPS-RX and disconnects the guard AF signal. Also the remote TPS-RX delivers 0dc voltage (command output) to remote RP while the AF guard signal is received. But if the remote TPS-RX receives command AF signals. Then the command output state change from 0vdc to 60vdc level with time prolongation, then the remote PR applies 110 or 60dc voltage for closing remote CB. If the AF guard signal and command AF signals not received or below the threshold value, the TPS-RX considering the communication missing and the alarm condition apply.

2.1.1. TPS-TX

The TPS TX contain three stages .In first stage the switch relays (SW) are controlled by the command input dc signal from four different PR's, The relays is passing the first input (COMM ON =1) or the third input (COMM OFF =0) based on the value of the second input (command input level). The second stage (logic block) is a truth table circuit which enables the AF signal F1, F2, F3, F4, F5, F6 and FG.

The table 2-2 and 2-3 are shown the input (b1, b2, b3, and b4) and the output (F1, F2, F3, F4, F5, F6, FG) of truth table. The product multiplications are used for enabling the assigned AF frequency as the last stage [18], [19], [20].

The AF frequency assigned with output power depends on bandwidth and operation mode as following rules:

- For signal tone: the power of command signal equal to guard signal power.
- For dual tone: the power of command signal equal to half guard signal power. Because the total power of dual tone will equal the guard power
- The frequency spacing should be distributed equally in band.

If we consider the all above rules we will have the following values

Signal Tone

$$B = \sqrt{10^{\frac{L}{10}} * .001 * R_o} * \sqrt{2}$$

where B is peak voltage for command and guard AF signal, L is Tx level in dBm and Ro is Output impedance.
And frequency schema as following:

Half Frequency spacing d= (Upper band Frequency -Lower band Frequency)/16;

F1= Lower band Frequency +2*d, F2= Lower band Frequency +6*d

F3= Lower band Frequency +10*d, F4= Lower band Frequency +14*d

AF for guard will assign depend on communication channel.

Dual Tone

$$B = \sqrt{10^{\frac{1}{10}} * .001 * R_O} * \sqrt{2}$$

where B is peak voltage for guard AF signal, the peak voltage for command AF is equal to B/2, L and Ro are same parameters for single tone. And frequency schema as following

Half Frequency spacing d=(Upper band Frequency -Lower band Frequency)/24;
F1= Lower band Frequency +2*d, F2= Lower band Frequency +6*d,
F3= Lower band Frequency +10*d, F4= Lower band Frequency +14*d,
F5= Lower band Frequency +18*d, F6= Lower band Frequency +22*d.
AF for guard will assign depend on communication channel.

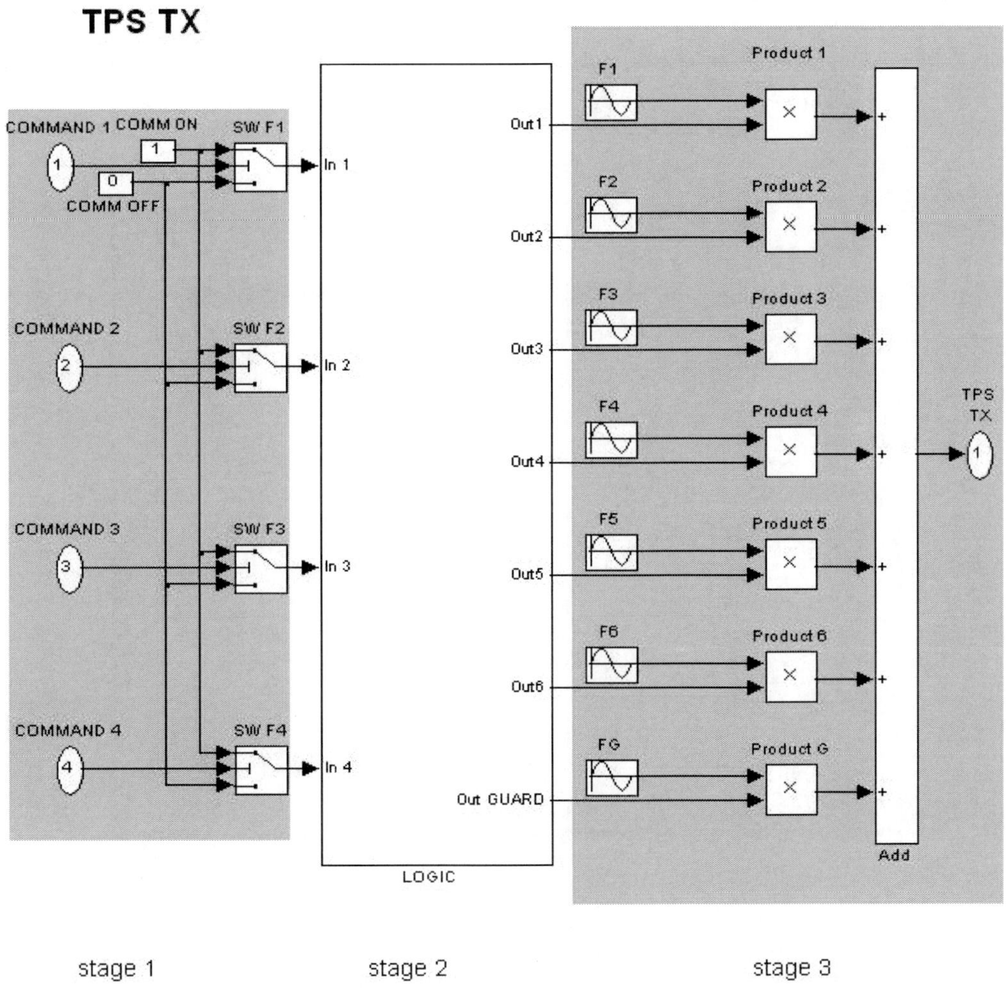

Figure 2.1. TPS-TX.

Operation mode: Type of mode (Single tone or Dual tone).

Guard frequency in Hz: Specifies the guard frequency depend on communication media.

TX level in dBm: Specifies the transmitter level in dBm, Depend on communication channel attenuation.

Output impedance: Type of output impedance (300 ohm or 600 ohm). Depend on communication media.

Lower band Frequency in Hz and Upper band Frequency in Hz: Specifies the transmitter bandwidth. Depend on communication channel attenuation.

Command input level in V: Specifies the threshold voltage level for command input, Depend on protection relay specification and local station charger.

Parameters and Dialog Box for TPS-TX

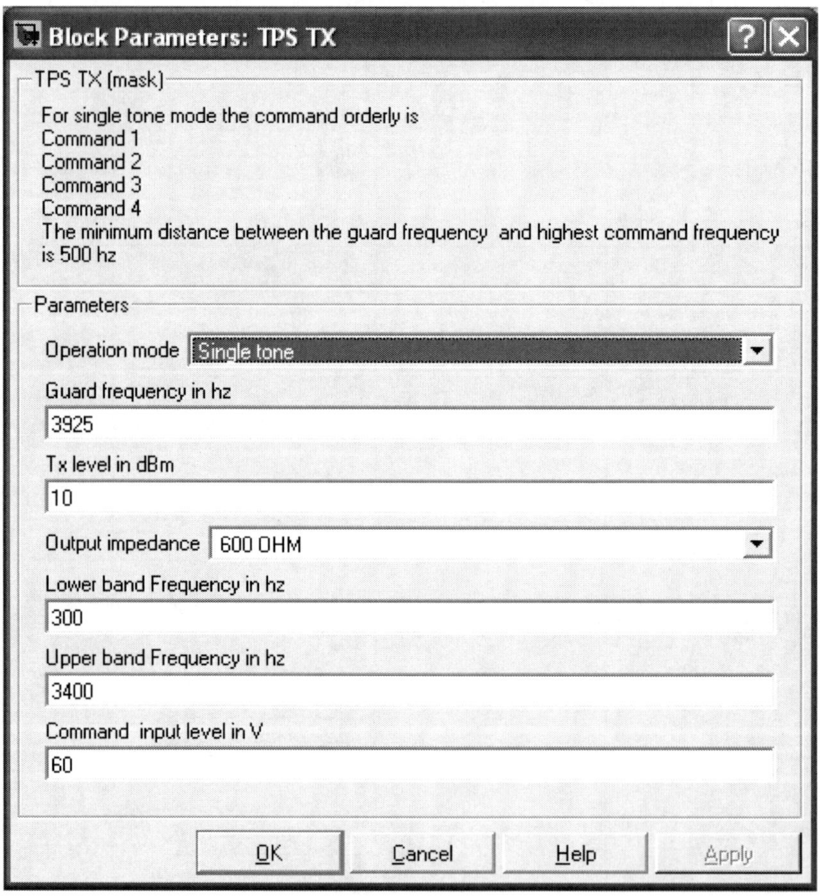

Figure 2.2. Parameters and Dialog Box for TPS-TX.

2.1.2. TPS-RX

The functions of the receiver are checking status of the communication link and operation condition. The TPS RX contains three stages. In first stage the voltage level for commands and guard frequencies are checked by applying band pass filter. Then the voltage peak is calculated for each frequency, also all voltage peaks added for calculating the receiver power in dBm. If the receiver power less than the threshold level the system will block all commands output and energize the alarm relay. The second stage is command checking block as Figure 2-4. The output of 'level comparing block' is enabled if the maximum level is feeding from first port. The logic block is simply truth table circuit, if we consider with the receiving frequencies as input and the selecting command as output. The last stage is enabling the output ports with prolongation using internal clock.

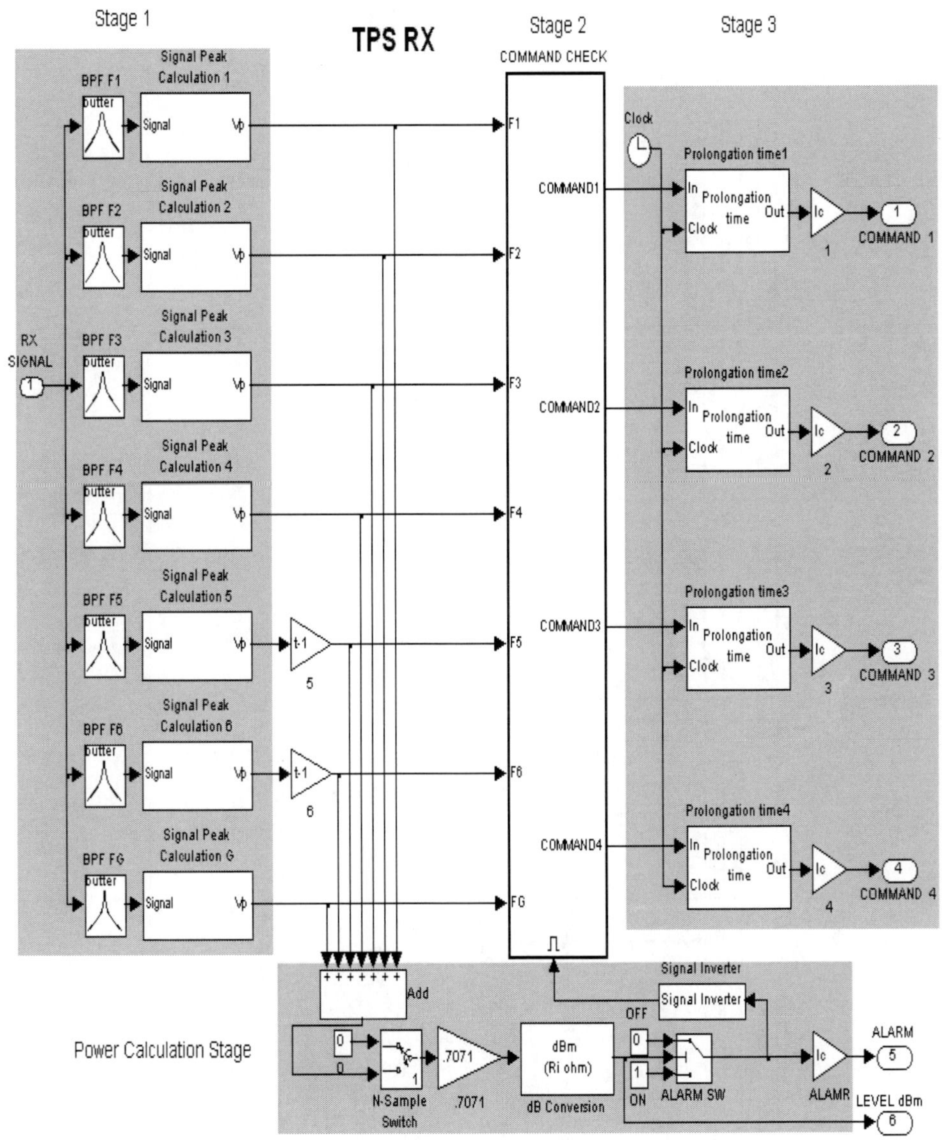

Figure 2.3. TPS-RX.

Parameters and Dialog Box for TPS-RX

Operation mode: Type of mode (Single tone or Dual tone).

Guard frequency in Hz: Specifies the guard frequency this depend on communication media.

Threshold level in dBm: Specifies the minimum receive level in dBm, you should select the level depend on communication channel attenuation.

Output impedance: Type of output impedance (300 ohm or 600 ohm). Depend on communication media.

Lower band Frequency in Hz and Upper band Frequency in Hz: Specifies the transmitter bandwidth, you should select the value depend on communication channel attenuation.

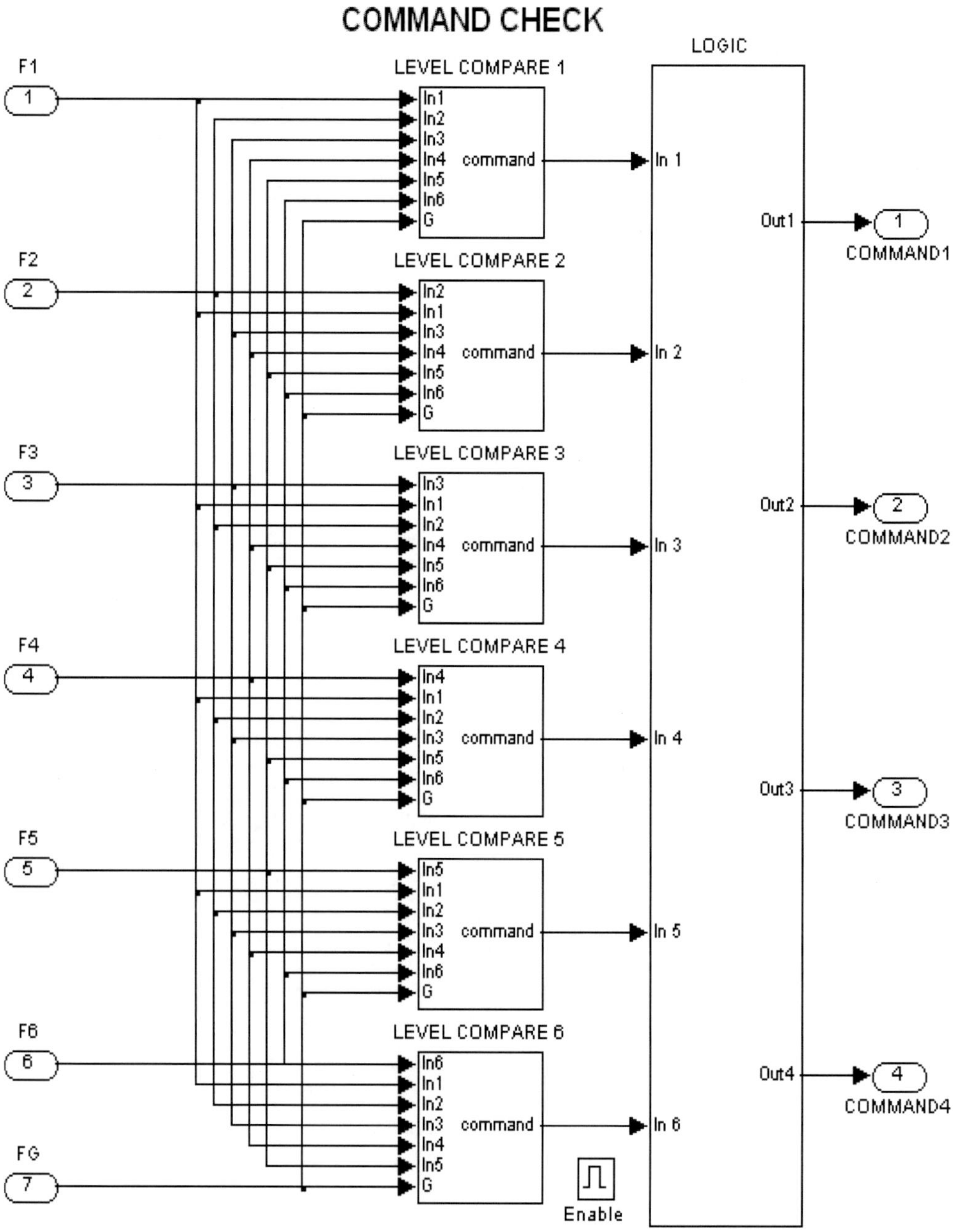

Figure 2.4. Command Check.

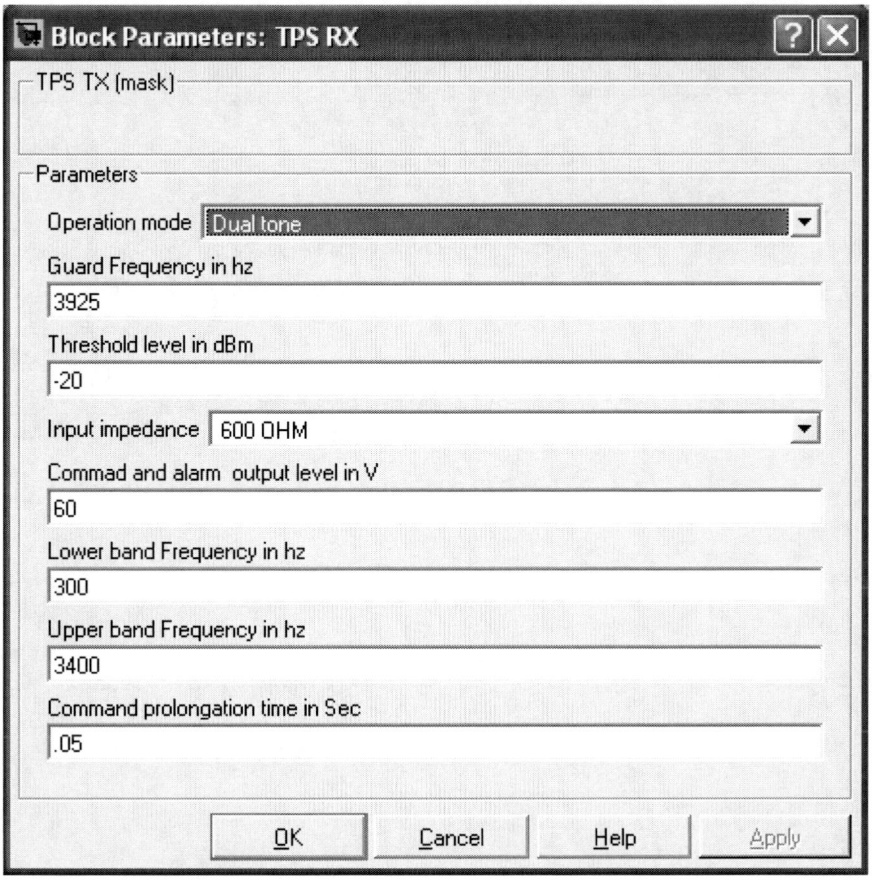

Figure 2.5. Parameters and Dialog Box for TPS-RX.

Command and alarm output level in V: Specifies the threshold voltage level for command and alarm level, you should select the value depend on protection relay specification and local station charger.

Command prolongation time in Sec: Specifies the prolongation time for command duration, you should select the value depend on protection relay and scheme.

2.2. Library Model

The library model contains all the blocks, which will be used in the project and is seen in Figure 2.6. To use a block simply drag and drop it onto the matlab files. Each block has specific parameters associated with its functionality. Once the block is placed onto the file, the blocks can be connected by signal wires. Simply drag and hold the mouse over one output port to another input port.

After connecting the blocks onto the file, double-click it to view the parameters. After all connections have been made and all parameter values are valid simulate the system file by Press the black play button on the top toolbar to begin the simulation.

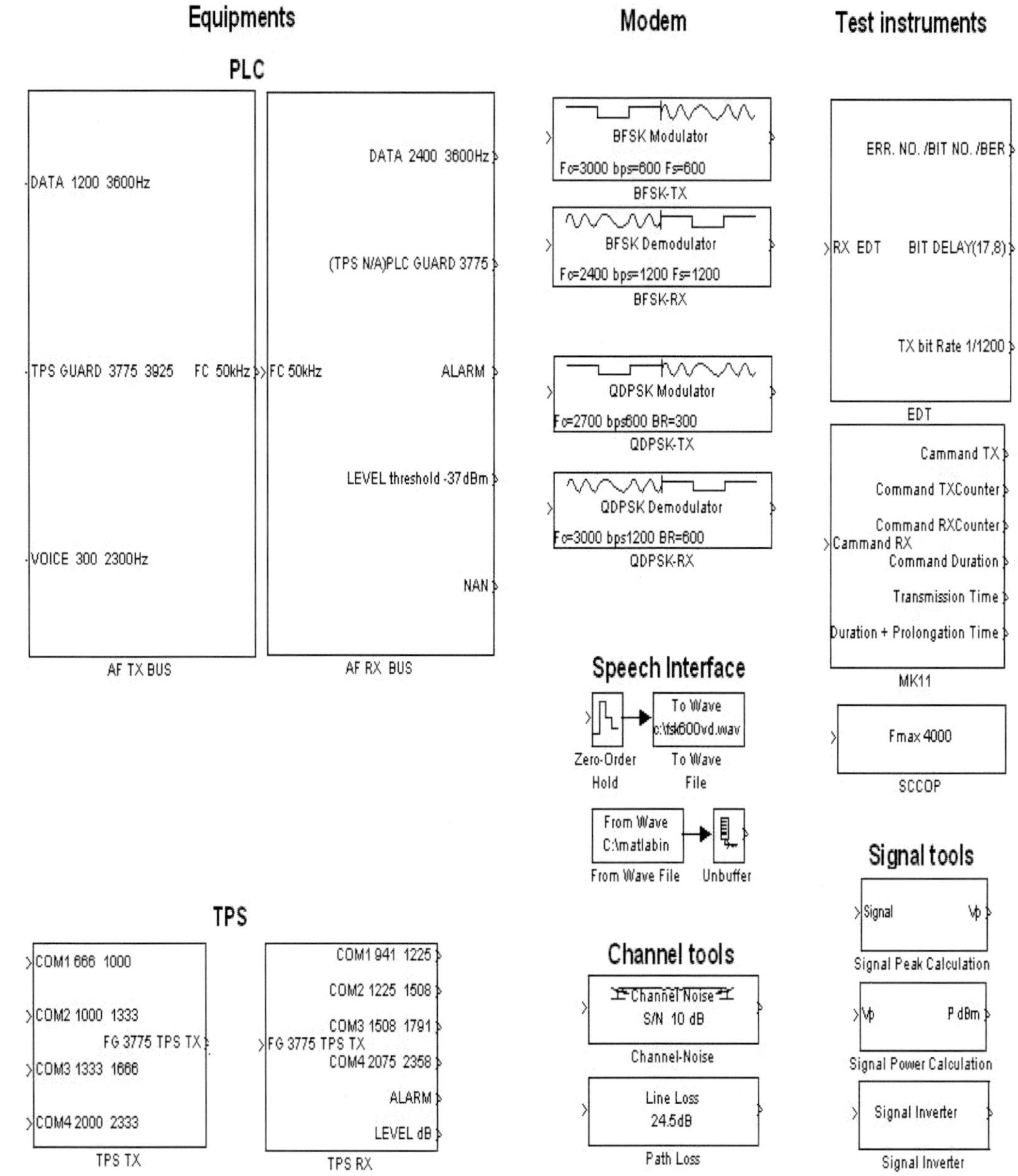

Figure 2.6. Library Model.

2.3. TPS Analysis

The table 2-1 is shown the transmission time, command prolongation time and trip counters values of the protection commands.

The modeling system is added around 5 ms delay. This is due to the modulation/demodulation jobs and timing of switching relays. In the real system the value is near to 12 ms delay.

Table 2.1. TPS Test

Transmission Media	Transmission Time <20ms	Command Prolongation Time setting 50 ms	Counters Values
Dual Tone TPS integrated in PLC	12.45 ms	44.9 ms	8/8
Single Tone TPS integrated in PLC	12.4 ms	46.75 ms	1/1
Dual Tone TPS Local loop	7.2 ms	48.28 ms	8/8
Single Tone TPS Local loop	7.15 ms	50.36 ms	1/1

CONCLUSION

3.1. Summary of Key Findings

This research is based on molding analog TPS. The main advantage of this study is used the existing equipments, which simplifies the implementation. We have been successful in implementing a working TPS model. During the design of this system we were faced with a number of difficulties due to the combination of modulations for TPS and programming interface. Due to this reason we had to redesign and test a number of blocks. However at the end of rigorous testing we were able to come up with a working TPS model.

We have succeeded in establishing the fowling applications:

- Single tone tele protection.
- Dual tone tele protection.

The required tests were performed, and the presented experiments show that duel tone TPS are designed for secure applications.

3.2. Suggestions for Further Research

It should be noted that even at this stage the model still only represent a system in single line diagram. Realistic modeling of protection relay and channel noise should be considered in the future to improve the models and to make it more feasible for real life situations.

REFERENCES

[1] Usta, O., Bayrak, M., Redfernand M. A. A New Digital Relay for Generator Protection against Asymmetrical Faults. *IEEE Power Engineering Review*.Volume: 21, Issue: 11, 2002.

[2] Bo, Z.Q., Weller, G. Lomas, T.Redfernand M.A. Positional protection of transmission systems using Global Positioning System. *IEEE Transactions on Power Delivery*. Volume: 15, Issue: 4, 2000.

[3] Apostolov, A., Vandiver, B. Automatic test system advances transformer protection. *IEEE Computer Applications in Power*. Volume: 13, Issue: 2, 2000.

[4] J.C.M. Vieira, W. Freitas, Wilsun Xu and A. Morelato. Performance of frequency relays for distributed generation protection. Power Delivery, *IEEE Transactions on. Volume:* 21, Issue: 3, 2006, Page(s): 1120 – 1127.

[5] Sinha, A.K., Yadavand A.K. *Design of HV transmission line protection and control system using wavelet transform for microcontroller application*. Power India Conference, 2006 IEEE.

[6] Ito, H., Shuto, I., Ayakawa, H., Beaumont, P.,Okuno,K. Development of an improved multifunction high speed operating current differential relay for transmission line protection. *IEE Seventh International Conference on Developments in Power System Protection,* 2001.

[7] Webb, A.C., Webb, M. Automated testing of power system protection relays. *Power Engineering Journal*. Volume: 2, Issue: 6.

[8] Baotao Shan, Dongmei Zhao, Xu Zhang, Fenghua Guan andZhe Liu. (2009). *Research on relay protection setting expert system for main equipment in power plant.* International Conference on Sustainable Power Generation and Supply, SUPERGEN '09, 2009.

[9] Guzman, A., Tziouvaras, D.A., Schweitzer III, E.O., Martin, K. Local-and Wide-Area Network Protection Systems Improve Power System Reliability. Power Systems Conference: *Advanced Metering, Protection, Control, Communication, and Distributed Resources,* PS '06, 2006.

[10] Adibi, A., Aghajani, A., Riahi, G.H. (1993). A comprehensive digital protection method in very high voltage power systems. *Proceedings of the 36th Midwest Symposium on Circuits and Systems*, 1993, Page(s): 382 - 384 vol.1.

[11] Zhang Qing, Shi Junwei, Liu Tianying and HouChenwei. (2007). System of Setting Calculation and Simulation of Relaying Protection on Local Power Network. *8th International Conference on Electronic Measurement and Instruments,* 2007. ICEMI '07. Page(s): 2-433 - 2-436.

[12] Jian Wu, Dongmei Zhao and Xu Zhang. Design and Development of Relay Protection Setting Calculation Expert System in Power Plant. International Conference on Energy and Environment Technology, 2009. ICEET '09. 16-18 October 2009, Guilin, Guangxi, China. Proceedings Volume: 2.

[13] Protection Relay System Technical Committee, "out-of-step and fault extension prevention relay technologies", *The In.rtitufe of Electrical Engineers in Japan*, Technical Report, No. 801, ISSN 0919-9195, October 2000 (in Japanese).

[14] *IEEE C37.2-1996 (Reaffirmed 2001)* – IEEE Standard Electrical Power System Device Function Numbers and Contact Designations.

[15] *IEEE Power System Relaying Committee, Working Group C3, Processes, Issues, Trend and Quality Control of Relay Settings*, Final Version 7.3, March 2007. http://www.pes-psrc.org/c/.

[16] *IEEE Standard 1613-2003*, IEEE Standard Environmental and Testing Requirements for Communications Networking Devices in Electric Power Substations.

[17] Kasztenny, Whatley, Udren, Burger, Finney, and Adamiak, IEC 61850: A Practical Application Primer for Protection Engineers, Georgia Tech Protective Relay Conference, 2006.

[18] Skendzic, V. and Moore, R., Extending the Substation LAN Beyond Substation Boundaries: Current Capabilities nd Potential New Protection Applications of Wide-Area Ethernet, Western Protective Relay Conference, 2006, or available at http://www.selinc.com/techpprs.htm

[19] Pozzuoli, R., Ethernet in Substation Automation Applications – *Issues and Requirements*, available at http://www.ruggedcom.com/whitepapers.html

[20] Serizawa, Y. Imamura, H. Kiuchi, M. Performance Evaluation of IP-based Relay Communications for Wide-area Protection Employing External Time Synchronization, *IEEE Power Engineering Society Summer Meeting 2001*, 15-19 July 2001.

In: Knowledge Management and Communication … ISBN: 978-1-62808-250-0
Editor: Rafiqul Islam

Chapter 20

THERMAL RUNAWAY ANALYSIS FOR THE SAFETY OF A CHEMICAL REACTOR

Badrtamam Benamrane[1,2*], Noureddine Bourmada, and Yahya Chetouani[3]
[1]Institute for Industrial Health and Safety,
[2]Laboratory of Research in Industrial Prevention,
LRPI, University of Batna, Algeria
[3]Department of Chemical Engineering,
University of Rouen, France

ABSTRACT

This paper deals with an analysis of different scenarios of thermal runaway in a chemical reactor. The HAZOP method is chosen for the analysis of risk generated by a chemical reaction, while FMEA method is used for the risk analysis linked to the operation of the chemical reactor. It is indicated in this paper the existence of interactions between the chemical reaction and the behavior of the chemical reactor. However, this leads to develop a new analytical approach to this specific type of system, which is based on a combination of (FMEA, HAZOP, Karnaugh Table) methods. A simplified electrical schematic and a simple equation which describes the occurrence of different thermal runaway scenarios are obtained for this chemical reactor.

1. INTRODUCTION

The analysis of past accident data indicates that sever accidents were caused by the release of the decomposition of products which being more hazardous than those present in normal operating conditions (Cozzani and Zanelli, 1999). A thermal explosion is the result of an uncontrolled increase of the temperature of the exothermic reaction (Gustin, 1993). The runaway scenario can serve as a basis for the assessment of thermal process (Eissen et al.,

* E-mail; benambad@yahoo.fr.

2003). Accident scenarios are selected in the frequent events of the emergency response plane (Dongwoon et al., 2003). Eissen et al. (2003) have described an experimental and its background for demonstrating TMRad by presenting the catalytic decomposition of hydrogen peroxide with potassium lode. Also, Dongwoon et al. (2003) have developed systematic method of generating accident scenario and implemented a soft-ware called yAGAS which produced automatically a list of accidents event and of generic hazardous situations. Papadaki et al. (2005) have examined the scenarios of runaway of the excess of hydrogen peroxide used during the N-oxidation of alkypiridine.

The contribution of this paper is to propose an approach for scenarios of thermal runaway and their analysis in a chemical reactor. This approach is based on a combination of three methods, namely FMEA, HAZOP, and the Karnaugh Table. The FMEA is an inductive method which starts from basic failures of components to deduce the faults impact of a system on its functions and its environment. The HAZOP method relies on the use of targets applied to the parameters (temperature, pressure ...) for a systematic analysis of deviations of the process. The HAZOP method is used in the chemical and pharmaceutical industries. It can also be applied to new processes with high hazard potential (Walkeling, 1988). The method of Karnaugh table presents the state of a logic function from the truth table. Each cell of the table corresponds to a combination of input variables. If a cell of the table changes, only one state variable changes (Kendall, 2003).

The originality of the present work concerns the use of the Karnaugh Table which is generally used for the functional analysis of a system. This table is used for the analysis of system faults. Therefore it allows to represent these scenarios as a simple wiring diagram and a simple logic equation.

2. Methods for Analysis of Risks

The proposed approach describes the combination of methods FMEA, HAZOP and Karnaugh Table for risk analysis of the system. FMEA is a method of risk analysis in components, the cause of failure, its effects and criticality. (Villemeur, 1988). The HAZOP method is particularly applicable for analyzing the risks of thermal hydraulic parameters such as flow, temperature and pressure (Salvi, 2003).

Interactions exist between components of these two subsystems FMEA, HAZOP and Karnaugh Table which is used to obtain the simple logical expression of the Boolean function (Jourlin, 2008). This approach gives a model of different thermal runaway scenarios and an electrical circuit diagram which is translated by a simple logic equation**.** The combination permits the complementarities between the results of both methods. The existence of interactions, however between the two subsystems and which are not taken into account by combining the results FMEA/HAZOP reflects the insufficient of this combination. The combination of the FMECA-HAZOP-Karnaugh Table allows to configure the failures of the subsystem 1 (chemical reactor), and the subsystem 2 (chemical reaction), but also the various interactions between the two subsystems which lead to the thermal runaway.

3. EXPERIMENTAL RESULTS

3.1. Experimental Device

The chemical reactor (Figure 1) is a glass-jacketed reactor with a tangential input for heat transfer fluid. The main aim of this input mode is to allow to give to the cooling a more important speed in the direction of the flow. It is equipped with an electrical calibration heating and an input system. It is also equipped with Pt 100 temperature probes. The heating-cooling system, which uses a single heat transfer fluid, works within the temperature range of -15 /+ 200 °c. Supervision software allows the fitting of the parameters and their instruction value. It displays and stores data during the experiment as well as for its further exploitation.

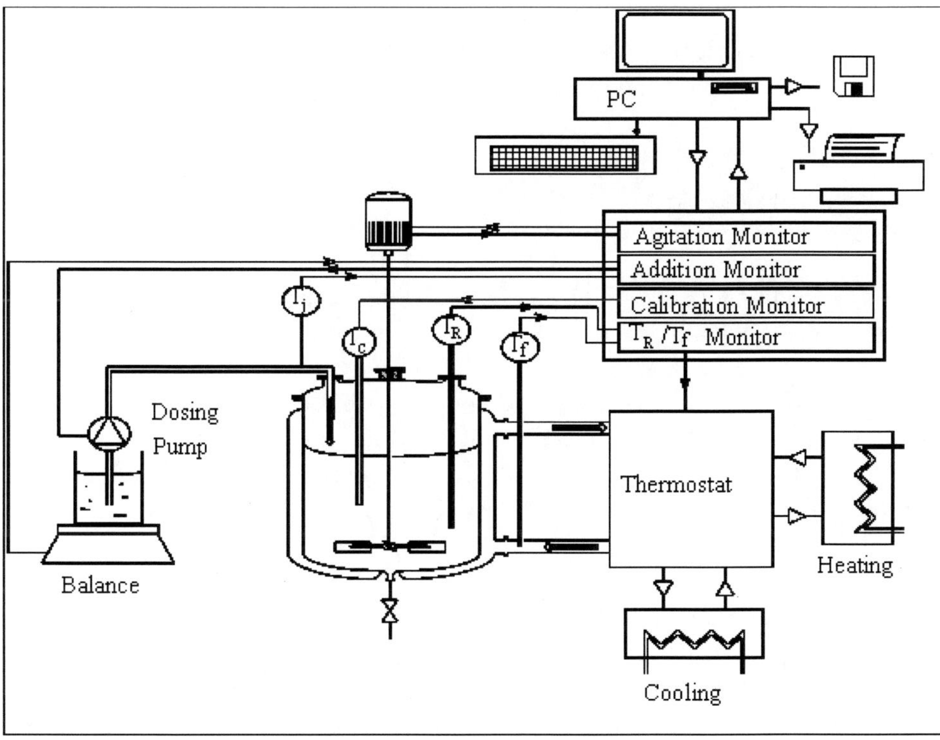

Figure 1. Experimental device: A chemical reactor.

A chemical reactor is composed of the subsystem 1, it consists also of various elements such as the cooling and stirring systems. The subsystem 2 is characterized by several parameters such as temperature, concentration and reaction rate.

3.2. Application of FMEA to the Subsystem 1 (Chemical Reactor)

The application of the FMEA method to the chemical reactor shows that the runaway scenario may be caused by the failure of some components such as the stirring, the cooling system. The application of the HAZOP method on the chemical reaction demonstrates that runaway scenario can be envisaged on the basis of drift parameters such as temperature, concentration.

3.3. Interactions Analysis

At stage of this study, the existence of different interactions can be established between the two subsystems, increase of the temperature (T) of the reaction environment (OR), increase in pressure (P) of the reaction environment (OR), inappropriate loading of products (R) and the stirring fault (A). On the other hand, each disturbance may be the combination result of the reactor failure components and the drift parameters of the reaction, it is explained by the following considerations:

T = Cooling system failure (a) AND increase the amount of heat of chemical reaction (b).
P = Failure of the pumping system (c) AND accumulation of gas in the course of reaction (d).
R = Inappropriate loading of reactants (e) AND presence of tow incompatible reactants (h). The variables (a), (c) and (e) are the failures related to the chemical reactor components. However (b), (d) and (h) are the drift of chemical reaction parameters.

Some causes leading to a disruption can be added to a cause leading to further disruption and thus increase the probability of its occurrence.

T = cooling system failure (a) AND load control system of reactant (e). T = increase of amount of heat of chemical reaction (b) AND accumulation of gas in the course of the reaction (b). P = accumulation of gas in the course of reaction (d) AND failure of pumping system (c) AND failure of the cooling system (a).

3.4. Application of Karnaugh Table

In electricity, a lamp (L) which is controlled by a contact (a) is on (L = 1) if the contact is closed (a = 1), it is off (L = 0) if (a) is open (a = 0). If L is controlled by two contacts (a) et (b) in series circuit, so L = 1 if a = 1 AND b = 1. If L is controlled by tow contacts (a) and (b) in parallel circuit, so L = 1 if a = 1 OR b = 1.

The AND function is translated by the series circuit, while the OR function is translated by the parallel circuit. In this work, the runaway phenomenon is likened to a lamp L, L= 1, i.e. the presence of the runaway, L = 0, i.e. the runaway phenomenon is not present. The events, increase in temperature T, increase in pressure P, inappropriate loading of reactant R, and a stirring fault A, are treated as bobbins A, P, Q, A.

The faults such as failure of the cooling system (a), increase the heat of reaction (b) are likened to contacts which control A, P, Q, A. The output T is a function of variables (a, b, e, and d). To establish the truth table of T we have 4 variables, so $2^4 = 16$ combinations (T = a and b, T = a and e, T = b and d).

$$T = ab + ae + bd \tag{1}$$

To establish the truth table of P we have 3 variables (a, c, d) i.e 8 combinations are possible (P = c and d, P = a and c and d). It gives the following equation:

P (a, c ,d) = c d (2)

Table 1. Truth table for T(a,b,e,d)

a	b	e	d	T(a,b,e,d)
0	0	0	0	0
0	0	0	1	0
0	0	1	0	0
0	0	1	1	0
0	1	0	0	0
0	1	0	1	1
0	1	1	0	0
0	1	1	1	1
1	0	0	0	0
1	0	0	1	0
1	0	1	0	1
1	0	1	1	1
1	1	0	0	1
1	1	0	1	1
1	1	1	0	1
1	1	1	1	1

Table 2. Simplification of T(a,b,d,e) equation

a b / e d	00	01	11	10
00	0	0	1	0
01	0	1	1	0
11	0	1	1	1
10	0	0	1	1

To establish the truth table of R, 4 combinations are possible in function of (e, f).

R (e, f) = e f (3)

A is a function of variables (g, h). To establish the truth table of A, 4 combinations are possible in function of (g, h).

A (g,h) = g h (4)

Figure 2 shows that the runaway occurs (L=1) if the temperature increases (T = 1) i.e the bobbin T is excited (T = 1) then it closes the contact T1, then lit the lamp L (L = 1), which reflects the accuracy of the thermal runaway. Similarly the Lamp (L= 1) if the pressure increase (P = 1), OR a stirring fault (A = 1), OR an inappropriate load control system (R = 1).

L = T + P + R + A (5)

On the other hand, the application of the Karnaugh table allows to find the respective equation of the bobbins T, P, R, A. Therefore, the final logic equation can be established reflecting the occurrence of the thermal runaway in the studied system:

$$L = (a . b + a . e + b . d) + (c . d) + (e . f) + (g . h) \qquad (6)$$

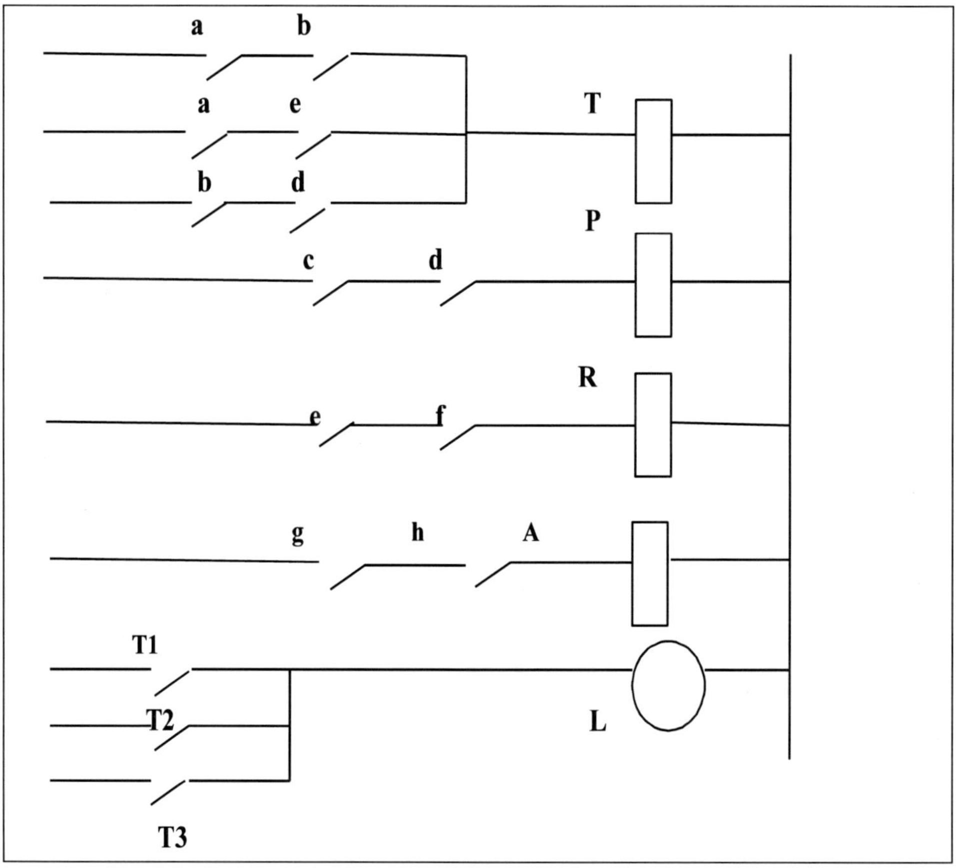

Figure 2. Model representing the scenarios of thermal runaway.

CONCLUSION

In this study the problem of runaway scenario was discussed. The proposed approach allows to present the various scenarios of the thermal runaway under the form of electrical circuit diagram which is translated by a simple logic equation. We emphasize that the study of various faults related to the chemical reactor or those related to the chemical reaction was not complete because there is a multitude of faults and failures that could generate other scenarios. The increase of the faults number led to an increase in the variables number. Therefore, it complicated the development of the runway analysis methods.

REFERENCES

Cozzani V., and Zanelli S, 1999, Precursors of dangerous substances formed in the loss of control of chemical systems, Eds. *Journal of Hazardous Materials* 65, 93-108.

Dongwoon K., Moon I. and Youngsoon Y., 2003, Automatic gerneration of accident scenarios in domain specific chemical plants, Eds. *Journal of Loss Prevention in the Process Industries*, 16, 121-132.

Eissen M., Zogg A. and Humgerbuhler K., 2003, The runaway scenario in the assessment of thermal safety: simple experimental access by means of the catalytic decompsition of H_2O_2, Eds. *Journal of Loss Prevention in the Process Industries*, 16, 289-296.

Gustin J., 1993, Continuous flow reactions and selection of security processes and control, Eds. *Chemie Technical Ingineer*, 65(4), 415-422.

Jourlin P., 2008, *Introduction to assemblor programation*, Eds. UNISCIE.

Kendall K., 2003, *Systems analysis and desing,* 5th edition Pearson Education Asia, 353-369, Delhi.

Papadaki M., Marqués D.E., Jun G. and Mahmud., 2005, Catalytic decomposition of hydrogen peroxide in the presence of alkylpyridines: Runaway scenarios studies, Eds. *Journal of Loss Prevention in the Process Industries* 18, 384-391.

Salvi O., 2003, *Risk analysis tools generated by industrial installation*, DRA-35.

Villemeur A., 1988, *Safety functioning of industrial installation, reliability, human factor, informations*, Eds. Eyrolles, Paris, France.

Walkeling H, 1988, *Preventing major chemical and related process accident, the institution of chemical engineering*, symposium serie N$^{\circ}$ 110, 13-23.

In: Knowledge Management and Communication … ISBN: 978-1-62808-250-0
Editor: Rafiqul Islam

Chapter 21

A Markovian Model for Economic Production Lot Size of a Process with Imperfect Production and Machine Failures

***Muhammad Al-Salamah*[*1] *and Ghufran Ahmed*[2]**
[1]Mechanical Engineering Department, College of Engineering,
Majmaah University, Majmaah, Saudi Arabia
[2]Systems Engineering Department,
King Fahd University of Petroleum and Minerals, Dhahran, Saudi Arabia

Abstract

In this paper, a single-machine single-product manufacturing system is considered; the system is assumed to operate under constant production rate that exceeds customer demand rate, customer demand rate is constant and certain. The machine is not perfect and is prone to failure, upon which it needs repair. The machine is also not perfectly efficient and it manufactures non-confirming and imperfect items. The possibility of failure and imperfect production are combined and modeled into a single model. The machine times-to-failure, the repair time, and the item manufactured being non-confirming are random. When the machine is under repair, the demand is backordered till a maximum backorder level is reached and after that the demand is considered to be lost.

The state of the system changes in discrete time and the process is modeled as a Markov chain which is used to find the cost model. The Economic Manufacturing Quantity of a machine which is prone to failure and also may manufacture non-confirming items is derived. Modeling the process in a Markov chain is very beneficial in incorporating backorders and failures into the process.

The results clearly indicate that the total cost is convex with respect to the maximum inventory and the total cost increases when the probability for the machine not to fail decreases. The results also show that the repair success has an effect on the total cost. The

[*]Email: m.alsalamah@mu.edu.sa.

total cost is higher when the probability of repair success is low than when the probability is high.

Keywords: Economic production lot size; Markov chain; Machine failure; Imperfect production; Inventory control

NOMENCLATURE

p: Production rate
d: Demand rate
I: Inventory level
B: Maximum backorder level
(i, j): State of the process
$V_{i,j}$: Visit to the state (i, j)
K: Expected setup cost per cycle
h: Inventory holding cost per unit per unit time
T: Expected length of the cycle
π: Cost of backordering one item
$\hat{\pi}$: Cost of backordering one item per unit time
s: Cost per unit of unmet demand (lost sales cost)
m_f: Cost incurred per failure
m_t: Maintenance cost per unit time of repair

1. INTRODUCTION

Economic production Lot Size (ELS) or Economic Manufacturing Quantity (EMQ) was developed from the Economic Order Quantity (EOQ) by modifying the inventory replenishment rate to be finite and also greater than the demand. The inventory level increases until the production is stopped. Machines involved in production systems are not ideal and are prone to failures. The times-to-failure and the repair times have to be included in the model. A single machine system that manufactures a single product is considered. The machine is prone to failure and therefore the failure rate and repair times of the machine are random and are considered to be geometrically distributed. Whenever the machine is under repair, the demand is met by the on-hand inventory if it is available, or else the demand is backordered up to a maximum limit that is selected by producer, and beyond this limit, the demand is considered to be lost. Also, the machine may produce non-confirming items. The system is modeled as a Markov chain and a total cost equation with the manufacturing quantity as a decision variable is developed which is used to find the optimal manufacturing quantity.

A Markov chain is a non-empty collection of states which are represented as vertices. The transition to a next state is represented by arrows and it has a probability. This is known as the one-step transition probability and this is determined entirely by the current state and is not dependent on the past states and these probabilities are fixed over the time. A process can remain in the same state as well. A probability transition matrix gives the probable transitions from one state to another in the chain.

The objectives of this study is developing a model for the EMQ in a single machine manufacturing process when the machine is prone to failure and, when in working condition, the machine might produce nonconforming items. Assuming constant and known production and demand rates, the economic maximum inventory level must be determined; after which production is stopped. Machine failure and repair are assumed random.

The paper is organized as follows. In Section 2, the literature related to unreliable manufacturing-inventory processes is presented. Section 3 describes the model and the Markov process. The cost model is presented in Section 4. A numerical example and sensitivity analysis is shown in Section 5, and the paper is concluded in Section 6.

2. Literature Review

Markov models are widely used for manufacturing system modelling, in a variety of environment such as system repair and failure aspect [1]; exponential service, failure, and repair process of machines [2,3]; exponential failure and repair process for machines and material handling robot [4]; and performability analysis of flexible manufacturing systems under machine failure, and spare inventory and repair processes [5].[6] proposed a model which had the decision variables as maximum inventory level at which the machine is switched off and the buffer inventory level below which the machine is started again. With no backorders allowed and considering the failure of the machine to be Poisson and the repair time as exponential, the cost function was determined.In addition, an analysis was conducted which assumed that the cost rate when machine is under repair is equal to the cost when the machine is under operation.An extension to this model was allowing backorders and considering the repair time constant which was presented by [7]. The assumptions of this model were that there is no machine breakdown when demand is backordered and the average backorder levels are minor in comparison with average inventory levels. Also, [8] proposed a model with no machine breakdowns but with imperfect production and rework.

[9-11] developed a model and a cost function when the time-to-failure of the machine was a general distribution. Repair time of the machine was taken to be insignificant and when a failure occurs, the whole inventory is consumed and then a new production cycle is started. [12] extended their previous model in which the repair times followed a general distribution and also added a safety stock level to the model which would be used when the machine failed and a lost sales case when the safety stock is depleted is also considered. [13] and [14] also considered the same model but the times to failure was considered to be Poisson distribution and the repair time a general distribution. Lost sales case was also considered.

[15] used the Markov chain modeling to model a system with two parallel input machines which manufactured parts for an assembly machine. The assumption made here is that the two parallel machines may fail and therefore a buffer level is also set between these two machines and assembly machine. The other assumption was that repair times and failure times were geometrically distributed.

[16] presented a model in which preventive maintenance and safety stock were considered. [17-20] considered the case where the production lot size was determined with

inspection schedules of processes that deteriorate with time and are prone to failure. [21-22] presented an inventory model in which there were machine failures and when the demand followed a compound Poisson process. [23] presented a model in which the economic production quantity and the preventive maintenance schedule are determined for a process which is subject to failures and imperfect maintenance. [24] considered a randomly failing manufacturing process but is directed towards perfect and imperfect maintenance of the machine.

3. PROCESS DESCRIPTION AND MARKOV MODEL

A single machine manufacturing process is considered. The rate of production is constant at p items per unit time and the demand rate is also a known constant$d$, where$p > d$. The notations used to develop the model are adopted from the model developed by [25]. The machine is put into the idle state when the inventory level reaches a prescribed level I and the next production level is started after all the on-hand inventory is consumed by the demand. As mentioned earlier, the machine is not perfectly reliable and is prone to failures or breakdowns and this may happen during a production run. These are represented as small circles on the axis of time. Whenever the machine fails, it receives immediate service. The times-to-failure and repair times are both random and are mutually independent. While the machine is in the state of repair, the on-hand inventory is consumed to satisfy the demand until shortage occurs and the demand is then backordered. Moreover, the number of items backordered is restricted to a maximum of $B$and after that the demand is considered to be lost. The cycle completes only when the machine manufactures items till the level I and then goes to idle state and all inventory is consumed and the inventory is zero before the machine is started again. The expected time for one cycle is the time to start from zero inventory level then reach the maximum inventory level and then again reach zero inventory level. As seen in Figure 1, which is a modification of Abboud work [25], the 1st cycle is a perfect cycle in which there are no failures of the machine, whereas in the next cycle it can be seen that there are two failures at the points which are denoted by a circle on the time axis. There is no manufacturing till the machine is repaired and so, the demand is being met by the on-hand inventory. If the machine is still not repaired and the entire on-hand inventory is depleted, then the demand is backordered. The backorders go to a maximum of B and then demand is lost after that because the machine is still under repair. When the machine is repaired it starts manufacturing and reaches the level I and after that the machine is switched to idle state. Then again the on-hand inventory is consumed to satisfy the demand till the inventory level reaches zero where the 2nd cycle gets completed. Nonconforming items can be produced at random with known and constant probability; nonconforming items cannot satisfy demand, therefore, nonconforming items are disposed of.

The following probabilities will be defined:

α =Pr{the machine will not fail | the machine is in operation}

$(1-\alpha)$ =Pr{the machine will fail | the machine is in operation}

β =Pr{the item manufactured is non-confirming | the machine is in operation}

$(1-\beta)$ =Pr{the item manufactured is confirming | the machine is in operation}

γ =Pr{the machine will not be repaired | the machine is down}
$(1-\gamma)$ =Pr{the machine will be repaired | the machine is down}

The state of the process is given by (i,j) where i is the inventory position and j is the status of the machine which is given by 0 – Repair, 1– Operation and 2– Idle state.

The process is now expressed by a Markov process as: $S=\{(i,j):-B\leq i\leq I, j=0,1,2\}$.

Define $P_{(i,j)(k,l)}$to be the one step state transition probability where the process is present in state (i,j) and will be in the state (k,l).

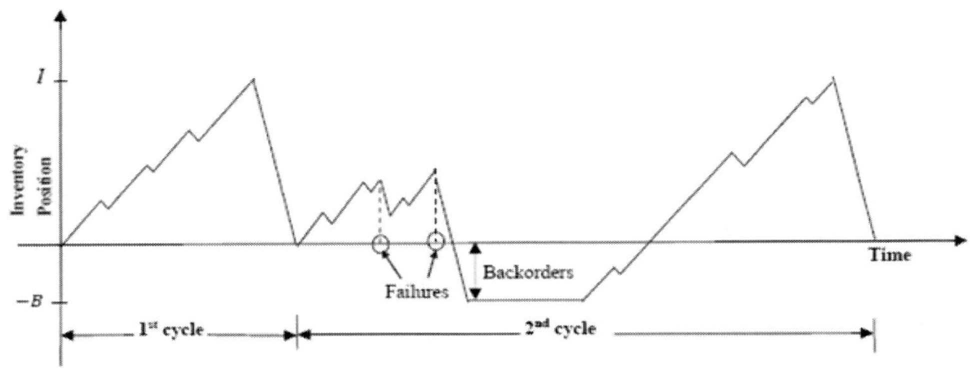

Figure 1. Inventory curve of a process with machine failure and nonconforming items.

The Markov chain developed to model the process is shown in Figure 2 which is modified from Abboud work [25]. In the Markov chain, the top row represents the states of the process with machine status as 1, which is that the machine is under operation, and then the final state I in the chain corresponds to maximum inventory level and the machine going to idle state. The process remains in the same row when the machine does not fail and it can go into the bottom array only if it fails. The bottom row corresponds to the states of the process where the machine has suffered a breakdown and repair has to be done on it. The process remains in the same row till it is repaired, and upon repair it moves to the top row.

The probability that the machine does not fail and the item manufactured is confirming is given by $\alpha(1-\beta)$. The probability that the machine does not fail and the item manufactured is non-confirming is $\alpha\beta$. The probability that the machine fails at the end of the discrete time step and the item manufactured is non-confirming is $(1-\alpha)\beta$. The probability that the machine fails at the end of the discrete step and the item manufactured is confirming is $(1-\alpha)(1-\beta)$. These probabilities are explained in Table 1.

Table 1. Probability transition table

Item / Machine	Confirming	Non Confirming
Not fail	$\alpha(1-\beta)$	$\alpha\beta$
Fail	$(1-\alpha)(1-\beta)$	$(1-\alpha)\beta$

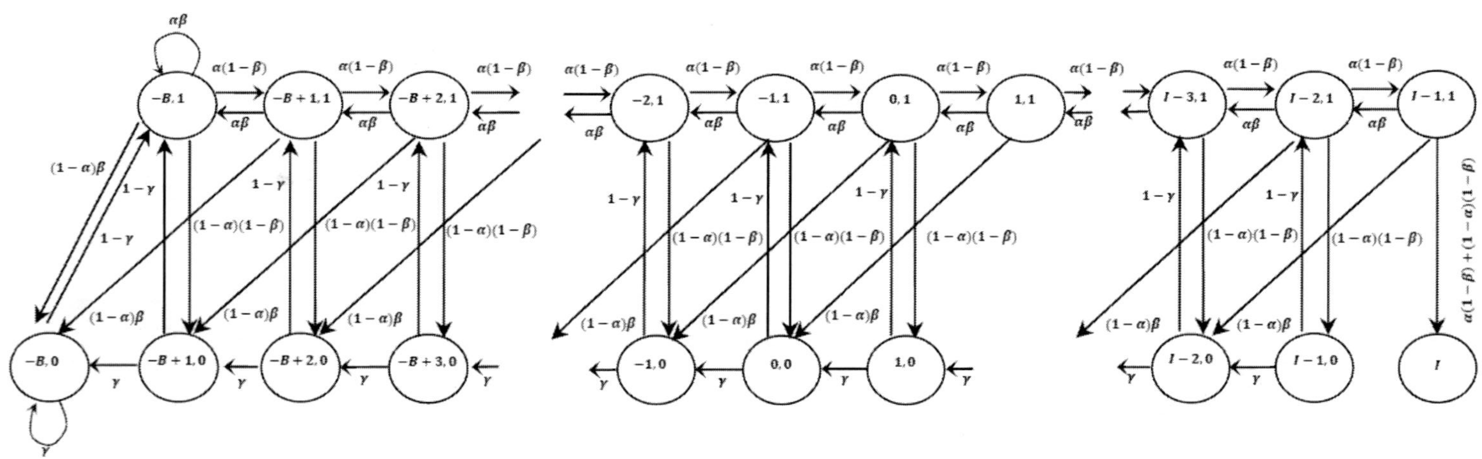

Figure 2. State transitions of the Markov chain of the process.

This table gives the transition from one state to the next state for the upper array of the Markov chain. For example, the process, if at present is in state $(0,1)$, then at the next discrete step it may go to $(1,2)$ with a transition probability of $\alpha(1-\beta)$ or it may go to $(-1,1)$ with a transition probability of $\alpha\beta$. These are the two states that the process can go while the machine is in operation and does not fail. The two other states it can go to are to fail at the end of the step with a conforming or non-confirming item. So, it can either go to $(1,0)$ state or to $(-1,0)$ state with the transition probabilities of $(1-\alpha)(1-\beta)$ and $(1-\alpha)\beta$.

The state $(-B,0)$ cannot go to a state $(-B-1,0)$ due to the constraint applied on the maximum amount of backorders and; so, if the machine is still not repaired in the state$(-B,0)$, then all demand is lost after this stage till the machine is repaired. This is depicted in the Markov chain by a self-transition probabilityγ, which is the probability that the machine is not repaired. Also, when the process is in the state $(-B,1)$, then even if the machine manufactures a non-confirming item, the process cannot go into state $(-B-1,1)$ and it remains in that state which is denoted by the self transition probability in the Markov chain as$\alpha\beta$. Each state when the machine is working can go to four different states in the Markov chain except at the boundaries where the constraints are set. Also, each state when the machine is under repair can go to only two states which are either it gets repaired or it does not get repaired.

So, $P_{(i,j)(k,l)}$ for different states is given by:

$$P_{(i,1)(i+1,0)} = (1-\alpha)(1-\beta) \text{ for } -B \leq i \leq I-1 \tag{1}$$

$$P_{(i,1)(i-1,0)} = (1-\alpha)\beta \text{ for } -B+1 \leq i \leq I-1 \tag{2}$$

$$P_{(i,1)(i-1,1)} = \alpha\beta \text{ for } -B+1 \leq i \leq I-1 \tag{3}$$

$$P_{(i,1)(i+1,1)} = \alpha(1-\beta) \text{ for } -B \leq i \leq I-1 \tag{4}$$

$$P_{(-B,0)(-B,0)} = P_{(i,0)(i-1,0)} = \gamma \text{ for } -B+1 \leq i \leq I-1 \tag{5}$$

$$P_{(-B,0)(-B,1)} = P_{(i,0)(i-1,1)} = 1-\gamma \text{ for } -B+1 \leq i \leq I-1 \tag{6}$$

$$P_{(1,2)(0,1)} = 1 \tag{7}$$

$$P_{(i,2)(i-1,2)} = 1 \text{ for } 2 \leq i \leq I \tag{8}$$

All other transition probabilities are assumed to be zero. The process becomes deterministic when it enters the state (I) and the state space can be reduced by eliminating the states $(i,2), i = 1,2, \dots I$.

If β (the probability that the machine produces a non-confirming item) is taken as 0, then the model developed here will return to the model developed by [25]. And also when α(the probability of machine not failing) is 1, then the model will be like the classical model.

An assumption is made here that there is no rework being done and also there is scrap but there is no cost associated with it.

Let $V_{i,j}$ be the expected number of visits to the state (i,j) before the process gets to the state I; and at this state, the process gets absorbed; and the states $(i,2)$, where $i = 1,2,\dots.I$, are eliminated. The discrete time step while the machine is in operation is $1/(p-d)$ and the time step when the machine has failed or is idle is $1/d$.

The expected visits are given by:

$$V_I = [\alpha(1-\beta) + (1-\alpha)(1-\beta)]V_{I-1,1} \tag{9}$$

$$V_{I-1,1} = \alpha(1-\beta)V_{I-2,1} \tag{10}$$

$$V_{I-1,0} = (1-\alpha)\beta V_{I-2,1} \tag{11}$$

$$V_{i,1} = \alpha(1-\beta)V_{i-1,1} + \alpha\beta V_{i+1,1} + (1-\gamma)V_{i+1,0} \tag{12}$$

$$V_{0,1} = 1 + \alpha(1-\beta)V_{-1,1} + \alpha\beta V_{1,1} + (1-\gamma)V_{1,0} \tag{13}$$

$$V_{-B,1} = \alpha\beta V_{-B+1,1} + (1-\gamma)V_{-B+1,0} + (1-\gamma)V_{-B,0} + \alpha\beta V_{-B,1} \tag{14}$$

$$V_{-B,0} = (1-\alpha)\beta V_{-B,1} + (1-\alpha)\beta V_{-B+1,1} + \gamma V_{-B,0} \tag{15}$$

$V_I$is the final state and it is only visited only once so the value of it is 1.The states are similarly calculated by moving backwards in the Markov chain.

Each time a state is visited, an inventory item is added or it is consumed. So, it takes into account the building up of the inventory and also the consumption of the inventory. These are then summed up to get the total inventory. The concept of visits to a state may also relate to the number of times it may be encountered in each time cycle.

4. Cost Model

Now, when the different $V_{i,j}$'s are calculated, the expected cost per unit time for a given level $I$is calculated. The cost equation is then used to find I^* which minimizes the total cost. The total expected cost per unit time $TC(I)$ is computed by employing the renewal reward theorem; and so, the expected total cost is divided by the expected cycle length. Define the effective production rate $p_e = p(1-\beta)$.

The expected length of the cycle is given by

$$E[T] = \sum_{i=-B}^{I-1}\left(\frac{V_{i,1}}{p_e-d} + \frac{V_{i,0}}{d}\right) + \frac{I}{d} + \frac{V_I}{p_e-d} \tag{16}$$

The second last term in the above equation refers to the states $(i,2)$ of the process, which is the time when the machine is idle and the inventory $I$is consumed by the demand. The last term refers to the final state of the state space and is the maximum inventory level.

The expected setup cost per cycle is written as

$$E[SC] = K \tag{17}$$

The expected lost sales cost is given by

$$E[LC] = sV_{-B,0} \tag{18}$$

The expected backorder cost is given by

$$E[BC] = \pi \sum_{i=-B+1}^{0} V_{i,0} + \hat{\pi} \sum_{i=-B}^{0} \left(\frac{iV_{i,1}}{p_e - d} + \frac{iV_{i,0}}{d} \right) \tag{19}$$

The expected inventory holding cost is given by

$$E[IC] = h \left\{ \sum_{i=0}^{I-1} \left(\frac{iV_{i,1}}{p_e - d} + \frac{iV_{i,0}}{d} \right) + \frac{I(I)}{2d} + \frac{V_I}{p_e - d} \right\} \tag{20}$$

Machine failures are due to many reasons and, therefore, are associated with costs. The expected cost due to failures per cycle is given by

$$E[MC] = m_f (1 - \alpha) \sum_{i=-B}^{I-1} V_{i,1} + m_t \frac{1}{d} \sum_{i=-B}^{I} V_{i,0} \tag{21}$$

The expected total cost equation is given by

$$E[TC] = \frac{E[SC] + E[IC] + E[BC] + E[LC] + E[MC]}{E[T]} \tag{22}$$

Substituting the equations (16)-(21) in equation (22), the total cost equation can be calculated. The total cost has to be minimized and the decision variable is the maximum inventory level I.The total cost equation is based on the renewal reward theorem.

5. NUMERICAL EXAMPLE

To examine and analyze the model developed, a numerical example is discussed in this section. The parameters selected are $p_e = 100, d = 60, B = 1, K = 500, h = 15, \pi = 45, \hat{\pi} = 30, s = 100, m_f = 75$ and$m_t = 25$.

The probabilities are taken as $\alpha = 0.9, \beta = 0.1$ and$\gamma = 0.3$. The system of linear equations represented in (9)-(15) is solved and then different expected cost terms represented by equations (16)-(22) are calculated while varying I. A graph of total cost versus the inventory level is plotted in Figure 3, which shows the variation of total cost against I.

It can be seen from the graph that the total cost decreases as the inventory level increases; and after a certain minimum of inventory level, the total cost again increases to give a convex curve. The minimum value or the optimal value of the inventory level at which the cost is minimized is of interest. The minimum cost is achieved at $I = 15$.

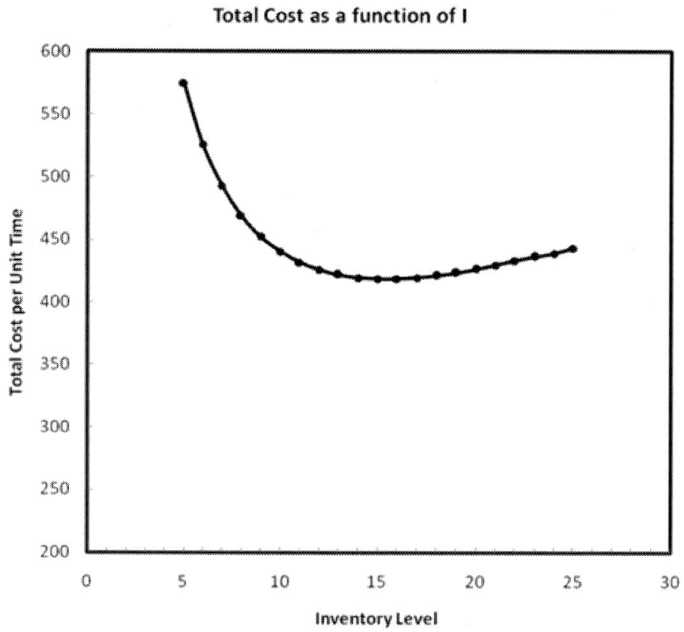

Figure 3. Total Cost as a function of Inventory Level.

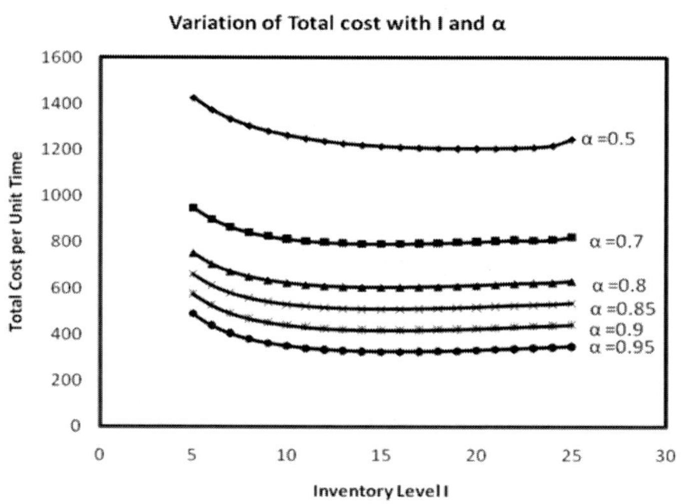

Figure 4. Total Cost as a function of α and I.

6. Sensitivity Analysis

To analyze the behavior and to validate the model, a sensitivity analysis is carried out with varying failure rate and repair rate. The probability for the machine not to fail α is varied and the behavior of the model is plotted in Figure 4. As the value of α is varied from 0.95 to 0.5, it is seen that the total cost increases; this result is expected because if the probability of the machine not failing is low then the machine fails often and so, the costs increase.

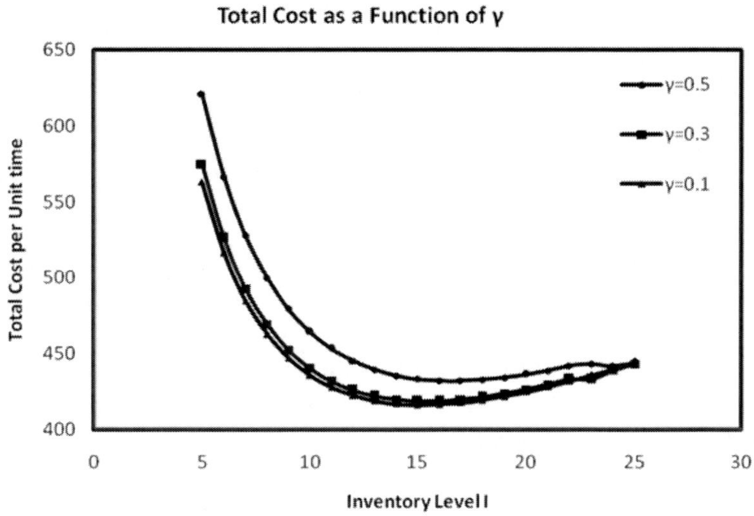

Figure 5. Total Cost as a function of γ and I.

Also, the probability of the machine not being repaired is varied to see the behavior of the model. The value of γ is varied and the total cost is calculated for three different values of γ to analyze its effect on the model. The total cost as a function of γ and I is plotted in Figure 5. The three different values of γ are $0.1, 0.3$, and 0.5. It is observed that the model behaves as expected because, if the probability of machine not being repaired decreases then the machine is repaired quickly and so the total cost is reduced.

CONCLUSION

The EMQ of a machine which is prone to failure and also may manufacture non-confirming items is derived. Modeling the process in a Markov chain is very beneficial in incorporating backorders and failures into the process. The machine failure and repair time are random and a constraint is applied on the maximum number of backorders. Imperfect manufacturing is combined with machine failure to derive a cost model. A numerical example is shown to prove the effectiveness of the model and sensitivity analysis is carried out to show the effect of failure rate and repair rate on the model. The main observations from the model are:

- The total cost is convex with respect to the maximum inventory;
- The total cost increases when the probability for the machine not to fail decreases;
- The repair success has an effect on the total cost. The total cost is higher when the probability of repair success is low than when the probability is high.

An extension in this model would be to add rework to the process for which the Markov Chain has to be re-modeled. Also, a scrap cost could be included in the model and its effect on the model can be tested.

REFERENCES

[1] Foster, J. W., III.,and Garcia-Diaz, A. (1983). Markovian models for investigating failure and repair characteristics of production systems. *IIE Transactions*, 15, 202–209.

[2] Gershwin, S. B., and Berman, O. (1981). Analysis of transfer lines consisting of two unreliable machines with random processing times and finite storage buffers. *AIIE Transactions,* 13, 2–11.

[3] Hong, Y., andSeong, D. (1993). The analysis of an unreliable two-machine production line with random processing times. *European Journal of Operational Research*, 68, 228–235.

[4] Suliman, S. M. A. (2000). A mathematical model for a buffered two-stage manufacturing cell with an unreliable transfer device. *International Journal ofProduction Economics*, 63, 69–81.

[5] Rupe, J., andKuo, W. (2001). Performability of FMS based on stochastic process models. *International Journal of Production Research*, 39, 139–155.

[6] Hopp, W.J., Pati, N., Jones, P.C., (1989). Optimal inventory control in a production flow system with failures. *International Journal of Production Research*, 27, 1367–1384.

[7] Moinzadeh, K., Aggarwal, P., (1997). Analysis of a production/inventory system subject to random disruptions. *Management Science*, 43, 1577–1588.

[8] Cardenas-Barron, L. E. (2009). Economic production quantity with rework process at a single-stage manufacturing system with planned backorders. *Computers andIndustrialEngineering,* 57(3), 1105-1113.

[9] Groenevelt, H., Pintelon, L. and Seidmann, A., (1992a). Production lot sizing with machine breakdowns. *Management Science*, 38, 104–123.

[10] Al-Tahat M. D., and Rawabdeh I. A. (2008). Stochastic Analysis and Design of CONWIPControlled Production Systems. *Journal of Manufacturing Technology Management,* 19(3),253-273.

[11] Al-Momani K. R., Al-Tahat M. D., Jaradat E. T. (2006). Performance Measures for Improvement of Maintenance Effectiveness: A Case Study in King Abdullah University Hospital (KAUH). International Conference on Service Systems and Service Management , IEEE.

[12] Groenevelt, H., Pintelon, L. and Seidmann, A., (1992b). Production batching withmachine breakdowns and safety stocks. *Operations Research*, 40, 959–971.

[13] Abboud, N.E., (1997). A simple approximation of the EMQ model with Poisson machine failures. *Production Planning and Control*, 8(4), 385–397.

[14] Abboud, N. E., Remeily, H. E. (1995). An approximate model to the production lot-sizing problem with Poisson machine breakdowns and low failure rate. Proceedings of the International Conference on Industrial Engineering and Production Management, 33–42.

[15] Simon, J.T., Hopp, W.J., (1995). Throughput and average inventory in discrete balanced assembly systems. *IIE Transactions*, 27, 368–373.

[16] Cheung, K.L., Hausman, W. H., (1997). Joint determination of preventive maintenance and safety stocks in an unreliable production environment. *Naval Research Logistics*, 44, 257–272.

[17] Hariga, M., Ben-Daya, M., (1998). Note: the economic manufacturing lot-sizing problem with imperfect production processes: bounds and optimal solutions. *Naval Research Logistics,* 45, 423–433.
[18] Lee, H.L., Rosenblatt, M.J., (1987). Simultaneous determination of production cycle and inspection schedules in a production system. *Management Science*, 33, 1125–1136.
[19] Makis, V., (1998). Optimal lot sizing and inspection policy for an EMQ model with imperfect inspections. *Naval Research Logistics*, 45, 165–186.
[20] Makis, V. and Fung, J., (1998). An EMQ model with inspections and random machine failures. *Journal of the Operational Research Society*, 49, 66–76.
[21] Liu, B., Cao, J., (1999). Analysis of a production-inventory system with machine breakdowns and shutdowns. *Computers and Operations Research*, 26, 73–91.
[22] Posner, M.J.M., Berg, M., (1989). Analysis of a production-inventory system with unreliable production facility. *Operations Research Letters*, 8, 339–345.
[23] El-Ferik, S. (2008). Economic production lot-sizing for an unreliable machine under imperfect age-based maintenance policy. *European Journal of Operational Research*, 186(1), 150-163.
[24] Liao, G.-L., Sheu, S-H. (2010). Economic production quantity model for randomly failing production process with minimal repair and imperfect maintenance. *International Journal of Production Economics*, 130(1), 118-124.
[25] Abboud N. E., (2001). A discrete-time Markov production-inventory model with machine breakdowns. *Computers and Industrial Engineering*, 39(1-2), 95-107.

In: Knowledge Management and Communication … ISBN: 978-1-62808-250-0
Editor: Rafiqul Islam

Chapter 22

APPLICATION OF ARTIFICIAL INTELLIGENCE METHODS IN UNMANNED AERIAL VEHICLE CONTROL

Vladimir Grishin*
Space Research Institute (IKI) of the Russian Academy of Sciences
Moscow, Russia

ABSTRACT

During the last 50 years, Artificial Intelligence (AI) made colossal progress. Beginning from the simple heuristic search in space now complicated methods of pattern recognition, planning, control and learning are developed. It seems that we are moving towards the creation of AI control system with a high degree of autonomy. Such systems are capable of efficient operation in unpredictable varying and hostile environment, including conflict situations. More and more practitioners in engineering will be involved in the process of AI control systems development. Unfortunately only the small part of up-to-date AI methods and algorithms are applicable to this task. The aim of this article is to list AI directions which, in author's opinion, are suitable for AI control system development. We will consider the control systems of unmanned aerial vehicles (UAVs). Such vehicles are characterized by the high dynamics and complicated system of restrictions on control parameters. Nevertheless, the main statements of this paper are applicable to control systems of other mobile objects based on AI principles.

INTRODUCTION

A subject of this article is to mark the possible ways of AI control systems development. This short article does not pretend to be a survey. Present-day AI includes too much direction for this. Simultaneously, this article does not pretend to make deep analysis and proof. Such an analysis and proof require a weighty book, rather than the small article.

* E-mail: vgrishin@iki.rssi.ru.

What is the Artificial Intelligence (AI)? There are many definitions of AI. We will use the following one. The AI is a complicated form of the information processing and control which promotes successful mission fulfillment and UAV survival. Mission – this is what should have to do. Mission can include series of the tasks. As the task can be reconnaissance, search of the defined objects (stationary or mobile), fight, or different demonstration actions aimed for misinformation of an opponent. Note that we will consider autonomous AI systems. It means that the AI system can solve effectively tasks and makes all necessary decisions in unpredictable varying and hostile environment without any remote control from the operator.

The functions of AI control system are:

- Analysis of surrounding environment. This environment may be defined only partly, or the environment is nonstationary and unfriendly.
- Creation of action plans. These plans may include research actions, which are intended for decreasing a current situation ambiguity.
- Implementation of these plans.
- Detection of deviation of the results from the planned ones and re-planning.

At present, there are a lot of papers and books, which are devoted to applications of AI methods. In [1] different applications of AI in robotics are presented. The development of AI pilot is described in [2]. This AI pilot is intended for games and flight simulators. Intelligent flight control is discussed in [3]. Different obstacles which make difficulties for creation of fully autonomous AI pilot are discussed in [4]. The problems and risks associated with autonomous control are discussed in [5]. Two approaches to natural and artificial intelligence are presented in [6, 7]. It is interesting to note that man-made AI control systems and control systems of living creatures are appropriate to be analyzed from the unified position.

Those, which are faced with the necessity to develop UAVs control system with usage of AI, should choose suitable methods. Some methods, that are convenient for development of practical on-board control systems, are listed in this paper. Although this paper deals with UAV control systems, this approach is appropriate for development of control systems for other mobile objects based on AI principles.

This paper contains some ideas of the course "Methods of artificial intelligence" that the author delivers for the final-year students of Bauman Moscow State Technical University during the last five years.

Autonomous Artificial Intelligence

It is necessary to mark the peculiarities of perspective control systems' development at present. On the one hand, the development of remote-controlled UAVs uncovers problems, which are connected with this way. In particular, it is the necessity of operators training for steadily growing UAV fleet. Pilots do not want to become the operators of UAVs. At the same time, the training of operators requires a significant time and money to study the UAVs application, tactics, peculiarities of the UAV dynamics, safe flight modes and so on.

On the other hand, remote control leads to the worse quality of control as a matter of principle in comparing with control quality of the manned aircraft pilot. This is conditioned by a smaller amount of information, which is obtained by UAV operator as compared with a pilot. Moreover, the operator has to use only visual information for control (i.e. perception channel with the minimal speed of information processing). On the contrary, the pilot in a cockpit uses all sense organs and the velocity of perception of other sense organs is much higher. All this seriously makes worse the quality of remote control. Considerable higher degree of UAVs accident rate as compared with manned aircraft proves this fact.

Note that the pilot takes into account the fact that the integrity of the aircraft substantially guarantees his own safety. So he will take efforts for airplane integrity in contradiction to UAV operator whose safety doesn't absolutely depend from the UAV integrity.

The next principal defect of remote control is the necessity of communication channel's usage. These channels are essentially vulnerable to countermeasures, and this vulnerability increases along with increasing of the distance between control center and UAV. Finally, it should be noted, that in any case, the human being is a very limiting element of the control system due to his low rate of information processing and decision-making.

Another principal restriction of remote control is the limitation of human being experience and knowledge. The ability to active learning and self-learning is typical not for every person. Moreover, this ability is losing more and more with human aging. On the contrary, AI system is capable of active learning and self-learning in a whole period of its usage. The huge flow of information provided by on-board sensors can promote the AI control system self-learning. Such huge flow cannot be processed by a human being. Accumulated knowledge can be easily transmitted to a more perfect control system during updating the hardware. Here, the point is that every AI system can accumulate *personal experience,* in particular, in navigation, in recognition of specified objects, in technique of pilotage and so on. All this information and knowledge, except for the set of basic skills, cannot be programmed into high degree perfection control systems. Nevertheless, it is possible to provide transfer of accumulated experience between different AI UAV control systems. The rate of information exchange between AI systems is some orders higher against the rate of information exchange between human beings or human being and AI system. Undoubtedly, the transfer of accumulated experience should be realized with mandatory verification of this experience with the aim of elimination of unfriendly reflexive control. The reflexive control will be discussed later.

Information (personal experience and knowledge) accumulated by AI system can be periodically backed up due to possible unexpected aircraft lost. In particular, the cloning of the system can be easily performed in the same hardware or in compatible hardware platforms. The specified bases contain knowledge and individual experience, and these bases can be used in development and manufacture of the new AI systems.

In author's opinion, the most effective UAV usage is possible only with *autonomous AI control systems.* Such systems should independently produce all necessary decisions in the frames of the specified task or, in the case of specified task solution impossibility, more general task. Such AI systems become very similar to highly organized biological systems. The difference between them is the absence in AI system of searching for food and self-reproduction tasks. Another significant difference is that in AI systems the process of information exchange (learning) can be fulfilled much faster. At the same time, the formation of social structures among autonomous AI control systems should be strongly limited. Such AI system, as well as biological systems, should be capable of intensive self-learning, effective usage of reflexive control methods, including detection of unfriendly reflexive control attempts.

The last peculiarity is very important because well-timed detection of unfriendly reflexive control attempts reveals true intentions of the opponent and allows ruining his plans. Note that reflexive control can be used only against the systems, which have a high degree of evolution and are capable of analyzing the current situation and adapt to it. High risk of reflexive control is confirmed by the fact, that intrigues, which mainly consist of reflexive control, are far from being always successful.

AI system has to function in conditions of conflict. Such systems should realize fault-finding, estimate the corresponding decrease of own functionality and then change the flight control algorithms. The further mission plan should be changed accordingly. The most similar to such AI system are highly organized living creatures, which successfully solve such tasks every time.

At present time, the progress in development of AI systems allows us to accumulate work-background, which can be considered as sufficient for beginning of autonomous AI control system's practical realization [8]. Progress of electronics and development of more and more powerful processors creates preconditions for such system development. Practical realization of such a UAV AI control system should be commercially effective. The efficiency will be achieved by increasing the probability of the mission success and decreasing the probability of UAV damage.

APPROACH TO PRACTICAL REALIZATION

Who should develop AI control system for UAVs? There are two possible variants of development. Either AI specialists will master UAVs control methods or specialists on control and application of UAVs will master AI methods. The author is convinced, that only the second way is promising. The point is that AI control system should have very good interaction with the ordinary (classical) aircraft control system. AI control system is building over the classical control system of an aircraft. Classical aircraft control system is built on the base of all well-known control methods approved by the practice. Only highly qualified and experienced designers of classical aircraft control systems are capable of guarantee such a seamless integration of AI control systems and algorithms of flight control. It is possible only on the second way of AI control system development. Another argument is, in the author's opinion, that the overwhelming majority of AI specialists are incapable of mastering the basic methods of control system development, including the aircraft control algorithms.

UAV as the object of control is the very complicated object. The behavior of UAV is described by aerodynamic laws. The behavior of UAV engine is described by gas dynamic's laws. Object of control is nonlinear with the complicated system of restrictions on phase variables [9]. The phase space used for description of the object behavior is many-dimensional. The development of aircraft control algorithms is much more complicated task as compared with the development of chess game programs or any other application of AI methods. So both researchers and engineers, who are experienced specialists in the area of aircraft control and specialists in the area of aircraft usage (application), must be involved in development of such AI control systems.

On the other hand, the presence of a complex system of restrictions on phase variables in many cases seriously decreases the number of variants to be analyzed and compared. Moreover, the critical lack of time in a real time control system does not allow performing the deep analysis. Therefore, classical search algorithms in AI state space are unlikely applicable to control tasks. In development of classic AI methods most of the effort was spent on search tasks' solution. Choice should be made on the base of subjective preferences. They are formed, on the one hand, during development of a control system, on the other hand, by individual experience. This individual experience is formed accordingly to analysis of available precedents and their estimation from the point of view of their effectiveness.

The starting point for autonomous AI control system development should be the classical control theory and practice. One can consider AI control systems as evolution of the classical control systems. In this case, a scheduled mission can be considered as the input reference value which should be worked out (executed) by AI system. In other words, AI system should decrease the difference between the current state and the reference. AI control system performs continuous monitoring of external environment. This process is similar to the measurement of parameters of a controlled object. Difference from the classical control system is the dimensionality of space, which describes the state of environment. The space dimensionality may be very high. Some of the measured or estimated parameters cannot be expressed in the numerical form. There is complicated system of restrictions on phase and control variables. Criteria of control quality are multidimensional and depend on the current situation and task.

The aim of AI control system is to transform the environment state to the reference state. In other words, it is necessary to form some state space trajectory, which allows transformation of current state to the new (reference) state. For example, we can consider a reconnaissance mission. If opponents will try to prevent to this mission, we can consider this as external disturbance. To decrease the influence of this disturbance, we can plan the sequence of actions aimed to decrease the probability of our detection. Another way is demonstration actions in another zone. Such actions will divert opponent resources and decrease his opportunities to prevent our mission. In this case, we should consider some sequences (possibly highly complicated sequences) of actions as “control”, which can be reflected as a hierarchical structure in different levels of generality.

As the object of control and its interaction with the environment are highly complicated, preliminary testing of actions on the model of the aircraft and the environment (external objects) is required. It is necessary to create such a plan of actions, which is capable of decrease undesirable reaction of environment on our actions or to use this reaction accordingly with our own aims.

The construction of control actions sequences is made with the account of both dynamic characteristics of UAV and external restriction, for example, stealthy character of actions. Taking into account of these restrictions, guarantees AI system effective functioning. All control actions are considered in the space of some qualitative or quantitative parameters. For instance, there can be:

- Integrity or minimum possible damage of the UAV.
- Required expenses of resources (time, fuel, and so on).
- Possible or probable disturbances to performance of target mission.

- "Leakage of information" to an opponent about one's plans and intentions, operational characteristics, resources.
- Accumulation of information about environment and possible algorithms of actions (self-learning).

Accordingly, control action's sequences are constructed in dependence of external conditions, scheduled mission (the mission goals) and the current state of UAV and its resources. Priorities of these parameters can be changed depending on the current situation. For instance, the priority can consist of unconditional fulfillment of the mission with desirable keeping UAV integrity and minimum expenditure of resources. In some cases, more priority can be assigned to the task of "leakage of information" minimization. In other words, UAV should design action sequence in such a way that opponents can receive minimal information about air route of flight, mission plan, UAV technical parameters, on-board knowledge bases and so on. As constraints in this case can be the probability of detection and tracing of this UAV by opponents. In other cases, it is necessary on the contrary to create "leakage of information" for reflexive control objectives (i.e. opponent misinformation).

As the background task can be considered the accumulation of information aimed to increase the control effectiveness. The data processing of accumulated information requires many computational resources and is performed mainly during the interval of time when UAV is not on a mission (i.e. UAV is "on recreation").

METHODS OF ARTIFICIAL INTELLIGENCE

At present, the huge number of methods, which usually are attributed to the area of AI, is accumulated. Essays devoted to 50 years of AI development are presented in [10]. Good reviews of AI present state are presented in [11, 12, 13]. The more understandable exposition of this material can be found in the book [14]. The books [15, 16] are addressed to the engineers and scientists.

At the same time, engineers, who involved into AI practical usage, face with a very complicated selection of principles, which should be laid in the base of control system development. However, the overwhelming majority of AI methods are of little avail or absolutely useless for such AI control systems.

In particular, the classical formal logic systems seem of little avail for realization of autonomous AI systems, except some particular cases. Classical formal logic is not agreed with a situation of a complicated world which can be changing during AI system function (dynamic character of knowledge). Classical logic is unsuitable for real-time control tasks. Formal approach to knowledge representation reduces possibilities of solution process speeding-up. Moreover, the changing and addition of new axioms which describes a world is a problem in autonomous mode without intervention of a human being. Presently different extensions of classical logic appear. There is spatial-temporal logic, abduction logic, probabilistic logic, non-monotonic logic, modal logic and so on [17]. Such modern types of logic seem more suitable for control tasks. On the other hand, it is very doubtful that a cat or an eagle which hunts on a mouse uses formal logic of any kind. At the same time, both cats and other beast of prey quite successfully solve the problems, which are posed by the life.

Formal logic is quite applicable in some simple areas, which can be completely described. Such areas can be a database querying, distribution of computational resources between processes in real time, fault diagnosis and so on. A few words should say about fuzzy logic. The problem of fuzzy logics is creation membership function. If there is the method to form correct membership function - then all is well. Otherwise the results of fuzzy logic calculations are described by the rule "garbage in – garbage out." Small utility of state search algorithms in the presence of a complicated system of restrictions has been mentioned above.

At the same time, semantic networks, frames, scripts can be successfully used for AI control systems [11, 12, 14]. Although there are the different forms of information and knowledge representation, these techniques will converge to approximately the same results during adaptation under concrete task, increasing effectiveness of knowledge and data representation.

The interest to the autonomous agents is continuously increasing during the last 15-20 years [1, 18, 19]. In [13] concept of "intelligent agent" is used for presentation of the all AI area. Autonomous agent has two properties – autonomy and embodiment. Embodiment means that agent is separate from surrounding environment. The term "embedded" means that agent is included in a social environment. Autonomous rational physically embodied agents are the most interesting for the UAVs control. Here "physical" means that agent is realized as a physical system which exists in the physical world. "Rational" means that agent can optimize the own behavior. Bio-inspired agents relate to the agents whose prototype was the living creatures. Natural agent refers to any living creature. Multiagent systems consist of interacting agents. The results of research in the area of autonomous agents are quite applicable to the development of autonomous AI control systems.

Let's consider such as the important concept as "precedent." We will use term "precedent" for the designation of the previous experience. Raw precedent consists mainly of the information which is fixed by onboard UAV sensors. This information includes all objects which were detected in the environment, their coordinates and velocities, coordinate and velocity of the UAV, decisions which have been made by UAV control system, internal parameters of UAV and so on. Raw precedents can be used for generalization for current and possible future tasks. Precedents can be used for the reasoning, planning, control, self-learning, etc.

Storing of precedents requires large volume of memory. However, as the base of precedents is filling, it is possible to establish the correspondence between the new precedents and previously stored ones. So instead of storing all information *only new information* should be stored in precedent base. The pieces of already known information in new incoming precedents are replaced by references to such information of previously stored ones. As far as progress in the environment studies the velocity of precedent base volume growing will decrease and then base volume will practically be stabilized in a stable environment. Such processing of precedents requires many computational resources and it appropriately to be fulfilled during the time intervals when UAV is not carried out a mission (UAV is "at rest"). There is an opinion that during night sleep, human brain fulfills similar work.

Precedents are the information of low level of abstraction and generalization. During precedent processing, it is possible to form more and more abstract representation requiring much smaller storage space for precedents storing. This abstract representation allows in one's turn to form bases of heuristics, frames and rules for reasoning, planning and control. So we can see that precedent storing and processing is the very powerful method of *self-learning,* which is appropriate to use for increasing of efficiency of control and solution of posed problems. Control efficiency depends on efficiency of precedents handling procedures.

Self-learning can be used for search more effective maneuvers, their optimization and updating corresponded knowledge bases. Furthermore, self-learning can be used for widening or narrowing the areas of applicability of available in knowledge bases maneuvers. Another area of precedents' application is detection of new tactical techniques and maneuvers of an opponent. Possible tasks are:

- development of counteraction methods to these opponent maneuvers;
- tracking of changing of the opponent technical opportunities;
- development of opponents models and models of their behavior.

There are terms case-based control and case-based reasoning. The difference between "case" and "precedent" is that "precedent" is "raw" while "case" is a more abstract level for information representation. Reviews of case-based reasoning are presented in [20, 21]. Review of similarity-based reasoning, analogy-based reasoning and case-based reasoning is presented in [22].

The integration of case-based reasoning with rule-based reasoning, solving the constraint-satisfaction problem and model-based reasoning are discussed in [23]. The more detailed description of case-based reasoning is presented in [24].

Many papers are devoted to constraint satisfaction problem and constraint-based control [25, 26]. Since the UAV is the real physical object, it has to obey to the physical laws. The physics laws are presented by mathematical models (frequently as a form of differential equations) from the one hand and the system of constraints on state variables from the other hand.

Satisfaction of these constraints guarantees normal operations of a controlled object. Unfortunately, the most part of researches, which are devoted to a constraint satisfaction problem not concerns to such a complex physical system such as UAV or other dynamic objects. Only small part of publications are devoted to complex physical systems. As an example we can refer to the article [27] where an efficient motion planning algorithm for stochastic dynamic systems with constraints on probability of failure is considered. For creation of effective methods for UAV control, the further development of constraint reasoning, scheduling, planning and control methods for physical dynamic systems in real time is needed.

Real tasks require performing flight with using extreme modes on stability and controllability. For the survival, short-term flight behind of stability and controllability borders is sometimes needed (tricks like Pugachev's Cobra and the like). Some situations require the optimal control for the UAV. Fortunately, now a great number of numerical optimal control algorithms are known.

The rule-based systems have wide applications in different areas. Wide spread of rule-based expert systems is conditioned by its effectiveness. Additionally, mention should be made that rule-based systems can operate in a partly defined world and can be formed automatically from precedents.

The nature of rule-based systems is adequate to state-space control methods, including real-time control. The aggregate of rules can be easily modified or extended. And what is most valuable, this modification can be done without human being intervention. Such AI systems are appropriate to development of practical systems.

The rules like IF <A> then <B> are formed during the system development. Here <A> is a condition of <B> applicability. <B> is the action which should be made. Such rules can be formed from precedents base during self-learning. Production systems are similar to the classical method of control where control is formed on the information of current system state. Introduction on production systems is presented in [11, 14].

COOPERATION AND CONFLICT

We shall consider the case of a small group of UAV (not more than 10). The centralized structure of control is effective for such a group. More precisely, the commander function is realized by one UAV and other UAVs operate under the centralized control. In the case of the commander failure, another UAV will be designated as the commander. Algorithm of appointment is known to all members of the group. The possibilities of UAV group strongly depend on the presence of communication between members of the group and throughput of communication channels. In the case of total absence of communications between members of the group, the principle of the swarm control is used. In other words, every UAV forms its own control on the base of common aim. Any group member operates avoiding counteraction to other group members. When the communication is available, the group forms the structure of information transmitting network and then returns to the centralized control. The principle of control can be defined by two conditions: firstly, by the mission which should be fulfilled; secondly – by the available communication channels between the members of the group. In general, the commander can set a task without specification. In this case, it is possible to switch to the group distributed control.

Only centralized control is capable of realize accurate control of group maneuvering and other actions. Such high coordination of group operations will realize highly effective actions which, in principle, beyond the capabilities of manned aircraft. High actions' coordination in time and space provides significant perspectives for development of UAV tactic. Such tactic gives undoubted advantages in effectiveness as compared with the groups of aircraft operating under control of a human being.

It should be marked that high-capacity communication channels not only assist to effective group control, but guarantee the transmission of information obtained from radars and television cameras. In particular, such information transmission contributes to effective information processing, what allows the considerable increase of distance and reliability of opponent's detection. This effect is achieved due to complex information processing of all aggregate of obtained information, which is collected by such a distributed sensor system. The joint information processing can be done on the possible powerful cluster computation system consisting of on-board processors of UAV group.

Contemporary trend is the usage for mission fulfillment the group of minimum size or even a single aircraft. Exception maybe is some reconnaissance or search missions. Small size of the UAV group seriously simplifies its coordination. In this case, the communication channels and used methods of communication, protocols of information exchange and the appropriate hardware are determined in advance during the UAVs design. Certain limitations on information exchange are defined by the mission nature. So during the mission fulfillment the sensory information distribution, joint information processing and group coordination should be realized. A good example of practical implementation of multi-agent cooperative system is Soviet and Russian naval anti-ship cruise missiles “Granit” (SS-N-19 Shipwreck) [28, 29].

The books [30, 31] are devoted to principles of multi-agent systems' functioning. Planning for multiple autonomous vehicles in dynamic uncertain environments is introduced in [32]. Results on cooperative control for inexpensive fixed-wing UAVs in remote-sensing operations are presented in [33]. Short review and taxonomy of multi-robot systems is proposed in [34]. In [35] online planning for operation of the unmanned air vehicle group in a hostile environment is presented. The agent approach is used for operation planning.

Let’s consider a conflict. Conflict is the way of complicated system existence. In the most cases conflict is characterized by very large uncertainty in a large number of fundamentally important parameters. In other words, the environment of conflict is ambiguity.

For UAV, we shall consider the narrow interpretation of conflict as counteraction and fight with “unfriendly” UAVs. Scenarios of conflict evolution and possible results of conflict are very various. The tasks associated with conflicts are traditionally referred to the game theory. Serious problem of conflict analysis caused by substantial incompleteness of information. For example, some or all the following parameters are frequently undefined:

- The aims (purposes) of the opponent can be known only very approximately. Frequently the aims of an opponent will be enough clearly revealed only when it is too late to undertake anything.
- The ways to achieve of these aims may be different and in advance are not quite obvious.
- The opportunities (resources) of the opponent are frequently hidden from observation or estimation.

To obtain all the unknown information AI control system should make reasoning from observations of opponent operations. So the opportunity to use or to become a victim of reflexive control appears. In that sense, the adaptability of AI control systems is the synonym of vulnerability to reflexive control, which can lead to catastrophic consequences. Such consequences are possible if AI control system is not provided with effective methods for detection of possible reflexive control from an opponent. However, if the attempt of reflexive control from an opponent was detected, so this immediately gives the advantage since opponent intention is disclosed.

It seems that applications of present-day game theory are appropriate for creation autonomous AI systems only at initial stage of their development. The development of the game theory applied to control tasks in conflict condition with the high degree of ambiguity is required. Another problem is that the ambiguity is interpreted in the game theory in the probabilistic sense. The average risk is estimated. Unfortunately, every conflict is individual and the statistical sample allowing to determine statistical parameters does not exist. For this reason very questionable estimations of average risk are used. Decisions based on such estimations are too cautious and hence ineffective. We can repeat once more that every conflict is individual and the theory of probability here is of little avail. So we need to develop the game theory for non probabilistic description of ambiguity. In [36] introduction in conflict theory is presented. Interesting direction of research is presented in [37] for reflexive agents.

The reflexive control plays a very important role in conflict. Reflexive control is the most complicated kind of activities in conflict. Certain information, in one way or another, is transmitted to the opponent. The way of delivery and content of information are chosen in such a way that to stimulate opponent actions in required for us a direction. Some possible ways of reflexive control are:

- To threaten with damage (real or pretended).
- To seduce with a benefit (real or pretended).
- To form false patterns of own behavior.
- To form false representation about own aims and resources.

Application of reflexive control frequently includes demonstration of different states and methods of behavior. The aim of these demonstrations is to stimulate the opponent to disclose his resources, intentions and aims, ways of actions or to undertake any actions, which allow us to get some advantages. The depletion of opponent resources for the solution of unnecessary tasks can serve as the advantage. Another advantage is the decreasing of opponent solution effectiveness for real tasks by stimulating the opponent to use inadequate algorithm's selection. Sometimes it is useful to demonstrate chaotic actions. The aim of such behavior is imitation of the panic or stimulation of an opponent impression that he does not control on the situation. Therefore, the opponent inclines to stop for some time active actions and wait for clearing of the situation.

In the monograph [38] the analysis of conflict and reflexive control is illustrated in application to radiolocation. Some approaches to development of such systems are shown in this book. Some information about reflexive control methods can be found in [39, 40 and 41]. The reflexive control task is closely interrelated with the adversarial reasoning (understanding the opponent's minds and actions) [42].

Let's look at the conflict between manned aerial vehicle and AI UAV. We should note that there are no data indicating that AI system should be very "intellectual". On the contrary, it is possible to give examples of conflict situations, which shows that very primitive systems (creatures) are capable of realize very effective counteraction to a human being. As such an example can be considered the conflict of biological systems – human and hungry shark or crocodile. A human being in this conflict has very small chances to win in spite of his undoubted intellectual superiority. Similarly, if an AI system is capable of estimate the current situation faster than the human beings, to plan its actions and to realize these actions faster than the human being, then AI system will get the victory.

So it is not necessary to develop very complicated and multi-functional AI system for obtaining the high effectiveness in conflict. It is enough to develop a system, which can effectively solve the very narrow class of tasks. It is quite possible because there are a number of important parameters in which AI systems have the fundamental superiority in possibilities over a human being.

It is known that with decreasing of time assigned for decision-making the quality of human being decisions seriously deteriorates. So in chess game even very primitive chess programs begin stably to win. Conflict interaction allows to impose to the opponent high rate of events, and consequently, the high rate of decision-making. This high rate can be adequate to possibilities of AI system, but substantially exceeds the physiologic possibilities of a human being. Some works on development of AI control systems for aircraft are performed [19].

FEATURES OF ARTIFICIAL INTELLIGENCE CONTROL SYSTEMS ARCHITECTURE

The dynamic flight models are absolutely necessary for simulation of planned actions if these actions and maneuvers are not included into standard knowledge bases of parameterized maneuvers. These models are critically important for planning of own actions and estimation of opponent capabilities for counteraction. For instance, the optimization is needed in the case of conflict, including a necessity to use flight modes, which are on the boundary of controllability, maximum admissible overloads or boundary of the turbojet engine stability. Taking into account all of these models is absolutely necessary for onboard AI control system.

In [43] model-based control and planning were used for the space autonomous system. In [44] environment model was considered for navigation purposes. Elaboration of rules for production expert systems for dynamic system simulation and control is described in [45]. The planning of dynamic object motion in conditions of uncertainty and limitation of failure probability is described in [27].

Control tasks are very integrated with the tasks of information processing [46, 47]. UAV can be equipped with the great variety of external information sensors. It can be radar, synthetic-aperture radar, television cameras, infra-red cameras and so on. The most important task of sensors and information processing systems is providing of situation awareness for control system. Sensor of the most interest is the television cameras. They differ by small size, weight and consuming power. Wide spread of vision among high organized living creatures on the Earth serves as reliable confirmation of high effectiveness of vision systems. At present, technical vision systems are capable of solving the great number of tasks, including angular stabilization (relative horizon line), navigation, self-direction and landing tasks [48].

Pattern recognition tasks are included into navigation and self-direction tasks. Additionally, pattern recognition is included into more general task of provision of information of AI system (situation awareness). In contrast to navigation and self-direction tasks in this case, the specification of classes of objects is needed. This division on classes is dictated by the required different reaction of UAV to these objects. The number of such class is not very large. However strict requirements are imposed on the precision and reliability of recognition in a wide range of aspect angle and distances, in the case of camouflage and imitation using, in a wide range of observation conditions changing. In the case of moving objects detection and tracking, we need to determine not only the distance but also the direction of velocity vector and spatial orientation of these objects. Such information is very important for control in conflict situations.

Pattern recognition tasks are fulfilled with the usage of special data bases, which store the description of specified classes of objects. Every class is presented with some number of types. Technical characteristics and geometric forms for each of these types are presented in such the databases. Initially, the preliminary recognition of aircraft class is carried out, then when the accumulation of information is done, more accurate definition of aircraft type is performed. In particular, both geometric form and dynamic characteristics of aircraft are used for recognition.

For preservation of high degree of awareness, the control system had to carry out the permanent monitoring and analysis of information, which is obtained from the on-board aircraft cameras. Since the aircraft can make active evolutions, then every obtained image has to be referred to the external Earth or internal coordinate system. The class and type of all detected objects are determined permanently. Images of all unrecognized objects are stored for further analysis. All recognized objects are categorized into the number of groups (set of groups): friendly / indifferent / not friendly, dangerous / not dangerous and so on. The checkpoints and targets are marked out. Additionally, the objects which require obeying to determined restrictions during the flight (for instance, building and wires during the flight at small altitude) are selected.

All these objects are tracking in space. As for an aircraft, the following parameters are measured: Its coordinates, velocity vector, spatial orientation and angle of attack. All these parameters are transmitted into the control system which analyses the current situation, makes the prognosis of the situation evolution and estimates possibilities of own disclosure by the opponent aircraft. Signs of disclosure and opponent attack are permanently monitored.

Control system defines the influence of a current situation on the current tasks and makes corresponding corrections of the action plans. Corrections may be very significant up to the cancellation of the whole tasks. It can be in the case when the UAV is attacked and transition to the solution of the defense task is promptly required.

Let's consider the choice of state space for situation description. This multi-level state space should be used for situation description, its analysis and planning of actions. Generally, a situation in high level of abstraction can be described with minimal details. After the selection the plan of actions it is translated to lower level of abstraction. The feasibility of all subtasks is verified. If all subtasks (aims) are reachable using available resources and restrictions are not broken, then transition on the next level down is accomplished and so on. This process is accomplished by synthesis of control programs for aircraft actuators control (aerodynamic control surfaces) and engine control. If on any level some subtasks are unreachable or any important restriction is violated, then planning process returns to the previous level up and the task is corrected. In reality, almost all control actions are the sequences of standard maneuvers in 3D space. These standard maneuvers have many free parameters, which can be tuned and optimized. These maneuvers are interconnected by transition maneuvers, which are necessary for satisfaction of boundary conditions for differential equations. Free parameters are determined by optimization of different aim functions.

Let's consider the methods which can be used for realization of a AI control system.

1. Upper level. The description of a situation is realized in extremely general form. Control is realized in conflict condition. Complicated analysis of a current situation is fulfilled. The analysis includes examination and comparison of different variants of situation evolution with the account of various opponent actions. Comparison of variants includes the account of the information, which can be obtained by opponent in different variants' situation evolution. At this level the game theory methods and

reflexive control can be used. Precedent bases can be used for variant of action creation. Selected variants of actions are tested on models. These models can be simple enough.

2. Middle level. More detailed definition of the variants defined at the upper level is fulfilled with the account of restrictions on dynamic, fuel and other resource. Similarly, specification of selected variants is performed with the account of information obtained by opponent. Production systems can be used on this level. Precedent bases can be used for specification and refinement. More precise and complex models are used for variant's testing.
3. Lower level. This level is more specified then a middle level. More detailed models are used. More constraints are taken into consideration. On this level different form of knowledge and data representations are used. Frames, semantic networks, scripts can be also used on this level. Precedent reasoning and planning can be used in this level too. Non-standard maneuvers can be synthesized on this level.
4. Lowermost (actuator) level. The sequences of maneuvers are formed at this level. The maneuvers are jointed by transitions. For the solution of the concrete task, the optimization of free parameters of these maneuvers is performed. Transitions guarantee smooth passing between maneuvers and are determined by allowable restrictions on overloads, angle of attack, sideslip angle, engine revolutions and so on. Algorithms which appropriate to these maneuvers are stored in the different libraries. If libraries off-the-shelf algorithms for any parameter are unsatisfactory new algorithms can be extracted from precedent bases and tuned to the specific task on this level.

A few words would be said about situational control. Situational control is very similar to control of the dynamic system in the phase space. For this reason, this method should be well-suited to UAV's control. There is the good theoretic foundation of situational control. One of the first applications of this method was control of real seaport operations. Unfortunately almost all books, which are devoted to this method, are in Russian. Some information about situational control can be obtained from [49, 50]. Applications of situational control are shown in [51, 52]. In [53] situational control is used for stabilization and control of the turbo-compressor engine.

One of the possible control system architecture is presented in [54]. This system includes four reasoning modules: Deductive, Inductive, Case-based and Associative. The very interesting 4D/RCS architecture of AI control systems is presented in [55, 56]. This architecture seems to be efficient for mobile vehicles of different classes.

CONCLUSION

Some AI methods which from the author's point of view can be promising for development of fully autonomous UAV are listed. All these methods can be presented as expansion of classical control theory. Such fully autonomous UAVs can be comparable in effectiveness with aerial vehicles under control of the pilot in the unpredictable varying and hostile environments without remote control in any form. The program complex which capable of realizing such AI control system is extraordinary complicated. Nevertheless, it seems that there is a sequence of stages, carefully combined with each other and allowing realizing such a project consistently, step by step. Each stage should get practically applicable results and justify the investment. This project should be realized as the superstructure over the classical aircraft control system.

ACKNOWLEDGMENTS

This work was partly supported by the Russian Foundation for Basic Research, project no. 09-01-00573. I want to express my sincere gratitude to Valery V. Afonin, for his help in improving my English in this paper. I am very grateful to the reviewers for their information about the 4D/RCS architecture.

REFERENCES

[1] Murphy, R. R. *Introduction to AI Robotics*; The MIT Press: Cambridge, MA, 2000.

[2] Laird, J. E.; Jones, R. M. *Building Advanced Autonomous AI Systems for Large Scale Real Time Simulations*; Proceedings of the Computer Game Developers Conference, Long Beach, CA, 1998; pp 365-378.

[3] Stengel, R. F. *Toward Intelligent Flight Control;* IEEE Trans. Systems, Man and Cybernetics. 1993, 23(6), 1699-1717.

[4] Kopp, C. *The UCAV Ascendancy: What are the Problem Issues?*; Proceedings of the UAV Australia Conference and Exhibition, Melbourne, Australia, 2001; pp 1-17.

[5] *Agents and Computational Autonomy: Potential, Risks, and Solutions*; Nickles, M.; Rovatsos, M.; Weiss G.; Eds.; Series: Lecture Notes in Artificial Intelligence 2969; Subseries: Lecture Notes in Computer Science; Springer-Verlag: Berlin, 2004.

[6] Hawkins, J.; Blakeslee, S. *On Intelligence*; *Times Books; Henry Holt and Company: New York, NY, 2004.*

[7] Zhdanov, A. A. *Autonomous artificial intelligence*; Series: Adaptive and Intellectual systems; Binom. Laboratory of knowledge: Moscow, Russia, 2008. (in Russian).

[8] Bowen, D. G.; MacKenzie, S. C. (2003). Autonomous Collaborative Unmanned Vehicles: Technological Drivers and Constraints. http://cradpdf.drdc-rddc.gc.ca/PDFS/unc105/p519508_A1b.pdf

[9] Grishin, V. A. *Roll and Yaw Stabilization of Hypersonic Vehicle;* Transactions on Scientific Session MEPHI-2004, MEPHI (Moscow State Engineering Physics Institute), Moscow, 2004; Vol. 12, pp 16-17. (in Russian).

[10] *50 Years of Artificial Intelligence. Essays Dedicated to the 50th Anniversary Artificial Intelligence*; Lungarella, M.; Iida, F.; Bongard, J.; Pfeifer R.; Eds.; Series: Lecture Notes in Artificial Intelligence 4850; Subseries: Lecture Notes in Computer Science; Springer-Verlag: Berlin, 2007.

[11] Luger, G. F. *Artificial Intelligence. Structures and Strategies for Complex Problem Solving*; 4th Edition; Addison Wesley: London, 2002.

[12] Jones M. T.; *Artificial Intelligence. A Systems Approach*; Infinity Science Press LLC: Hingham, MA, 2008.

[13] Russell, S. J., Norvig, P. *Artificial Intelligence. A modern Approach*; Second Edition; Prentice Hall, Pearson Education Inc.: Upper Saddle River, NJ, 2003.

[14] Coppin, B. *Artificial Intelligence Illuminated*; Jones and Bartlett Publishers, Inc.: Sudbury, MA, 2004.

[15] *Expert Systems in Engineering: Principles and Applications;* Gottlob, G.; Nejdl, W.; Eds.; Series: Lecture Notes in Artificial Intelligence; Subseries: Lecture Notes in Computer Science 462; Springer-Verlag: Berlin, Heidelberg, 1990.

[16] Hopgood, A. A. *Intelligent Systems for Engineers and Scientists*. Second Edition; CRC Press LLC: Boca Raton, FL, 2001.

[17] Vagin, V. N.; Golovina, E. Yu.; Zagoryanskaya, A. A.; Fomina M.V. *Exact and Plausible Reasoning in Intelligent Systems*; Fizmatlit: Moscow, 2004. (in Russian).

[18] *Embodied Artificial Intelligence*; Iida, F.; Pfeifer, R.; Steels, L.; Kuniyoshi, Y.; Eds.; International Seminar, Dagstuhl Castle, Germany, July 7-11, 2003. Revised Selected Papers; Series: Lecture Notes in Artificial Intelligence 3139; Subseries: Lecture Notes in Computer Science; Springer-Verlag: Berlin, 2004.

[19] Solinger, D. (2005). *Creating a Dogfight Agent: The design and implementation of a fully autonomous agent that can fly an airplane during a one-to-one dogfight*; Technical Report DKS-05-01 / ICE 10. Version 1.0, April 2005. Mediamatics Data and Knowledge Systems group. http://www.kbs.twi.tudelft.nl/docs/MSc/2005/Solinger_David/thesis.pdf

[20] Kolodner, J. L. *An Introduction to Case-Based Reasoning*; Artificial Intelligence Review. 1992, 6, 3-34.

[21] Aamodt, A.; Plaza, E. *Case-Based Reasoning: Foundational Issues, Methodological Variations, and System Approaches;* AI Communications. 1994, 7, issue 1, 39-59.

[22] Bao, D. (2008). A Survey on Similarity-based Reasoning. http://www4.ncsu.edu/~dbao/Analogy%20Survey_final.pdf

[23] Marling, C.; Sqalli, M.; Rissland, E.; Muñoz-Avila, H.; Aha D. *Case-Based Reasoning Integrations*; AI Magazine. 2002, 23, no. 1, 69-86.

[24] Pal, S. K.; Shiu, S. C. K. *Foundations of Soft Case-Based Reasoning*; John Wiley and Sons, Inc.: Hoboken, NJ, 2004.

[25] Mackworth, A. K. *Constraint-Based Design of Embedded Intelligent Systems*; International Journal "Constraints". 1997, 2, no. 1, 83–86.

[26] Zhang, Y.; Fromherz, M. P. J.; Crawford, L. S.; Shang, Y. *A General Constraint-Based Control Framework with Examples in Modular Self-Reconfigurable Robots*; Proceedings of IEEE/RSJ Conference on Intelligent Robots and Systems (IROS 2002), Lausanne, Switzerland, 2002; Vol. 3, pp 2163-2168.

[27] Ono, M.; Williams, B. C. *An Efficient Motion Planning Algorithm for Stochastic Dynamic Systems with Constraints on Probability of Failure*; Proceedings of the 23rd AAAI Conference on Artificial Intelligence (AAAI'08), 2008; Vol. 3, pp 1376-1382.

[28] Richardson, D. (2001). "Shipwreck" surfaces after 20 years. Jane's Missiles and Rockets. Anti-Ship and ASW. http://knygos.sprogmenys.net/knygos-2/Rocketry/Missiles%20and%20Warheads/Jane%27s%20Missiles%20and%20Rockets.pdf

[29] Russian/Soviet Sea-Based Anti-Ship Missiles. (2005). Defense Threat Informations Group. Special Report. http://www.dtig.org/docs/Russian-Soviet%20Naval%20 Missiles.pdf

[30] Vlassis, N. *A Concise Introduction to Multiagent Systems and Distributed Artificial Intelligence*; Series: Synthesis Lectures on Artificial Intelligence and Machine Learning; Morgan and Claypool Publishers: 2007.

[31] Liu, J.; Wu, J. *Multi-Agent Robotic Systems;* Series: International series on computational intelligence; CRC Press LLC: Boca Raton, FL, 2001.

[32] Pongpunwattana, A.; Rysdyk, R. *Real-Time Planning for Multiple Autonomous Vehicles in Dynamic Uncertain Environment;* Journal of Aerospace Computing, Information, and Communication. 2005, 1, pp 580-604.

[33] Ryan, A.; Zennaro, M.; Howell, A.; Sengupta, R.; Hedrick, J. K. *An Overview of Emerging Results in Cooperative UAV Control;* Proceedings of the 43rd IEEE Conference on Decision and Control, 2004; Vol. 1, pp 602-607.

[34] Iocchi, L.; Nardi, D.; Salerno, M. *Reactivity and Deliberation: A survey on Multi-Robot Systems;* In Balancing Reactivity and Social Deliberation in Multi-Agent Systems: From RoboCup to Real-World Applications; Hannebauer, M.; Wendler, J.; Pagello, E.; Eds.; Series: Lecture Notes in Computer Science / Lecture Notes in Artificial Intelligence; Springer-Verlag: London, UK, 2001; Vol. 2103, pp 9-34.

[35] Baltié, J. and others. *Multi-Vehicle Mission: Architecture and Algorithms for Distributed Online Planning*; In Artificial Intelligence for Advanced Problem Solving Techniques; Vrakas, D.; Vlahavas, I. PL.; Ed.; Series: Premier Reference Source; Information Science Reference: Hershey, PA, 2008; pp 1-22.

[36] Druzhinin, V. V.; Kontorov, D. S.; Kontorov, M. D. *Introduction in conflict theory*; Radio and Communications: Moscow, Russia, 1989. (in Russian).

[37] Lefebvre, V. *Sketch of Reflexive Game Theory;* Proceedings of the Workshop on Multi-Reflexive Models of Agent Behavior, Los Alamos, NM, 1999; pp 1-40.

[38] Druzhinin, V. V.; Kontorov, D. S.; *Conflict radiolocation*; Radio and Communications: Moscow, Russia, 1982. (in Russian).

[39] Thomas, T. L. *Reflexive Control in Russia: Theory and Military Applications*; Reflexive Processes and Control. International Interdisciplinary Scientific and Practical Journal. 2002, 1, no. 2, 60-76.

[40] Shemayev, V. N. *Cognitive Approach to Modeling Reflexive Control in Socio-Economic Systems*; Information and Security. 2007, 22, 28–37.

[41] Thomas, T. L.; *Russia's Reflexive Control Theory and the Military;* Journal of Slavic Military Studies. 2004, 17, issue 2, pp 237–256.

[42] *Adversarial Reasoning: Computational Approaches to Reading the Opponent's Mind;* Kott, A., McEneaney, W. M.; Ed.; Series: Computer And Information Science; Chapman and Hall/CRC, Taylor and Francis Group: Boca Raton, FL, 2007.

[43] Knight, R.; Chien, S.; Gat, E.; Starbird, T., Gostelow, K.; Keller, B.; Smith, W. *Integrating Model-based Artificial Intelligence Planning with Procedural Elaboration for Onboard Spacecraft Autonomy;* Proceedings of the International Conference on Space Operations (SpaceOps 2000), Toulouse, France, 2000. http://trs-new.jpl.nasa.gov/dspace/bitstream/2014/14438/1/00-0875.pdf

[44] Tani, J. *Model-Based Learning for Mobile Robot Navigation from the Dynamical Systems Perspective*; IEEE Trans. on Systems, Man and Cybernetics Part B: Cybernetics. 1996, 26, no. 3, 421-436.

[45] Aksoy, M. S. *Dynamic System Modelling Using Rules3 Induction Algorithm*; Mathematical and Computational Applications. 2005, 10, no. 1, 121-132.

[46] Patrick, C. H. (2006) *The Generation of Situational Awareness within Autonomous Systems – A Near to Mid term Study – Analysis*; Report No.: DSTO-TR-1896, DSTO Information Sciences Laboratory; Australia. http://dspace.dsto.defence.gov.au/dspace/bitstream/1947/4561/1/DSTO-TR-1896.PR.pdf

[47] Ollero, A.; Merino, L. *Control and perception techniques for aerial robotics*; Annual Reviews in Control. 2004, 28, 167–178.

[48] Grishin, V. *Computer Vision and Artificial Intelligence in Flight Control of Unmanned Aerial Vehicles;* Journal of Information, Intelligence and Knowledge (JIIK). Vol. 2, Issue 4, pp 327-338.

[49] Osipov G. S. (2000). *Origines of Applied Semiotics*; Proceedings of the 14th European Conference on Artificial Intelligence (ECAI 2000); Workshop "Applied Semiotics: Control Problems (ASC2000)". http://cool.iitp.ru/asc2000/ps/01_ORIGI.PS

[50] Gudwin, R.; Queiroz, J. *Towards an introduction to computational semiotics.* Proceedings of the International Conference on Integration of Knowledge Intensive Multi-Agent Systems, 2005; pp 393-398.

[51] Andoga, R.; Madarász, L.; Giertl, J. *Overview of New Scopes and Approaches in Situational Control of Complex Systems*; Proceedings of the 4th Slovakian-Hungarian Joint Symposium on Applied Intelligence (SAMI 2006), 2006; pp 126-133.

[52] Madarász, L.; Andoga, R.; Lazar, T. *Development, Perspectives and Application Possibilities of Situational Control of Complexes*; Proceedings of the Budapest Polytechnic Jubilee Conference, Science in Engineering, Economics and Education, 2004; pp 411-423.

[53] Andoga, R.; Madarász, L.; Karas, M. *The Proposal of Use of Hybrid Systems in Situational Control of Jet Turbo-compressor Engines*; Proceedings of the 3rd Slovakian-Hungarian Joint Symposium on Applied Machine Intelligence (SAMI 2005), 2005; pp 93-104.

[54] Nikitenko, A. *Knowledge-based Robot Control;* International Journal "Information Theories and Applications". 2008, 15, pp 126-137.

[55] Albus, J. S., et al. *4D/RCS Version 2.0: A Reference Model Architecture for Unmanned Vehicle Systems*, Internal Report (NISTIR) - 6910, National Institute of Standards and Technology, Gaithersburg, MD, 2002.

[56] *Intelligent Vehicle Systems: A 4D/RCS Approach*; Madhavan, R.; Messina, E. R.; Albus J. S.; Eds.; Nova Science Publishers Inc.: New York, 2006.

In: Knowledge Management and Communication … ISBN: 978-1-62808-250-0
Editor: Rafiqul Islam

Chapter 23

A Progressive Study on Integrated Patch Antennas

***Rakesh Kumar Yadav*[*1], *R. L. Yadava*[1], *and Yogesh Gupta*[2]**
Department of Electronics and Communication Engineering,
[1]Galgotias College of Engineering and Technology, Greater Noida, U. P. India
[2]KCC Institute of Technology and Management, Greater Noida, U. P. India

Abstract

This paper provides an over view of design, development and application of active integrated antennas. Active integrated antennas (AIA's) provide a new paradigm for designing modern microwave and millimeter-wave architecture with desirable features such as compactness, low cost, light weight, low profile, minimum power consumption, and multiple functionality. This paper reviews recent research related to this emerging technology with emphasis on its applications in high-efficiency radio-frequency (RF) front-end, filter, and phase arrays, as well as switching antenna. Optical controlling techniques for AIA's are also described.

Keywords: Active Integrated Antennas, Active Devices, Phased Array, Injection-locked oscillator, RF front-end Amplifier

Introduction

The Active Integrated Antenna means that the active devices and the passive antenna elements are integrated on the same substrate to enhance the performance of antenna. According to a microwave engineer an AIA can be assumed as active microwave circuit in which input or output port is free space rather than a conventional 50-Ω interface. Here antenna can provide some circuit function such as Duplexing, Resonating and Filtering inspite of its basic characteristics as a radiating element. While from antenna designer's point

[*] E-mail: er.rakeshyadava@gmail.com.

of view, the AIA is an antenna that possesses built-in signal- and wave processing capabilities like mixing and amplification.

The concept of using Active Antenna can be traced back to as early as 1928 [1]. An antenna of small size with an electron tube was used in radio broadcast receivers around 1 MHz at that time. The research on active antenna received much more attention and several pioneering works were reported [2]–[11] in the 1960s and 1970s when high frequency transistors were invented. There are several advantages like bandwidth enhancement, decrement in mutual coupling between array elements and improvement in noise factor when we integrate an active device on passive radiating elements [12]. As the of operation increases, the available power from the solid state devices decreases. So, power combining of solid state devices using quasi optical techniques in millimeter–wave region became an important issue [13]. There are two methods of quasi-optical power combining namely active antenna approach and grid approach. The coupled oscillator array to combine power spatially is an example of active integrated antenna approach while the other approach, the grid array is very useful one for high power applications [14]-[21].

The classification of active integrated antenna is based on their application and named as their counterparts in radio systems. Two basic categories are transmitter and receiving types AIAs. The others types with both functions of transmitting and receiving are transceivers, repeaters and so on. All these have a common feature: the integration of an amplifier and antenna elements. The other classification is based on the functions of active devices they integrate, according to which three basic groups, namely, the oscillator type [22]-[32], the amplifiers type [33]-[38] and the frequency conversion type [39]-[47].

Recently, numerous innovative designs based on the AIA's concept have been proposed and successively demonstrated. AIA technology has evolved to a point where practical implementation for use in the latest microwave and millimeter-wave system is considered feasible. It is currently pursued in a number of related fields such as power combining, beam steering and switching, retro directive arrays, as well as high- efficiency power-amplifier designs. These AIA-based designs are particularly attractive for millimeter-wave systems because they provide an effective solution to several fundamental problems at these frequencies, including higher transmission-line loss, limited source power, reduced antenna efficiency, and lack of high-performance phase shifters.

This paper reviews the latest research related to this technology with emphasis on its application in RF front end, filters, phase array, switching and optical antennas.

RF FRONT END AMPLIFIER

The RF front-end section consists of a large number of components assembled on a printed circuit board. It also offers a good value for both the manufacturer and the consumer with respect to price and complexity. A brief progress in the research of AIAs technology emphasizing its applications in RF FRONT END amplifier is presented in this section.

In 1997, T. Seki proposed a three dimensional active antenna configuration for a RF front end module of a high speed wireless communication system at millimeter and quasi millimeter wave frequencies. The active antenna is integrated with a MSA array on multilayer alumina ceramic and the individual MMIC chip including transmit –receive amplifier, time division duplex switches and filter on multilayer polyimide. The MSA with parasitic element for the fundamental part of the active antenna at 20 GHz band including the via patch construction is analyzed. It was also found that when the slot width around the 1mm x 1mm via patch was 300μm, the good radiation pattern and impedance matching condition were achieved [48].

However, the evolution from single band cellular phones to today's standard dual band or triple band phone engines have led to a quite complex antenna interface. For every additional RF band implemented in the phones an additional frontend preselector filter in the receiver section is required. It has been demonstrated that LTCC ceramic using non shrinkage technology is perfectly suited to fulfill the requirement of increased functionality and higher integration level. In the first level of integration the switching between different band and modes were comprised in antenna switch module (ASM). In the next step a front end module includes the surface acoustic wave filter and dozens of matching elements. LTCC is a low loss, high precision substrate allowing integrating a large number of passive devices in small volume. The typical performance requirements on the frontend of a GSM multiband terminal are summarized in Table II [49], for both Tx and Rx sections. FEMs and Duplexers based on LTCC will pick a significant share and play an important role for the RF front end of mobile phones for the next few years.

In year 2003, Y. Chug proposed a high efficiency and compact AlGaN/GaN hetrojunction field effect transistor power amplifier integrated with a microstrip antenna at 7.25 GHz for RF frond end application. In which a microstrip circular sector antenna is employed as both a radiator and frequency dependent output load. In this the measured antenna impedance is directly transformed to the optimum load impedance for maximum efficiency. The power amplifier having 1mm gate periphery shows 42% peak PAE and 30.3 dBm saturated output power with a linear gain of 8 dB. It is also found that more than 30 dB harmonic suppression has been achieved at both the 2nd and 3rd harmonic frequency in both the H and E plane (Figures 1, 2). It can successfully used to produce high efficiency power amplifier required for phase array radar and wireless communication systems.

However, W. M. Proposed a new FMCW front end module for the purpose of next generation automotive radar sensors. It has one transmit and 8 receive channel and operate between a multi beam antenna and an advanced signal processing unit. For parallel operation of all receive channels, cascaded buffer amplifier is used.

However, N. Khandelwal proposed a system in package which have cost and size benefit and with integrated antenna can help mitigate the feed network losses and phase error in antenna array application. Here a printed antenna is integrated with a multilayer BGA package using low cost laminate substrate. The measured and estimated gain in both the modes is comparable, thus showing that packaging does not affect the IC performance drastically (Figures 3, 4 and 5). By using central short package resonances are suppressed as well as isolation is improved [50].

In year 2005, H. Kim proposed a structure to obtain both high PAE and a compact RF front end. For this the direct integration approach between the power amplifier and an antenna can be achieved without any impedance transformer. It is found that the proposed transmitter front end provides high PAE of 67.5% and compact and integrated RF front structure by the size reduction of 43% compared with the conventional class F antenna for high PAE (Figs 10, 11, [51]).

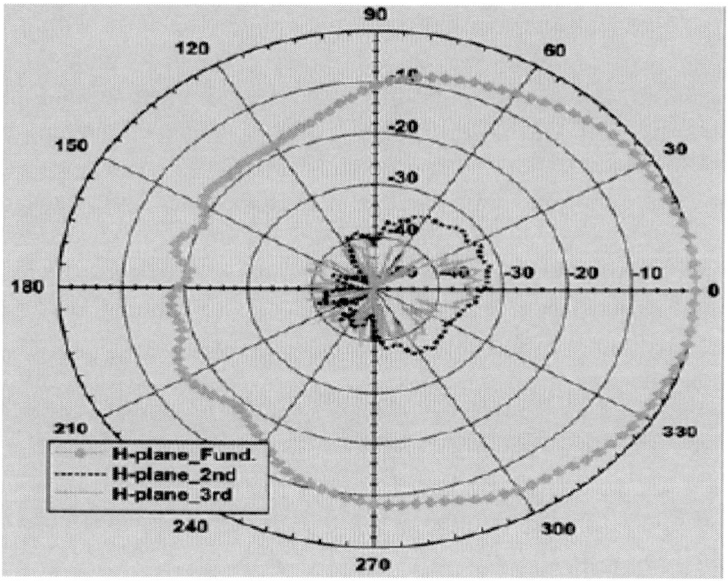

Figure 1. Measured H-plane radiation patterns at the fundamental of 7.25-GHz, second, and third harmonic frequencies for the AlGaN/GaN HFET power amplifier integrated with the microstrip circular sector antenna.

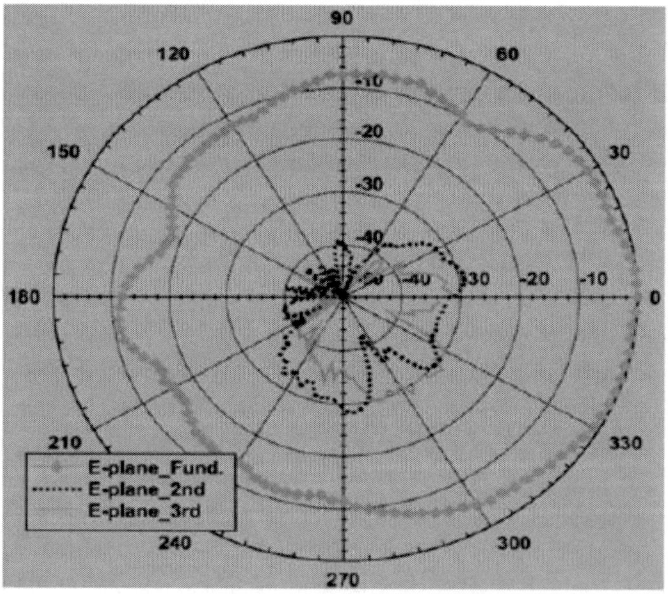

Figure 2. Measured E-plane radiation patterns at the fundamental of 7.25-GHz, second, and third harmonic frequencies for AlGaN/GaN HFET power amplifier integrated with the microstrip circular sector antenna.

A high performance synthesizer provides programmable frequency ramped waveform with high stability and linearity. All active millimeter wave function are realized with GaAs MMIC assembled in a hermitically sealed microstrip module which is incorporated into 3 Layer PCB stack on which IF amplification, frequency control and D.C. bias control is realized.

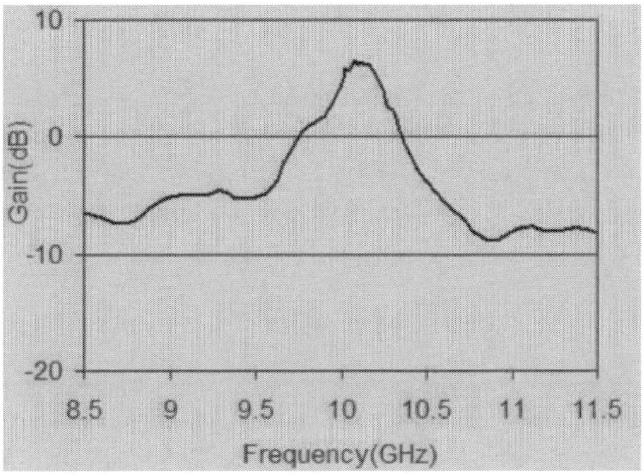

Figure 3. Gain vs. frequency for the passive antenna.

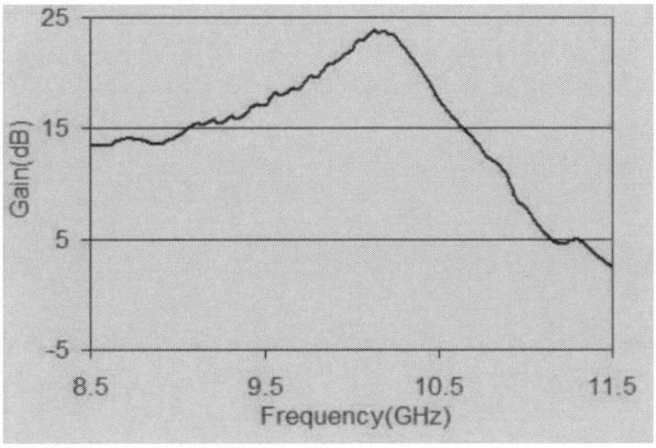

Figure 4. Gain vs. frequency for receive mode.

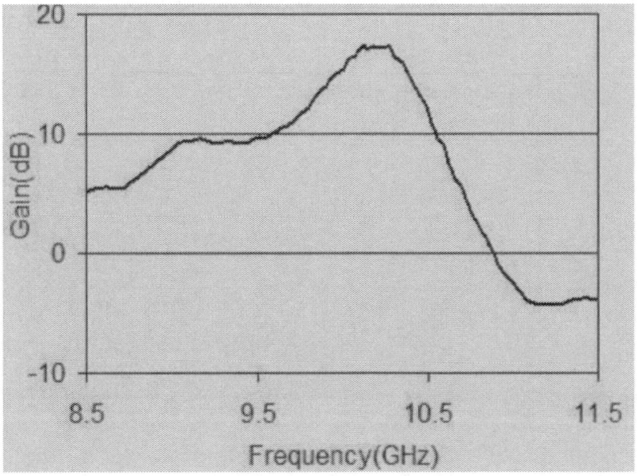

Figure 5. Gain vs. frequency for transmit mode.

Only a small selection of the most important key properties of the module is listed in Table 1 addressing transmits power and receiver sensitivity properties, as well as synthesizer performance data.

In year 2006, Y. Qin proposed the design of broad band high efficient CP active antenna as well as broad band high efficient CP active array antenna. The antenna is used to serve as radiator as well as a harmonic terminated load for class –E high efficiency PA integration. This active antenna achieves a 9 % bandwidth (1.84-2.01 GHz) for axial ratio below 3 dB and a 12 % bandwidth for PAE over 60%.

Table I. Typical Performance data of the front end module

Parameter	*Condition*	*Value*
Operating Frequency Range	taking into account spread, aging and environmental conditions	76.1..76.9 GHz
DC-Power Consumption	TX-off	5.1 W
	TX-on	10.2 W
Transmit RF Power		13.7 dBm
Phase Noise	$\Delta f = 10$ kHz	-77 dBc / Hz
	$\Delta f = 100$ kHz	-82 dBc / Hz
	$\Delta f = 200$ kHz	-83 dBc / Hz
RX Conversion Gain	$f_{\mathrm{if}} = 500$ kHz	23 dB
RX Noise Figure	$f_{\mathrm{if}} = 10$ kHz	$<$ 24 dB
	$f_{\mathrm{if}} = 100$ kHz	$<$ 18 dB
	$f_{\mathrm{if}} = 1$ MHz	$<$ 16 dB
RX Channel Isolation	adjacent channels	27 dB
	non adjacent channels	44 dB
Frequency Linearity	frequency deviations $\leq$ 400 MHz and slopes $<$ 2000 GHz / s	$\leq$ 0.5%
Settling/Retrace time	frequency deviation $\leq$ 400 MHz	$<$ 10 μs
Explanations:	Δf = offset frequency from carrier f_{if} = baseband IF output signal frequency (any of the eight channels)	

In addition, to form an active array, four antenna elements each integrated with broadside class-E PA are sequentially rotated. At 1.95 GHz the peak drain efficiency of 71.5% for class-E amplifier is measured. The active array has a peak effective radiated power of 39.7 dBm,

AR is below 3 dB over 27 % BW (1.72-2.26 GHz) and PAE is over 50% within a 22.6% BW (1.72-2.16 GHz) [52].

However, a novel wideband method to attain increased PAE bandwidth and transmitting power bandwidth of the fully integrated transmitter front end amplifier is described in [**53**]. Direct integration between power transistor output and antenna input can be achieved without any impedance transformer. The measured PAE bandwidth is 930 MHz (4.9-5.83 GHz) while transmitting power bandwidth is 1030 MHz (4.8-5.83 GHz) under the condition of PAE over 50% and transmitting power within 3 dB for the peak transmitting power. It provides excellent size reduction. The 2nd and 3rd harmonic radiations for the normalized peak power of the fundamental frequency are measured less than -30 and -40 dB in all direction respectively (Figures 6 and 7).

The 10% area of wireless communication devices are occupied by ICs while 90% by RF front end model. These powerful ICs enable effective implementation of MIMO digital processing but the problem is implementation of MIMO RF front end module in small devices such as handset still remains unsolved. Without solving this network provider is not able to deploy next generation wireless broadband network that can sustain tens of Mbps throughput with mobility. So an LTE MIMO air interface solution for handsets using metamaterial design, which offer small, low cost and low profile antenna printed directly on PCB for easy integration and manufacturing. The proposed antenna has centre frequency at 771 MHz of the LTE band 746-796 MHz. They also measure the different parameter of proposed antenna in two different configuration based on physical proximity and current distribution on the ground plane for the MIMO application in the LTE standard.

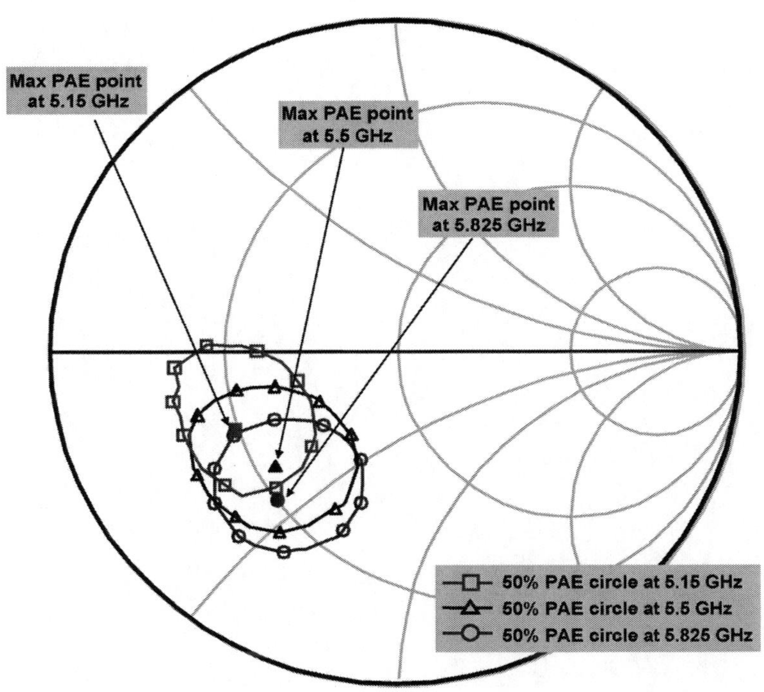

Figure 6. Load–pull simulation for PAE characteristics of the power transistor at each frequency.

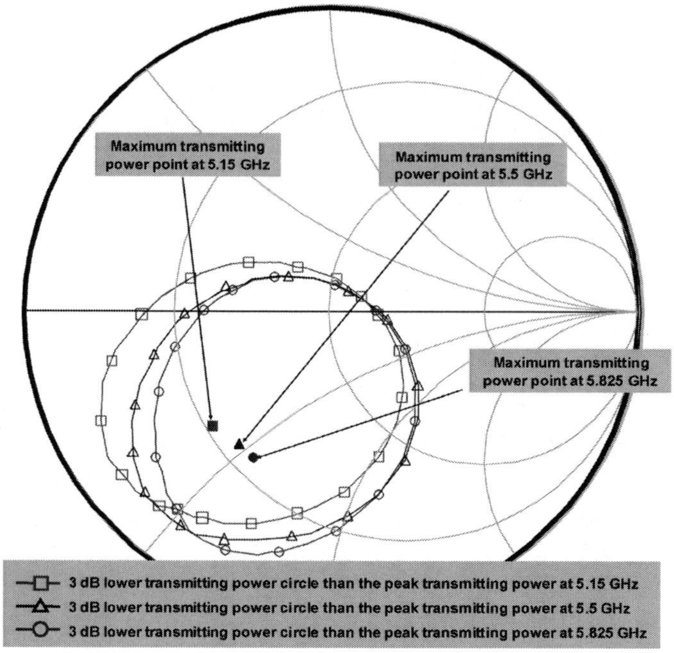

Figure 7. Load–pull simulation for transmitting power characteristics of the power transistor at each frequency.

Figure 8 shows the measured return loss and isolation for configuration 1 where both antennas are located at two corners on the short side of the PCB while Figure 9 shows the measured return loss and isolation for configuration 2 where the antennas are placed at the two corners on the long side of the PCB and oriented at 900 relative to each other. The physical separation between the antennas is 31mm (λ0/13) for configuration 1 and is 79mm (λ0/5) for configuration 2 [54].

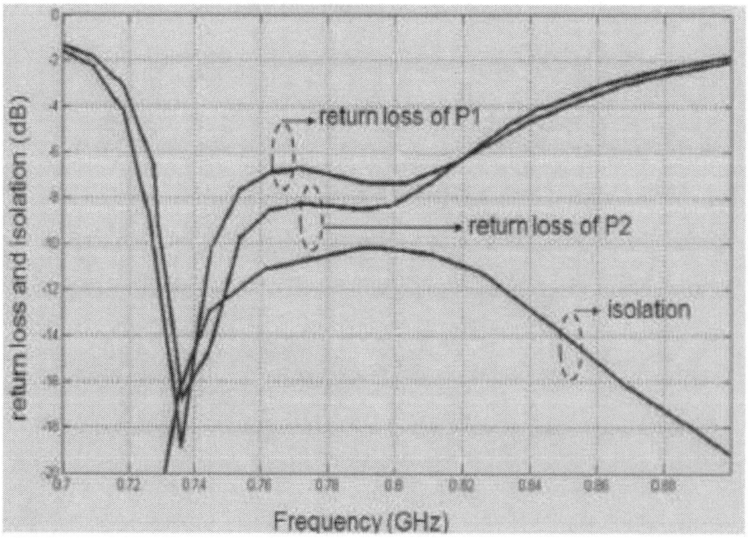

Figure 8. Measured return loss and isolation of Antenna array in configuration 1.

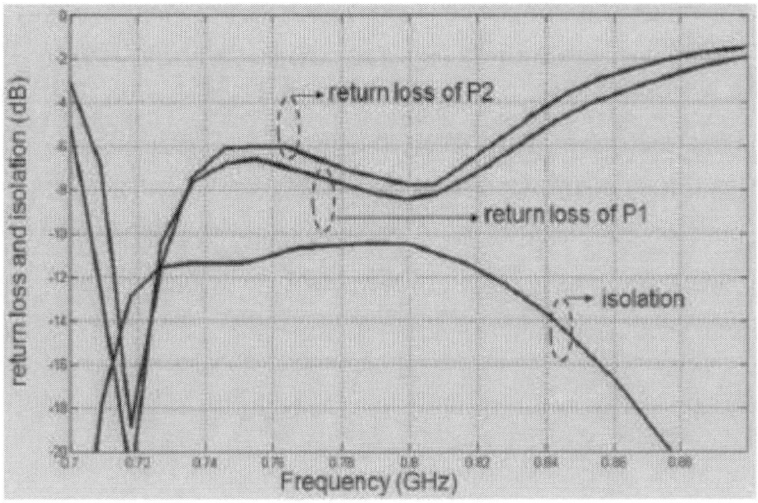

Figure 9. Measured return loss and isolation of Antenna array in conFigure 2.

FILTER

All AIAs have a common feature such as the integration of an amplifier/oscillator and antenna elements. In terms of technological applications, high frequency filters based on active devices are widely used in many microwave and millimeter-wave systems. The available literatures on the active antenna indicate that several useful results have been found. B.K.Kuanr proposed a novel band pass filter using nanoparticles as the active elements in microstrip geometry. The microstrip structures showing different layers and FCL structures are shown in Figure 10.

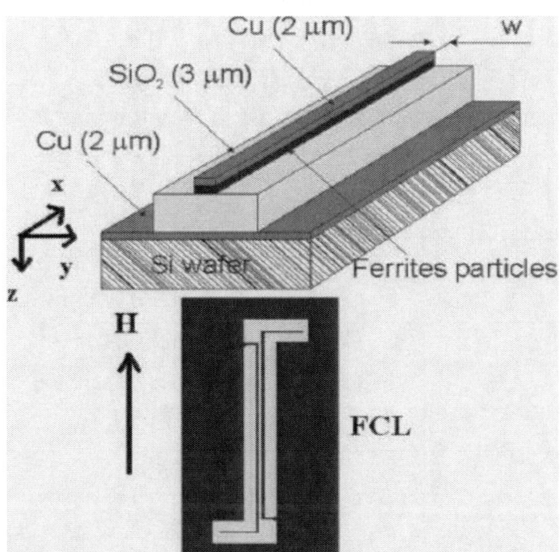

Figure 10. (a) (Top) Microstrip structures showing the different layers with their thicknesses. (b) (Bottom) FCL structure showing the two antennas separated by 5 μm.

This band pass filter shows tunability with the external field. Since this is microstrip based technique so these filter can be easily integrated with other MIC components in different planner microwave system. It was found that quality factor Qext= fc/Δf, increase from 4.5 to7.5 as the centre frequency increases from 3 to 13 GHz and bandwidth of the device is almost constant (1.2-1.6 GHz) over the field range.

However, K. Jin Koh and G. M. Rebeiz proposed a 0.18 μm SiGe BiCMOS technology based 5 bit 6-18 GHz active phase shifter. The performance is summarized in Table II. This proposed structure is suitable for high resolution on-chip phased arrays with multiple elements. Whereas, S.M. Han proposed a rectenna system (convert RF power to DC power) which is very useful for point to point RF power transmission and the space power satellite.

The efficient power conversion in rectenna can be achieved either by exciting high microwave power into rectifying diodes or by increasing the rectifying efficiency of rectenna. They measured the efficiency of rectenna for various resistive loads varying from 125Ω to 200Ω and found 56.7% with a 150 Ω load when incident power is 10 dBm. It has less complexity due to use of a circular sector antenna that has a built in harmonic rejection property.

However, R. Deshbashi proposed an AIA based rectenna whose block diagram is shown in Figure 11 and Figure 12. The band pass filter rejects the harmonics of the fundamental frequency produced by non linear diode to eliminate radiation of them and thus system efficiency increases. However, a trade-off exist between antenna size and the radiation gain but they found that by using C- shaped U-slot coaxial fed microstrip antenna size is compact and radiation gain is high. This rectenna is designed at 2.4 GHz and found that size is reduced by 32% while gain is increased from 2.677 dBi to 6.96 dBi. It is also observed from Figure 13 that the return loss of new antenna is very high at 2_{nd} and 3_{rd} harmonics.

Table II. Summary of the Phase Shifter Performance

Technology	0. 1 8 - μ m S i G e B i C M O S (Jazz SiGe12D, 1P6M)
Frequency band	6 - 1 8 G H z
Supply voltage	3.3 V
Current consumption	18.7 mA (phase shifter core: 3.2 rnA)
Chip area	1 .2x0.75 mm2 (phase shifter core: 0.45x0.35 mm2)
Input return loss	≤s -10 dB @ 11.2-15.2 GHz
Output return loss	≤s -12 dB @ 6-18 GHz
Power gain (ave)	16.5-19.5 dB @ 7.5-15.2 GHz
Phase resolution	5-bit
Gain error	< 1.1 dB (rms) @ 6-18 GHz
Phase error	< 5.60 (rms) @ 6-18 GHz
NF	4-5.7 dB @ 7.5-15.2 GHz
Isolation(output to input)	< -38 dB@6-18 GHz

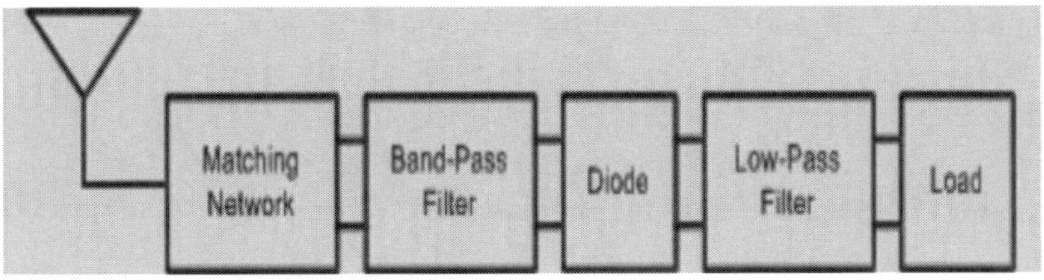

Figure 11. Configuration of a conventional Rectenna.

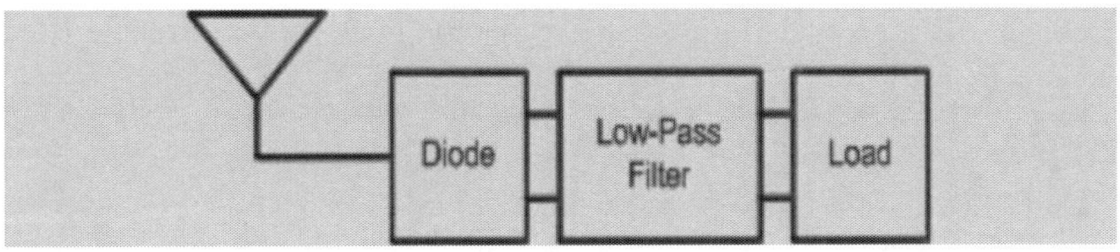

Figure 12. Configuration of the proposed AIA based Rectenna.

Fundamental 2nd 3rd Harmonics harmonics

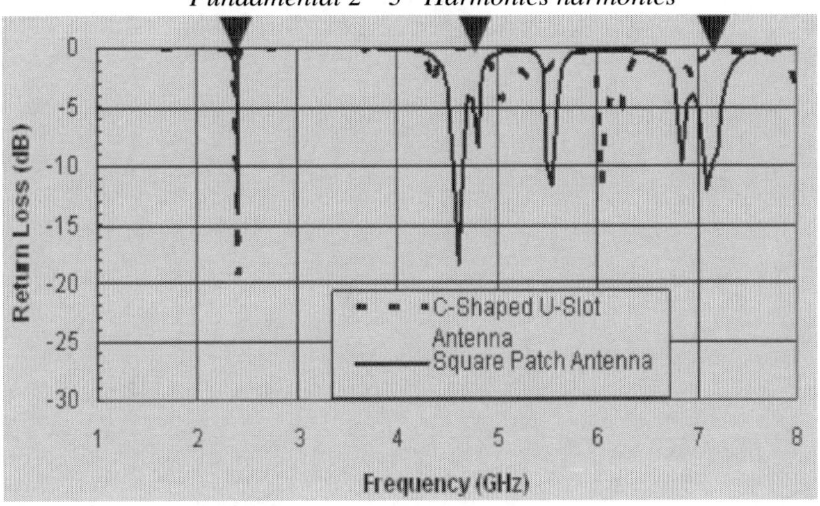

Figure 13. Return loss of the c-shaped u-slot and square patch antennas.

R. Malmqvist *et.al*, proposed a RF tunable front end filters and LNAs for wideband phase array using 0.25 μm ultra thin silicon CMOS technology. They found that a switched passive X-band filter as well as a tunable active c-band filter can achieve reasonable RF performances when implemented with this technology with high- Q passives. Figure 14 shows that filter with fixed frequency have minimum loss of 4 dB at 7.9 GHz while for switched filter loss is 7.7 dB and 8.1 dB at 7.9 GHz and 10.4 GHz respectively.

They also proposed wideband as well as tunable band pass LNA for multifunctional RF systems using a SiGe BiCMOS technology. The response are plotted in Figure 15 shows that the measured LNA gain is more than 10 dB between 0.7- 9.1 GHz while simulated gain is above 10 dB between 2 to 10 GHz. The tunable range of band pass LNA is between 5.23-5.51

GHz with a gain of 10.2 dB and 12.3 dB at the two centre frequencies and shown in Figure 5 [55]. However, K.J. Koh and G.M Rebeiz proposed a four bit active digital phase shifter for K band multiple antenna arrays with digital control circuitry in 0.13 μm CMOS technology. They developed a resonance based quadrature network to minimize loss and to increase operation bandwidth in the phase shifter exhibits 6.5-13° of rms phase error at 15-26 GHz and average voltage gain ranges from -4.6 dB at 15 GHz to -3 dB at 24.5-26 GHz. Measured phase and gain response are plotted in Figure 6 and Figure 7 [56].

M.G. Eldin et al. proposed an Ultra wide band diplexer using a Si-Cu technology, which can perform filtering, Impedance transformation and mode conversion. They found that diplexer has an insertion loss of 3.5 dB in both bands while common mode suppression of around 15 dB and 20 dB for low and high band respectively as shown in Table III.

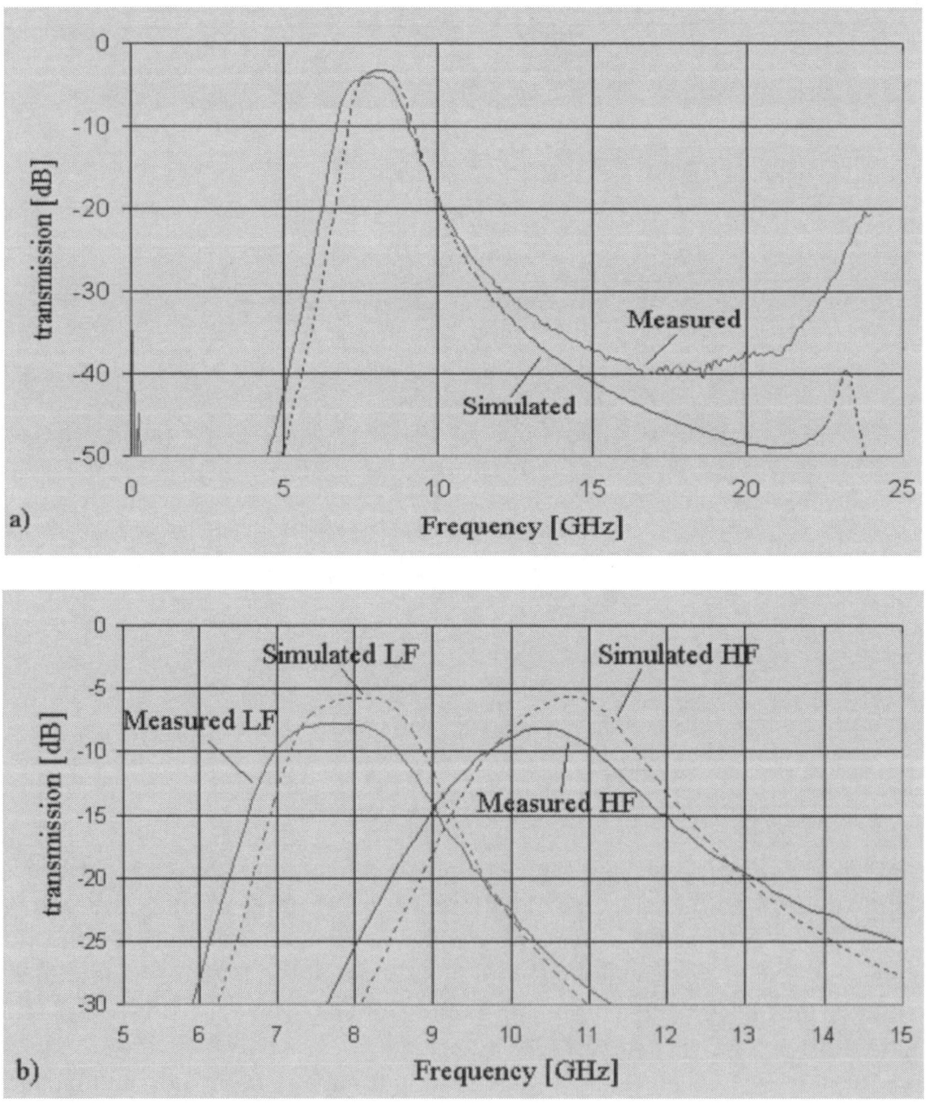

Figure 14. Measured and simulated transmission (s21) of two passive filters in 0.25 μm UTSi® CMOS a) with fixed frequency b) a switched filter-bank.

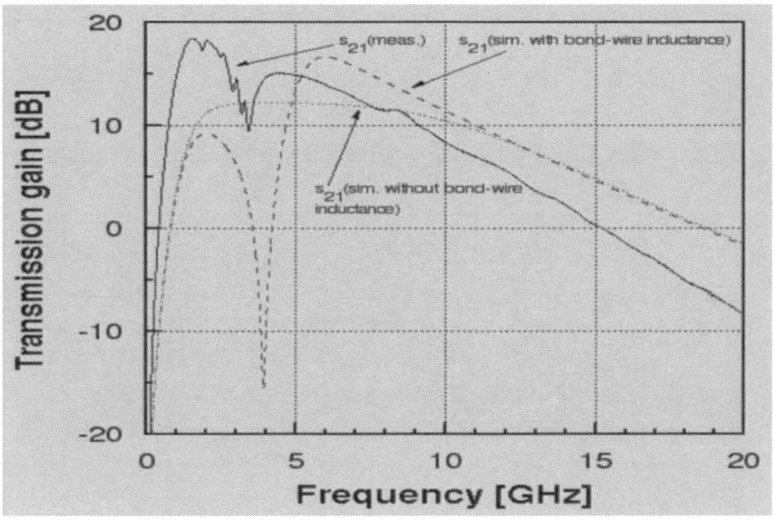

Figure 15. Measured and simulated s21 of the SiGe BiCMOS wide-band LNA. (ICC=10mA and VCC=1.8V).

Table III. Diplexer Target Specifications

Parameter	Freq. (GHz)	Target Spec
First Band		
Insertion loss	*3.182-4.738*	*3.5dB*
Return loss	*3.182-4.738*	*15 dB*
Attenuation	*0.1-2.5*	*25 dB*
Attenuation	*5.4-8*	*20 dB*
Second band		
Insertion loss	*6.4-9.0*	*3.5 dB*
Return loss	*6.4-9.0*	*15 dB*
Attenuation	*0.1-5.8*	*25 dB*
Attenuation	*9.5-12.0*	*20 dB*

The insertion loss, return loss, common mode suppression and group delay for the low band are plotted in Figure 6,7,8,9[57] and for high band in Figure 10,11,12,13[57].

Zhang Qi et al proposed active power filtering technology that is reliable and accurate method for monitoring the harmonics in power systems. They mainly analyze the three methods which are based on PWM, Harmonic decoupling control and generalized integral iterative. All these methods conclude that choosing appropriate active filter is the solution to mitigate harmonics and improve the power quality of electric prides. However, E. N. Onggosanusi et al proposed a decentralized space time multiuser detection scheme based on canonical space time co ordinates (CSTCs) for representing the received signal. The frame work is based on partitioning of the signal space into active and inactive CSTCs. The active CSTCs having the signal of desired user facilitate maximal diversity exploitation and minimal complexity interference suppression while the inactive CSTCs only contain MAI and can be included progressively to attain a desired level of MAI suppression at the cost of increased complexity. They designed a LCMV-based decentralized multiuser detectors and found that the receiver is near-far resistant as long as the total number of active and inactive co ordinates is greater than the effective rank of the MAI. The performances of different receiver are plotted in Figure 16.

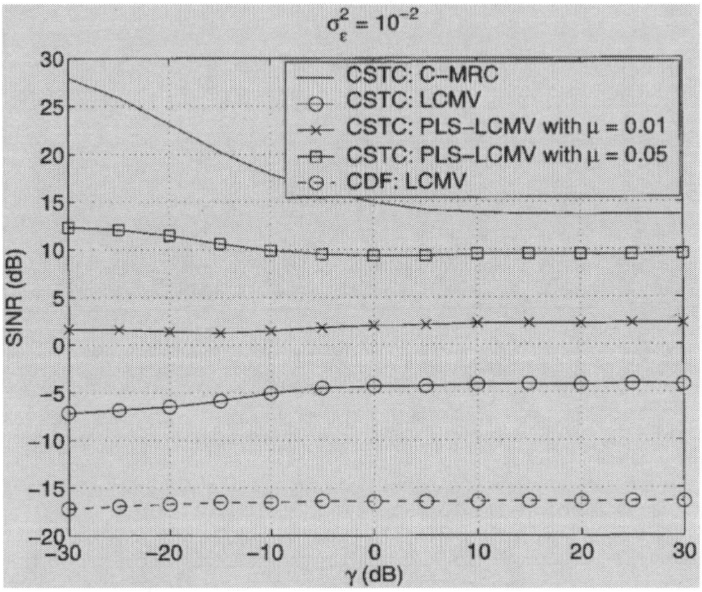

Figure 16. Performance of different receivers in the presence of channel estimation error for 1-sensor system. For CSTC-based receivers, NA = 5 and NI = 8.

PHASE ARRAY ANTENNAS

In this section design and development of active integrated antenna in phase array receivers emphasizing its application in SIGNALPROCESSING array has been reviewed. Retrodirective arrays, Reflectarray, multilayer Microstrip Array Antenna, circularly polarized active antenna and Injection locking have described in detail focusing on the characteristics; bandwidth and gain.

Retrodirective arrays are a type of special antenna arrays, which provides an omnidirectional coverage, while simultaneously maintaining a high level of antenna gain. This is the reason retrodirective arrays important in widely used in Radar transponders, self-steering antennas and in mobile communication systems [58-60]. Retrodirectivity is realized when each element in the array radiates an outgoing wave whose phase is conjugate to that of the incoming signal relative to a common reference [61]. The classical example of retro-directive array is the Van Atta array, in which the conjugated elements of a symmetric array are connected by transmission lines of equal length [62], however its limitations on symmetry of the array and uniformity of the phase front. To overcome these limitations, the phase conjugation with heterodyne mixing using a local oscillator (LO) hat having twice the RF frequency was proposed. In this technique, the lower sideband product has the same frequency as the RF, but the phase is conjugated. When combined with an antenna and placed in an array, the phase-conjugated signal from each antenna element will be radiated toward the source direction [63]. But this technique also has limitation of poor isolation, because of sharing of same frequency (RF/IF),Since MESFETs have the characteristics of providing conversion gain and mixing operation, they can be used in phase-conjugated circuitry [64, 65] and can allow an array to send amplified signals toward the source without amplifiers, resulting in compact profile and lower cost [66]. A prototype four-element retro-directive antenna array using the circuit topology has been designed and from the experimentation it was found that the array has excellent retro-directive performances. Such type of self tracking system can be used in advanced wireless applications such as RF ID tags and remote information retrieval [67, 68].

Later, in year 2003 T. Brouner et al proposed an aperture coupled active antenna integrated with low-noise amplifier (LNA) and tested it as a receiving antenna in the frequency range of 5 to 6 GHz, and found that the new structure offers the advantage of compact and robust adaptive antenna array. A thick low dielectric substrate (polyguide ε_r = 2.32, h = 3.175mm) was selected for designing. The measured re that the antenna gain is improved up to 17 dBi with cross-polarization around -27 dB. Figure 17 shows the gain variation of active and passive antenna [69].

R. W. Clark proposed a new element or unit cell incorporating the amplifier and an integrated microstrip reflectarray. The simulation result of return loss and isolation characteristics, indicate that the triangular slot causes the maximum coupling between the port 1 and 2 at 1.6 GHz, while square slot has the minimum. It was also found that the 1x4 array and single active element provides 23 and 13.1 dB normalized gain, over the passive component at 1.6 GHz. The close spacing between arrays is helpful to avoid grating lobes in receiving as well as transmitting patterns with minimum side lobes. The aperture efficiency and gain is also found to be increased as more number of elements is placed in small area. The application areas repeaters for indoor satellite communication as well as in spatial power combining [70].

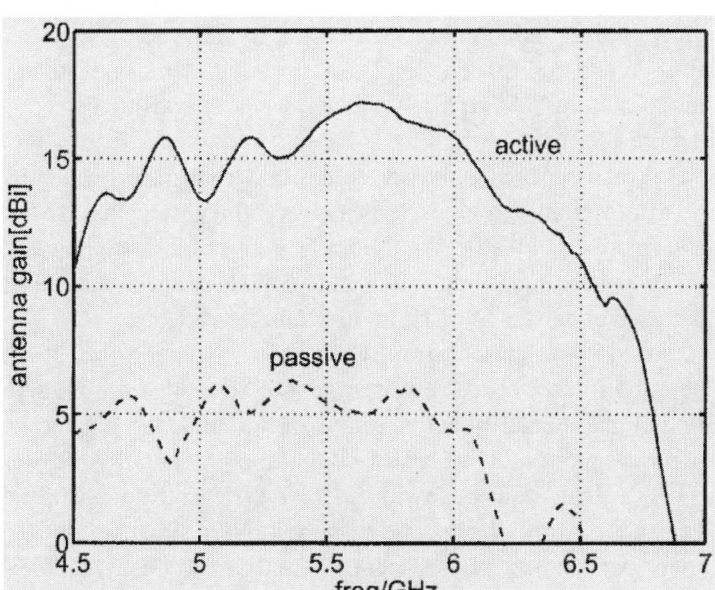

Figure 17. Gain of active (—) and passive (- - -) antenna.

However, H C Chang studied the Phase Noise and Phase Dynamics of N coupled PLL with N independent Oscillator and found that the near carrier the PLL element in the array can be reduced up to 1/N that of single independent VCO of the PLL element. free-running frequencies of the oscillators are 8.194 GHz and the injection power after applying the power divider and an attenuator in two coupled PLLs is P_{inj} = - 2.0 dBm. It is also found that the phase noise of the PLL element is reduced by approximately 3 dBc/Hz as compared to the free running Oscillator Phase Noise [71]. However, R C Ramirez proposed a highly integrated active antenna array module using simple and less expensive packaging techniques. An array of four cavity backed patch antenna is firmed by the above packaging process and then integrate it with down converter. The proposed antenna down converts a 39 GHz received signal [72].

Later in 2005, T Seki proposed a compact high gain multilayer parasitic microstrip array antenna which employs a multilayer low ceramic substrate that is very good for the assembly of MMIC chips. The maximum absolute gain for this antenna is 8.0 dBi at 60 GHz [73]. S-Band APAA with analog Phase Shifters that uses two DBMs. The phase and amplitude error during phase rotation were 0.5° rms and 0.3 dB respectively which is given in Figs. 4, 5, and 6 respectively[23] gain is more than 12 dBi, EIRP is more than 25.01 dBW and G/T is more than -14 dB/K for the proposed antenna. In addition, beam pointing resolution of 0.04° and beam pointing accuracy in the null-pointing of 0.4° rms. Application areas of the antenna are mobile vehicles to handle satellite communication. [74]. T. Seki also proposed a multi-layer parasitic MSA array that is constructed on a multi-layer Teflon substrate for millimeter wave system on package modules. It was that the radiation efficiency is greater than 91% and an associated antenna gain of 11.1 dBi at 60GHz, Figures 14 ,15 and 16 respectively [75].

M .Salehi proposed two element microstrip antenna arrays in which a dumb bell shape defect on the ground plane of the antenna is inserted between patches creating a band gap in the operation frequency band of antenna. The DGS antenna was analyzed using an FDTD method and a mutual coupling reduction of 18.28 dB was achieved by suppressing the surface waves as shown in Figure 5.[26] .The plots of broadband return loss and gain pattern indicates that radiation pattern have minimal change in the broadside direction but back lobe level is slightly increased.[76].

However, V. Douval proposed a novel monolithic millimeter wave two dimensional active conical horn array, suitable to be used as a quasi optical receiver at 95 GHz. For better reception, a low cost micro machined conical horn array was fabricated and placed on the top of active antenna. The pattern measured at 95 and 93 GHz are plotted in Figure 10 and 11, and reveals it can be used either for fixed beam or scanning. [77].

T. Yu proposed a 24 GHz phased array receiver implemented in 0.13 μm CMOS. The active phase shifter using the novel active vector generator results in a phase resolution up to 6 bit. The rms gain and phase error of active phase shifter are < 0.5 dB and < 2.8° at 23-24.4 GHz which is shown in Figure 6.[28] He also find that the phased array receiver has a gain of 14 dB, a NF of 6 dB, a 3 dB gain bandwidth of 4.7 GHz and wideband input and output match. The chip consumes 30 mA from a 1.5 V supply with dimension of 0.66 x 1.25 mm2 including pad. It has outstanding phase accuracy and is excellent candidate for satellite communication [78].

N. Yang proposed a fixed beam frequency tunable coplanar strip line antenna array which is able to obtain a wide range frequency tunings with good impedance matching. An array of 6 elements can give a gain of 9.3 dBi for bidirectional radiation and12.5 dBi for unidirectional radiation using a back reflector. To achieve the frequency tuning through a bias voltage control, shunt varactor diode are incorporated along the CPS structure. The variation of reflection coefficient of antenna with frequency, for different biasing voltage is shown in Figure 18.

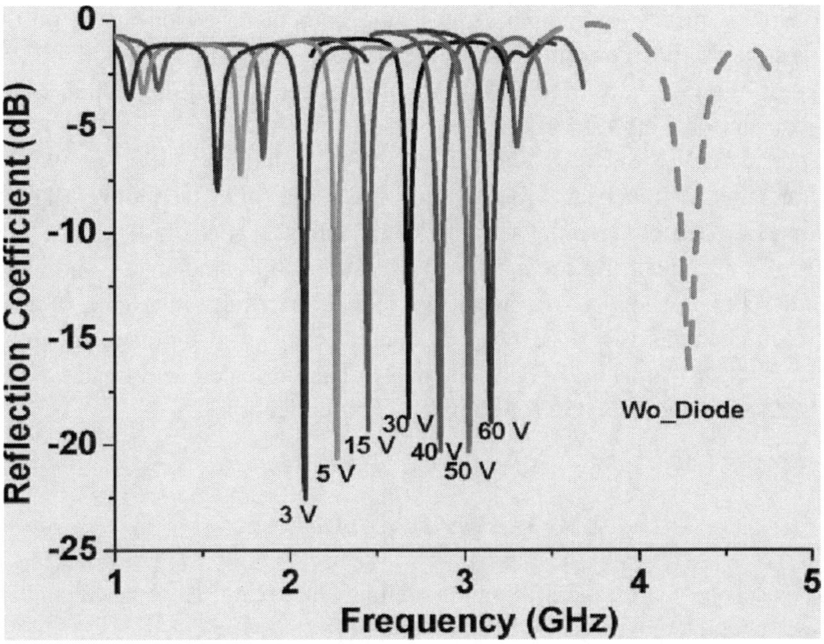

Figure 18. Measured reflection coefficient of the antenna array for different biasing voltages.

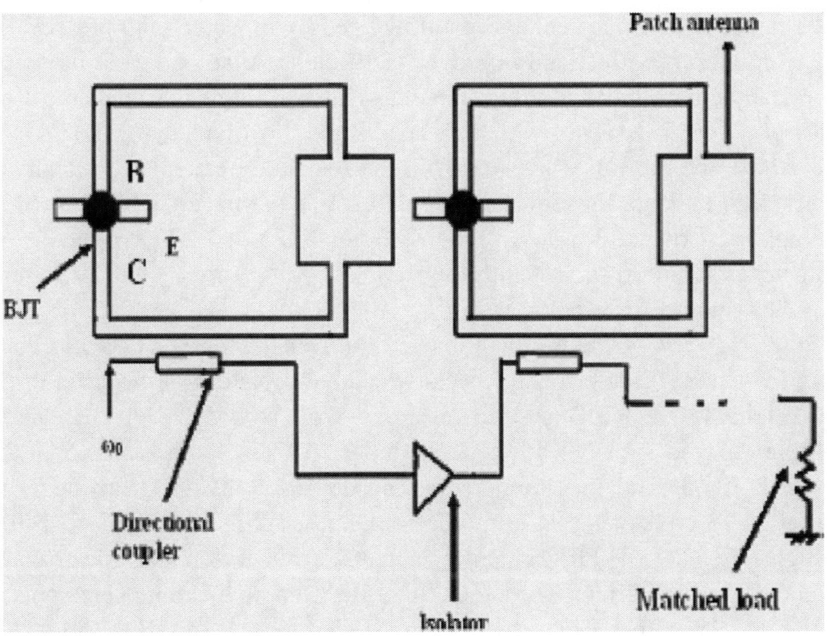

Figure 19. New scheme for injection locking.

The several advantages like uni-planarity, MIC/MMIC integration, efficient and stable frequency tuning of the antenna make it to be used in frequency agile communication system, measurement equipment and high gain RFID tags and readers [79].

Recently, L S Modur proposed a new method for the analysis of transient response to an external injection locking signal at circuit level through simulation shown in Figure 19.

They developed an active antenna array using element based on a two port microstrip patch which is suitable for translating to millimeter-wave frequencies with MMIC realization. In addition, they demonstrated a new scheme for injection locking which overcome the limitation of conventional circuit [80].

Very recently, D W Kang proposed a Ku band two antenna phased array receiver chip capable of simultaneous four beam in a 0.18μm Si Ge BiCMOS technology. The four beam chip has a gain of 4-6 dB per channel at 13-15 GHz, a noise figure of 10-11 dB, an rms phase and gain error of < 12° and 1.5 dB respectively.Measured S parameters are shown in Figure 7 [81]. When this chip is tested using two input signals it results in a gain of 9-11 dB and isolation between channels is > 40 dB. It can operate over an instantaneous bandwidth of > 1GHz at any frequency from 13-15 GHz. The design can be scaled to Ka band (35 GHz) or V band (60 GHz) using same structure for complex communication.

SWITCHING ANTENNAS

The AIAs also provide circuital function such as Duplexing, Resonating and Filtering in spite of its basic characteristics as a radiating element [82].

In year 2001, W. Choi proposed an Automatic On-Off switching repeater that is switched of automatically when there is no active user within its coverage. This has advantage of preventing the unnecessary noise enhancement induced by repeaters and protects the reverse link capacity from harmful effect and improves it. While, J. Vian and Z. Popovic proposed a high speed optically controlled T/R active antenna. This is designed as an element of 6 λ_o by 3 λ_o cylindrical active lens array with a focal distance to diameter (F/D) ratio of one, a directivity of 20 dB and a beam width of 10° in the focusing plane. The optically controlled SPOT switches used to route the signal in the active antenna have 0.31dB insertion loss, 36 dB isolation and -10 dB return loss from 8.36 to 10.8 GHz. The optically controlled microwave switches requires significantly less control energy than a MEMS, since energy distribution is permanently different.

In year 2009, M. Y.W. Chia et al . proposed a novel low power CMOS chip to enable beam steering for the RFID interrogator because radiation pattern is the main drawbacks of RFID. This is suitable for RFID system operating from 900MHz-2.4 GHz bands. It can provide multiple delayed signals with 5ps resolution. This will translate for a 2° scanned angle at 2.4 GHz band. The measured radiation pattern at 924 MHz using array of two antennas and at 2.412 GHz using an array of 4 antennas. While, S. Pajic and Z. B. Popovic proposed a high efficiency 10 GHz amplifier antenna array for spatial power combining of switching mode power amplifiers. They found that the 16 active antenna elements exhibit an output EIRP of 162 W with 70% average drain efficiency at 10.2 GHz. The 4 element sub arrays and the 16 element arrays have power combining efficiencies around 80%. In the combiner used here , all Pas saturate simultaneously as the input power is raised and the drain efficiency across the array remains constant maintaining the main benefit of switching mode high efficiency PAs.

F. Ferrero et al. proposed a novel azile micro strip patch antenna in which radiating patch is fed by a tunable quasi lumped coupler. By changing the operating mode of QLC the complete structure can be altered to radiate the electromagnetic wave with vertical linear, horizontal linear, right handed circular or left handed circular polarization. It has the advantage of compactness and bias simplicity with no additional capacitors or inductors. The frequency bandwidth is limited to 4% in circular polarization mode and 9% in linear polarization mode while maximum gain of 6.2 dBi in the broad side direction is found.

Figure 20 shows the 3D view the polarization – azile antenna and Figure 21 shows the radiation pattern at 3.5 GHz. While in year 2004, S. S. Zhong proposed a novel polarization - agile micro strip patch antenna which can be operated either in linear or circular polarization by electrically controlling with special FET phase shifter. He found that for circular polarization a bore sight axial ratio of 0.5 dB is achieved while isolation is better than 30dB over a 15% BW for the 16 element corner fed dual polarization array.

B. Poussot et al. Proposed reconfigurable antenna with a single input port which can combine polarization and pattern diversities. It is switched by means of 8 pin diodes. They found that under specific propagation condition the diversity gain is 12.9 dB with a combination of two patterns ($\theta = \pm 45°$) and two polarization for a diagonal orientation of the RAUT.

The diversity gains for various combinations of branches are given in Table IV. The improvement is approximately 3 dB compared to the space diversity scheme based on two monopoles separated by two wavelengths.

A. C. K. Mak proposed the design of two reconfigurable multiband antennas. The first one uses switched-feed technique in which by changing the feed location it can be operated within the GSM, DCS, PCS and UMTS frequency bands.

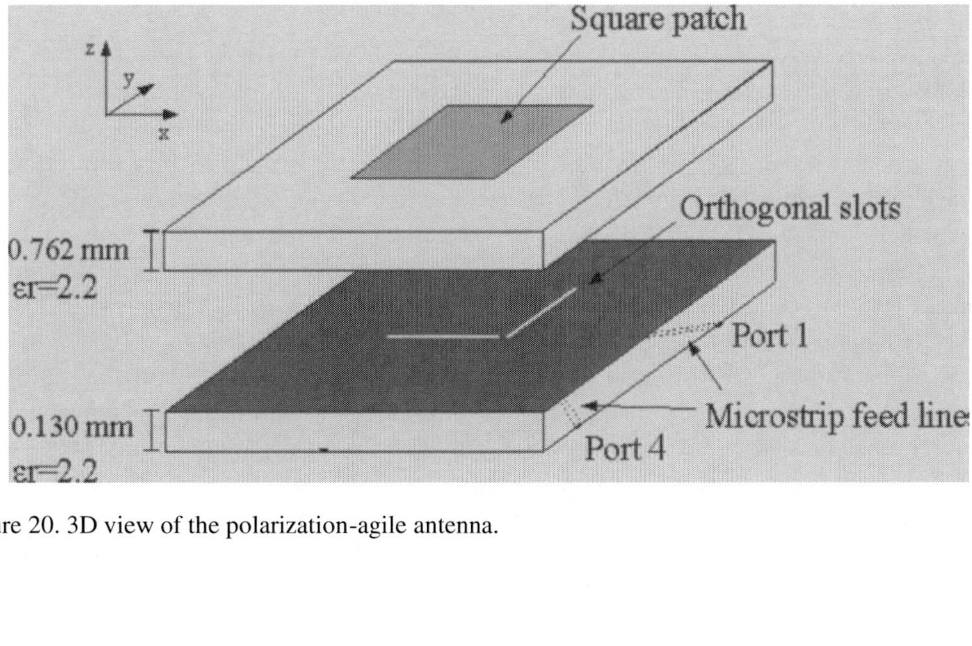

Figure 20. 3D view of the polarization-agile antenna.

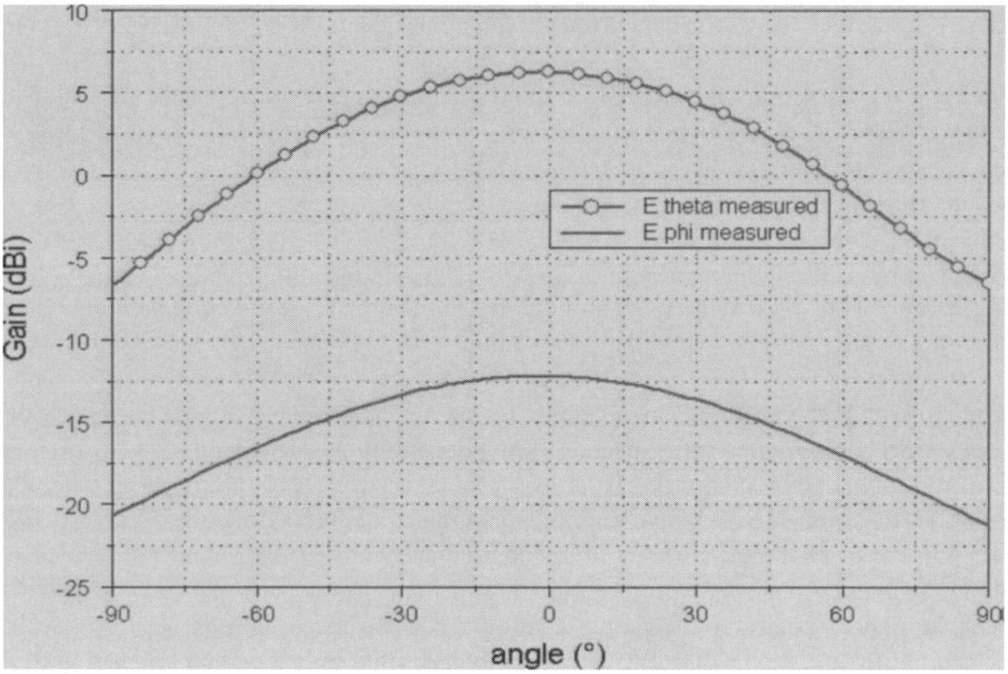

Figure 21. Measured radiation pattern of the antenna for Vr=15 V at 3.5 GH (uncoupled line mode).

Table IV. Diversity Gain for Various Combinations of Branches with a Diagonal Orientation of the Raut

Number of channels	DG at 1%
2: Pattern diversity with $\theta = \pm 45°$	8.2 dB
2: Polarization diversity only.	9.7 dB
4:Polarization diversity and Pattern diversity with $\theta = \pm 45°$	12.9 dB
6:Polarization diversity and Pattern diversity with $\theta = \pm 45°$ and $\theta = 0°$	13.4 dB
8: All Polarization and Pattern diversities.	13.9 dB

The second design uses a switched ground approach in which by connecting or disconnecting the antennas around the specific location different operating bands can be created which includes GSM, DCS, PCS, UMTS, Bluetooth and wireless LAN frequency bands. Figure 22 and Figure 23 show the plot of efficiency for switched feed and switched ground. Different active switches such as PIN diode, GaAs switch and MEMS were implemented into two reconfigurable antennas.

G.K Mohanty proposed a method based on real coded genetic algorithms that optimize the voltage excitation of the elements for the design of a reconfigurable array antenna with or without presence of a ground plane behind the array and the results are given in Table V and Table VI.

In this the maximum variation of the active impedances of the elements is also minimized when the active antenna switches between patterns without changing the geometry of the elements.

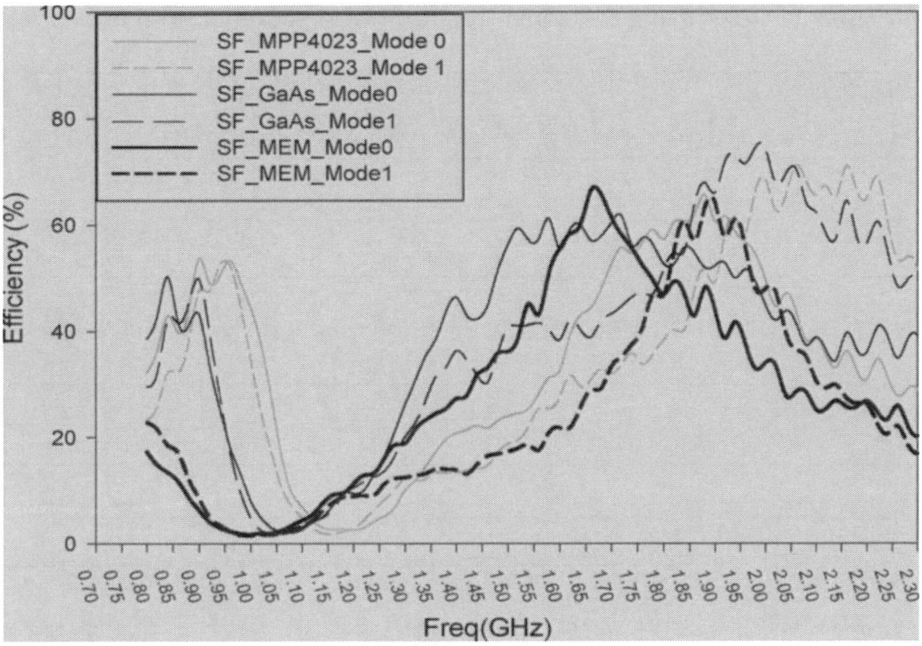

Figure 22. Switched feed using different active devices efficiency.

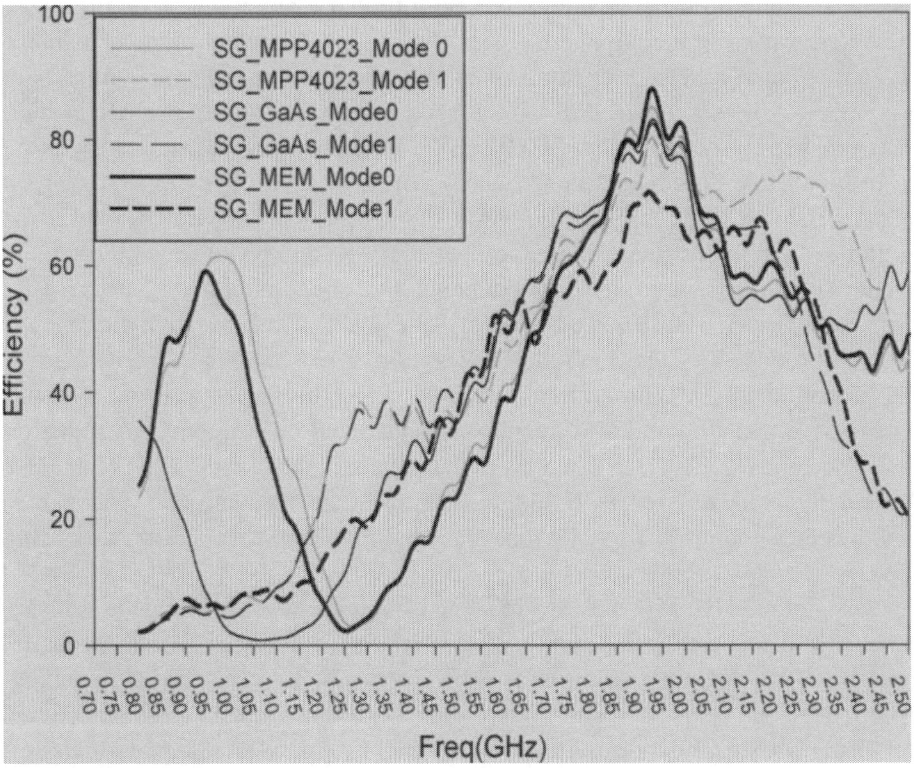

Figure 23. Switched ground using different active devices efficiency.

Table V. Desired and Obtained Results in Absence of Ground Plane

Design parameters	Pencil-beam		Flat top beam	
	Desired	Obtained	Desired	Obtained
Side lobe levels (dB)	-20.0	-21.49	-20.0	-21.06
Ripple (dB,-0.19≤u≤+0.19)	N/A	N/A	0.5	0.439

Table VI. Desired and Obtained Results In Presence of Ground Plane Λ/4 Behind

Design parameters	Pencil-beam		Flat top beam	
	Desired	Obtained	Desired	Obtained
Side lobe levels (dB)	-20.0	-22.82	-20.0	-21.85
Ripple (dB,-0.19≤u≤+0.19)	N/A	N/A	0.5	0.610

J. J. Maciel et al. Proposed 0.4 m^2 MEMS electronically steerable antennas containing 25000 MEMS which is integrated with AN/APG-67 radar system. This MEMS demonstrated radar system successfully detected both airborne and ground moving targets. The low cost, light weight and low power technology demonstrated can enable weight and power constrained platforms with electronic steering. They found that when a large radiant 0.4 m^2 MEMS ESA was interfaced with existing transmitter, receiver and display unit , the MEMS radar beam successfully scanned a 120° azimuth sector under control of the AN/ APG-67.

L. Petit proposed a model to design MEMS switched parasitic antenna array at 5.6 GHz as shown in Figure 24. This model is fast and accurate method of optimization of switched parasitic array, consisting of an active central element and a uniform ring of 5 parasitic antennas. In this system the beam switching is at ±25°. At any given time only one parasitic element open circuited, while the rest remains the short circuited. The array exhibit directional features as well as strong advantage of the beam switching capability.

C.W Jung proposed a reconfigurable rectangular spiral antenna with a set of MEMS which are monolithically integrated and packaged on to the same substrate. On activating these switches, the overall spiral arm length is changed and consequently its radiation beam direction is also changed. The two antennas radiates right handed circularly polarized and left handed circularly polarized for PCB and quartz substrate respectively. The gain of two antennas varies between 3-6 dBi, while axial ration is 3dB at their operating frequency band i.e. 10 GHz.

S. C. Panagiotoy et al. Proposed a broadband circular switched parasitic arrays (SPAs) for portable DVB-T receiver application in VHF/UHF bands. It can be also incorporated in an indoor portable DVB- the remaining two is short-circuited. By appropriately selecting active and short-circuited elements, a set of four radiation patterns can be formed, covering the horizontal plane alternately. Furthermore, a comparison between the measured reflection coefficients is plotted in Figure 25.

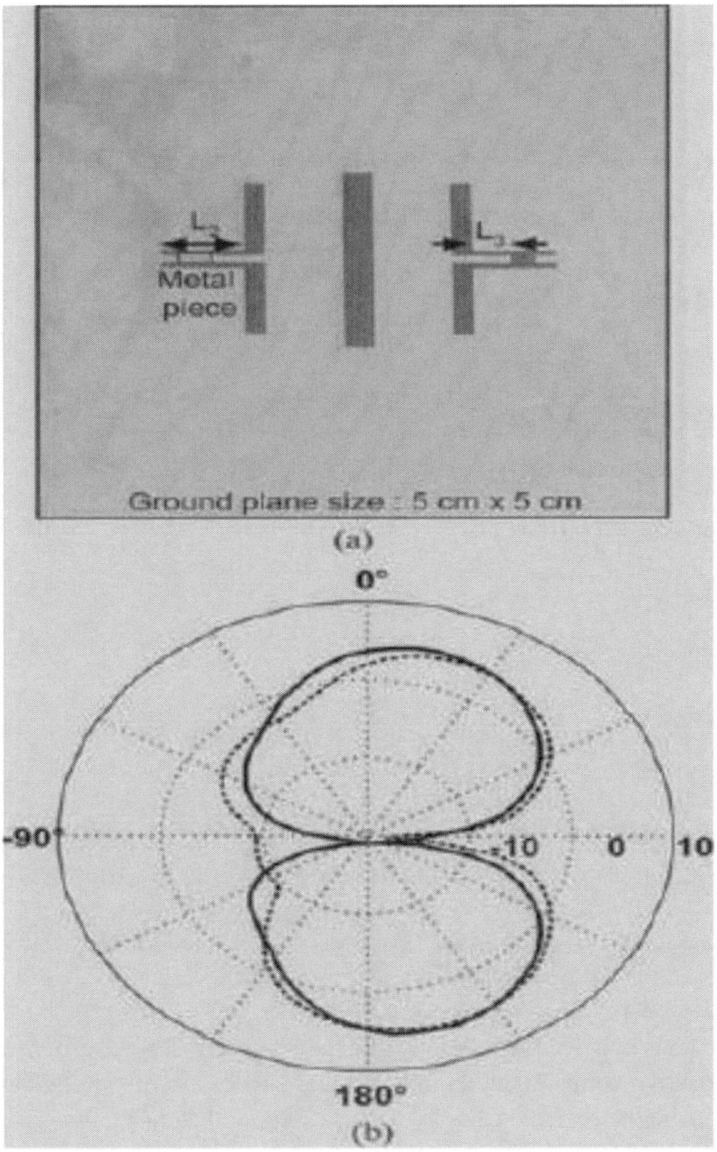

Figure 24. (a) Photograph of the passive prototype with metal connections for load selection (active slot: 2.2 X 1.7 mm, parasitic slots: 1.6 X 14.7 mm,d = 8mm (d/λ = 0:15)) and (b) simulated (solid) and measured (dashed) radiation patterns at 5.6 GHz of the passive prototype.

J.T.Aberle et al. proposed a reconfigurable antenna that can tune different frequency bands having enough efficiency and BW. They found that such tunable antenna technology is an enabler for software defined radios and RF front ends of which must be reprogrammable on the fly. It is also possible to cover a large range say 800 MHz to 2 GHz with A single T/R antenna pair. A good tuning circuit can be achieved by implementing a switched band of high Q capacitors with very low loss, which is the application of MEMS. The efficiency and measured (S_{11}) for the reconfigurable SPA is shown in Figure 4 and 9 respectively [83].

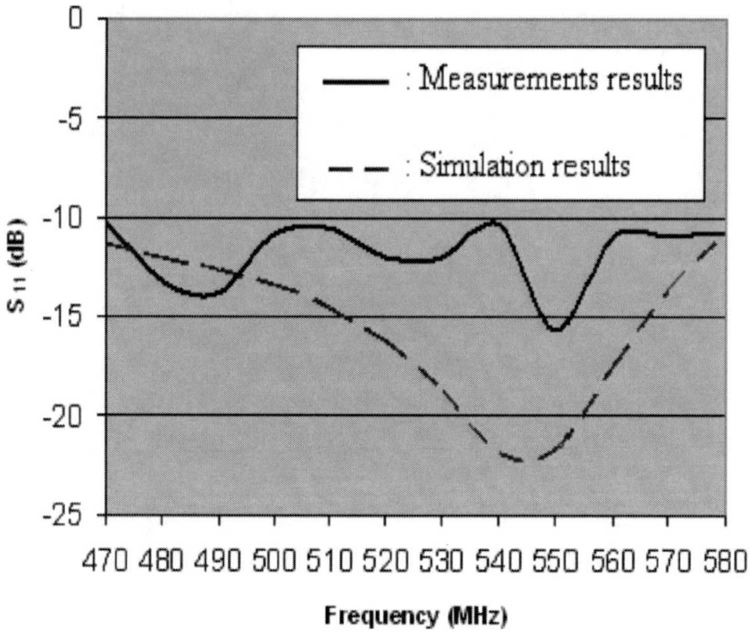

Figure 25. Simulation and measurement results for S 11.

Optical Antennas

Optical antennas are a solution to the mismatch between the small dimensions of nanoscale devices and the length scale associated with optical wavelengths. An optical antenna is a device that efficiently couples the energy of free-space radiation to a confined region of sub wavelength size. The optical antennas will be used for artificially enhancing the absorption cross-section or quantum yield of optoelectronic devices (e.g. solar cells), for efficiently releasing energy from nanoscale devices (e.g. LED lighting), and for boosting the efficiency of biochemical detectors relying on a distinct spectroscopic response. The available literatures on the active optical antenna indicate that several useful results have been found. To be compatible with the optical technology, it is essential that an antenna structure may also be integrated with other components, ***so*** that the cost, size and weight of the antenna can be greatly reduced. In present paper authors describe the progressive study of integrated optical antennas.

In year 1990, G. J. Simonis and K.G. Purchase proposed the optical generation and transmission of a microwave signal by mixing two high quality optical signals from diode laser pumped Nd: YAG ring laser.

Figure 1 [84] shows the optical mixing configuration. This signal has narrow microwave line width and broad microwave tunability (dc to 52 GHz). An III-V semiconductor waveguide with a doping super lattice active region is used to optoelectronically provides 20 dB of amplitude control and upto 5Π of phase shift. However, M. Righini et al. proposed novel optical trapping scheme that allows us to hold living E-coli bacteria for times of several hours using moderate light intensities. The trapping efficiency corresponding to incident wavelength of laser is plotted in Figure 26 for both longitudinal and transversal polarization.

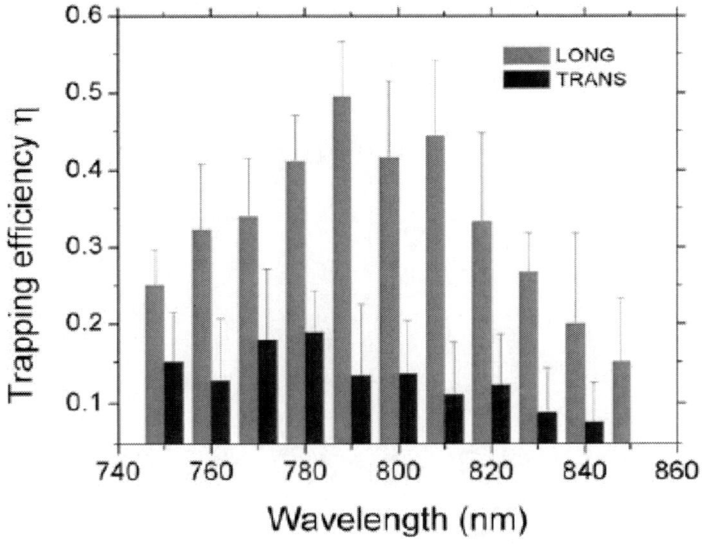

Figure 26. Resonant trapping Evolution with incident wavelength of the trapping efficiency η under both longitudinal and transversal polarization.

In year 1998, K. Hettak, G.Delisele and M. Boulmalf proposed a CPW aperture coupled antenna for the extremely high frequency band around a centre frequency of 37 GHz. The measured return loss and normalized input impedance as a function of frequency is shown in Figure 2 [85]. And it is found that the bandwidth at the resonance frequency is around 1 GHz for VSWR < 2. However, I. Pavlosoglou, R. Ramirez-Iniguez, M. S. Leeson and R. J. Green designed Warwick optical antenna for securing wireless communications in a combined radio and infrared environment.

For this purpose, deploying an infrared link at the physical entrance(s) to the location of the network enables the user to exchange a key using a secure line-of-sight channel, which they can in turn use for RF connectivity. The Warwick optical antenna represents such a front end to an optical receiver as shown in Figure 27.

Figure 27. The Warwick optical antenna enables for a wider acceptance angle in an optical front end.

The key advantage of this is a wider acceptance angle for securing wireless communications lies in the wider field of- view (FOV) of the optical front end.

H.Yang et al. proposed a new aspheric surface pre-collimation lenses system for the optical antenna of space laser communication as shown in Figure 1[86]. Since high gain is the main requirement for intersatellite laser communication, therefore Schmidt Cassegrain antenna having much higher gain is used. The transmit distances verses minimum power has been plotted in Figure 6[86] by calculating the parameter values from the equation

$$P_{rmin} = \left[\frac{a}{\omega(z)}\right]^2 P \tag{1}$$

$$\omega(z) = \omega_0\sqrt{1+\left(\frac{\lambda z}{\pi\omega_0^2}\right)^2} \tag{2}$$

where

P_{rmin} = The minimum power of receiver signals required by the system;
P = Emit power;
$\omega(z)$= Radius of Gaussian beam;
ω_0 = Minimum of $\omega(z)$;
a =Aperture of the optical antenna.

By (1), and (2), we can also obtained the transmit distance z satisfied with

$$z^2 = CP - D$$

where

$$C = \frac{1}{P_{rmin}}\left(\frac{\pi\omega_0 a}{\lambda}\right)^2 \quad , \quad D = \left(\frac{\pi\omega_0^2}{\lambda}\right)^2$$

So the values of the parameters are:

$$P_{rmin} = 1.0 \times 10^{-9}\,W \quad , \quad \lambda = 910nm$$

$$\omega_0 = 0.1\,m \quad , \quad a = 0.1\,m$$

The structure of Schmidt Cassegrain antenna is shown Figure 3 [86]. According to the measured value of rms for wave aberration: 0.041λ, which is less than λ/20, we can conclude that the optimum design optical antenna can be used to space laser system.

H. Wang, C. T. Chong and L. Shi suggested using bowtie aperture type surface plasmon antenna with finite outer boundary to generate nano-sized light source for 10 terabit/in^2 data storage. The size of the beam at 3 nm below the optical antenna is 10.4 nm in the x-direction and 9.7 nm in the y-direction, which is good enough for 10 terabit/in^2 optical data storage. Because of cross polarization it is very difficult to achieve such small field localization with other type of antenna. The finite outer boundaries introduce extra resonances to this optical antenna, which makes the localized field much stronger than that obtained with a conventional bowtie aperture.

P. Liao and R. A. York proposed wideband quasi-optical voltage-controlled oscillator. This VCO uses a varactor-tuned microstrip patch antenna and may be used as an element in quasi-optical power combining oscillator arrays. It was found that for both the case a tuning BW of 1.3 GHz was achieved with less than ± 1.5 dB of amplitude variation over the range. Figure 2 [87] illustrate a plot of the measured transmission through this tunable feedback circuit and shows that when the varactor capacitance is varied from 1.5 to 0.5 pF, the VCO should be tunable from 7.0 to 8.4 GHz. The measured DC-to-RF conversion efficiency of the VCO was 11 %, the effective radiated power of this VCO was measured to be 29.5 mW and phase noise of - 63 dB/Hz at 100 kHz.

H. S. Tsai, M. J. W. Rodwell and R. A. York proposed a broadband quasi-optical amplifier cell and 4x4 array using folded-slot antennas, suitable for monolithic power combining. The plot of EIPG with respect to frequency of single amplifier cell and of amplifier array is shown in Figure 2 and Figure 4 [88] respectively. The peak effective isotropic power gain in the transmission mode is 11 dB at 4.3 GHz with 10% bandwidth for the single cell, a factor of 10 improvement in bandwidth over the same amplifier cell using patch antennas, and 32 dB at 4.24 GHz with 8% bandwidth for the array.

J. Tong et al. proposed a photonic true time delay cell with two waveguide couplers and five semiconductor optical amplifiers. The block diagram of the time delay measurement is in Figure 28.

It was found that 6-dB average losses associated with fibre-to-waveguide coupling will be reduced with integrated waveguide architectures in the next-generation devices. With the amplifiers off the signal are blocked with extinction ratios of more than 10 dB and optical fibre provides a delay of 20 ns. The device has application in very agile phased array antenna, in optical switching and routing, and in optical filtering. However, K. C. Ho, Y. K. Kim and S. H. Hong proposed design criteria to investigate the radiation characteristics of photonic antennas, which transfer the energy of a beam out of an optical waveguide. As applicable examples, they evaluate gratings having symmetric or asymmetric groove shape.

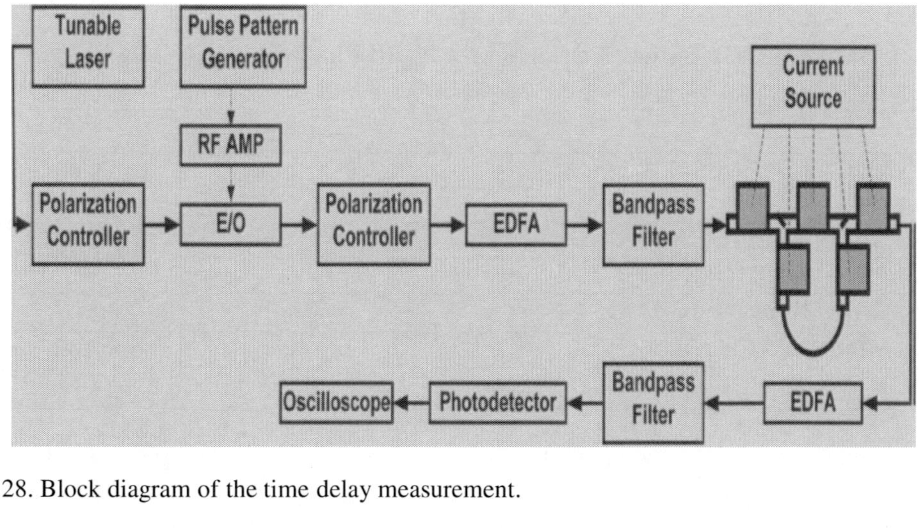

Figure 28. Block diagram of the time delay measurement.

For symmetric grating profiles, the radiation beam of the incident surface-wave coupled to the outgoing direction is equally partitioned between the cover and substrate regions if only a single diffracted order is leaky out, while for the case of asymmetric grating has preferentially coupled the surface-wave into either the cover or the substrate region with an efficiency of over 80% or about 99% at a specific condition as shown in Figure 6 [89]. Furthermore, the analytic perturbation model transmission line theory (P-MTLT) can readily apply to determine the design parameters of photonic antennas with the grating profiles.

R. A. Flynt, Lu Fan, Julio A. Navarro and Kai Chang proposed a FET and a Schottky mixer diode integrated within an inverted patch antenna for transceiver applications. The position of the mixer diode on the patch was optimized to achieve a minimum conversion loss. It was found isotropic mixer conversion loss is 5.5 dB at 6 GHz for a 200 MHz intermediate frequency. The FET acts as both the transmitter and the local oscillator at 5.8 GHz. Figure 10 and Figure 11 [90] shows the H plane and E plane pattern of the integrated antenna transceiver. A two-way communication system using these transceivers can be designed. However, A. K. Gupta, E.V. Korpinen, A. D. M. Chen and D. S. Matthews proposed a novel, high-yield, FET gate fabrication technology. The advantage of this technology is that it permits fabrication of sub-half micron gate-length devices using standard optical photolithography without recourse to deep UV or electron-beam lithography. Exceptionally good gate length control, typically 10% for a 0.4 μm long gate is described. For a typical 10-wafer lot, the standard deviation in gate length for all wafers is less than 0.025μm. This technology is used to fabricate a 17.5-GHz, 3-bit phase-shift receive MMIC of moderately high complexity. The feature of 3 bit phase shifter MMIC is given in Table VII.

C. W. Hicks, Huan-Sheng Hwang, Michael B. Steer, James W. Mink and James F. Harvey designed a two-dimensional hybrid dielectric slab–beam open and closed waveguide spatial power combining systems suitable for the design of planar quasi-optical integrated circuits and devices. An open system consisting of an active E-plane amplifier array having Vivaldi-type antennas with MESFET and MMIC devices was investigated. The amplifier gain for a unit cell amplifier with MMIC chips are shown in Figure 10 [91]. It was found that the 4 x 1 MESFET amplifier array generated 11 and 4.5 dB of amplifier and system gain, respectively, at 7.12 GHz while the cascade MMIC Vivaldi-type antenna produced 24 dB of amplifier gain at 8.4 GHz.

Table VII. Salient Features of A 3 -Bit Phase-Shiftemr Mic

Features	Units	Value
NOMINAL GATE LENGTH	μm	0.4
NUMBER OF FETS		28
TOTAL GATE PERIPHERY	mm	6.62
NUMBER OF RESISTORS		57
NUMBER OF MIM CAPACITORS		24
TOTAL MIM CAPACITANCE	pF	85
NUMBER OF VIA HOLE		20

K. Horikawa, Y. Nakasuga and H. Ogawa proposed novel self-heterodyning optical waveguide beam forming and steering network integrated on a Lithium Niobate (LN) substrate. Figure 2 [92] shows the experimental set up for LN-BFN. The LN-BFN contains an optical frequency shifter and optical signal processing circuit, and is totally functional circuit in conjunction with a light wave source, a PD array, and connecting optical fibres. This integrated LN-BFN can simultaneously control phase and amplitude feeding distribution for eight array elements. A fully integrated LN-BFN connected by optical fibres can steer beam direction with the sensitivity of 1.5 degrees per volt.

H. S. Tsai and R. A. York proposed CPW-fed multiple slot antennas for active array and integrated antennas which are suitable choice as the radiating elements for quasi-optical system because of their relatively wider bandwidth. They are also preferable for MMIC's since no backside processing (via holes) are required for integration with devices and they can be integrated directly with off-the shelf MMIC's. The calculated and measured frequency response of a triple slot antenna is shown in Figure 1[93] .The concept of engineering the antenna for a self resonant 50Ω input impedance are verified using a three slot antenna on ε_r=2.2 substrate and for a five slot antenna on ε_r= 9.8 substrate.

S. C. Ortiz, John Hubert, Lee Mirth, Erich Schlecht and Amir Mortazawi proposed a high-power K_a band quasi-optical amplifier array. They found the radiated power level of 44 dBm at 34 GHz with a 3-dB bandwidth of 800 MHz .An estimated 25 W powers is radiated when the amplifier is used as an antenna feed, otherwise 13W is collected into waveguide. Figure 29 shows that the antenna provides a 10-dB return-loss bandwidth of greater than 1 GHz at a centre frequency of 34 GHz and simulated radiation efficiency of 73%.

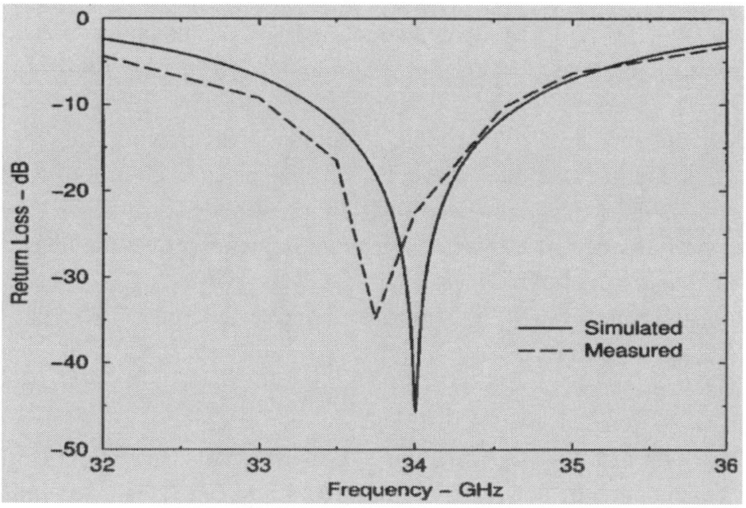

Figure 29. Measured return loss of the microstrip patch antenna compared with simulated results from Agilent HFSS.

CONCLUSION

As presented above, the AIA is an interesting subject of study with many examples presented above. The topic is rather interdisciplinary in nature and, hence, is challenging, but provides many opportunities. We have tried to highlight some of the major areas of active research in this review. However, it should be pointed out that potential applications of the AIA will far exceed what we can describe or even predict here. With the steadfast progress in solid-state devices, integrated circuits, antennas as well as novel fabrication technologies such as microelectromechanical systems (MEMS), more innovative designs based on the AIA concept are expected to emerge in the coming years.

ACKNOWLEDGMENTS

The authors would like to thank Dr. Sushrut Das. ISM, Dhanbad for providing early literature of active antennas and valuable suggestions.

REFERENCES

[1] H. A. Wheeler, “Small antennas,” *IEEE Trans. Antennas Propagate.,* vol. AP-23, pp. 462-469, July 1975.

[2] J. R. Copeland, W. J. Robertson, and R. G. Verstraete, “Antennafier arrays,'' *IEEE Trans. Antennas Propagate,* vol. AP- 12, pp. 227-233, Mar. 1964.

[3] H. H. Meinke, “Active antennas,'' *Nachrichtentech. Z.,* vol. 19, pp. 697-705, Dec. 1966.

[4] P. Anderson, W. S. Davies, M. M. Dawoud, and D. E. Galanakis, “Notes on transistor-fed active--array antennas,” *IEEE Trans. Antennas Proptrgat.,* VOI. AP-19, pp. 537-539, July 1971.

[5] M. M. Dawoud and A. P. Anderson, “Calculations showing the reduction in the frequency dependence of a two-element array antenna fed by microwave transistors,” *IEEE Trans. Antennas Propagate.,* vol. AP-20, pp. 497499, July 1972.

[6] ------, “The performance of transistor fed monopoles in active antennas,” *IEEE Trans. Antennas Propagate.* Vol. AP-21, pp. 371-374, May 1973.

[7] M. I. Kontorovich and N. M. Lyapunova, “Active antennas,'' *Radio Eng. Electron. Phys.* vol. 19, pp. 126-127, 1974.

[8] T. S. M. Maclean and P. A. Ramsdale, “Short active aerials for transmission,” *In?. J. Electron.,* vol. 36, pp. 261-169, Feb. 1974.

[9] M. M. Dawoud and A. P. Anderson, “Experimental verification of the reduced frequency dependence of active receiving arrays,” *IEEE Trans. Antennas Propagate.,* vol. AP-22, pp. 342-344, Mar. 1974.

[10] P. K. Rangole and S. *S.* Midha, “Short antenna with active inductance,” *Electron. Lett.,* vol. IO, pp. 462463, Oct. 1974.

[11] J.-P. Daniel and C. Terret, “Mutual coupling between antennas optimization of transistor parameters in active antenna design,” *IEEE Trans. Antennas Propagar.,* vol. AP-23, pp. 513-516, July 1975.

[12] B. Grob, *Basic Electronics,* 6th ed. New York: McGraw-Ell, 1959, ch. 8.

[13] J. W. Mink, “Quasi-optical power combining of solid-state millimeter wave sources,” *IEEE Trans. Microwave Theory Tech.,* vol. MTT-34, pp. 273-279, Feb. 1986.

[14] W. W. Lam, C. F. Jou, H. Z. Chen, K. S. Stolt, N. C. Luhmann, Jr., and D. B. Rutledge, “Mil1imete:r-wave diode grid phase shifters,” *IEEE Trans. Microwave Theory Tech.,* vol. MTT-36, pp. 902-907, May 1988.

[15] -----, "Millimeter-wave diod'e grid frequency doubler," *IEEE Trans. Microwave Theory Tech.,* vol. MTT-36, pp. 1507-1514, Nov. 1988.
[16] Z. B. Popovic, R. M. Weikle I]., M. Kim, and D. B. Rutledge, "A 100- MESFET planar grid oscillator," *IEEE Trans. Microwave Theory Tech.,* vol. 39, pp. 193-200, Feb. 1991.
[17] R. M. Weikle 11, M. Kim, J. B. Hacker, M. P. De Lisio, Z. B. Popovic, and D. B. Rutledge, "Transistor oscillator and amplifier grids," *Proc. IEEE,* vol. 80, pp. 1800-1809. Nov. 1992.
[18] Pance and M. J. Wengler, "Microwave modeling of 2-D active grid antenna arrays,'' *IEEE Trans. Microwave Theov Tech.,* vol. 41, pp. 20-28, Jan. 1993.
[19] M. Kim et al., "A 100-element HBT grid amplifier," *IEEE Trans. Microwave Theov Tech.,* vol. 41, pp. 1762-1771, Oct. 1993.
[20] L. B. Sjogren et al., "A monolithic diode array millimeter-wave beam transmittance controller," *IEEE Trans. Microwave Theory Tech.,* vol. 4 1, pp. 1782-1790, Oct. 1993.
[21] K. D. Stephan, P. H. Spooner, and P. F. Goldsmith, "Quasi-optical millimeter-wave hybrid and monolithic PIN diodes switches," *IEEE Trans. Microwave Theory Tech.,* vol. *41,* pp. 1791-1798, Oct. 1993.
[22] D. B. Rutledge, D. P. Neikir,k and D. P. Kasilingam, "Integrated circuit antennas," in K. J. Button, Ed. *Infrared and Millimeter- Waves.* London: Academic Press, vol. 10, 1983, p. 25.
[23] H. J. Thomas, D. L. Fudge, and *G.* Moms, "Gunn source integrated with microstrip patch," *Microwave RF,* pp. 87-91, Feb. 1985.
[24] N. Camilleri and B. Bayraktaroglu, "Monolithic millimeter-wave IMPATT oscillator and active antenna,'' *IEEE Trans. Microwave Theory Tech.,* vol. MTT-36, pp. 167g1676, Dec. 1988.
[25]] R. A. York, R. M. Martinez, and R. C. Compton, "Hybrid transistor and patch antenna element for array applications," *Electron Lett.,* vol. 26, pp. 49U95, Mar. 1990.
[26] J. A. Navarro, K. A. Hummer, and K. Chang, "Active integrated antenna elements," *Microwave J.,* vol. 34, pp. 115-126, Jan. 1991.
[27] R. E. Miller and K. Chang, "Integrated active antenna using annular ring microstrip antenna and Gunn diode," *Microwave Optical Tech. Lett,* vol. 4, no. 2, pp. 72-75, Jan. 20, 1991.
[28] J. Birkeland and T. Itoh, "Two-port FET oscillators with applications to active arrays," *IEEE Microwave Guided Wave Lett.,* vol. 1, pp. 112-1 13, May 1991.
[29] ------, "A circularly polarized FET oscillator active radiating element," *IEEE MTT-S In?. Microwave Symp. Dig.,* vol. 3, June 10-14, 1991, pp. 1265-1 268.
[30] J. A. Navarro, Y-H. Shu, and K. Chang, "Broadband electronically tunable planar active radiating elements and spatial power combiners using notch antennas," *IEEE Trans. Microwave Theory Tech.,* vol. 40, pp. 323-328, Feb. 1992.
[31] J. Birkeland and T. Itoh, "An FET oscillator element for spatially injection locked arrays," in 1992 *IEEE MTT-S Int. Microwave Symp. Dig.,* vol. 3, June 1-5, 1992, pp. 1535-1538.
[32] M. Vaughan and R. C. Compton, "Resonant-tee cpw oscillator and the application of the design to a monolithic array of MESFET's," *Electron. Lett.,* vol. 29, no. 16, pp. 1477-1479. Aug. 1993.
[33] D. J. Roscoe, A. Ittipiboon, and L. Shafai, "The development of an active integrated microstrip antenna," *IEEE AP-S In?. Symp. Dig.,* vol. I, June 24-28, 1991, pp. 48-51.

[34] R. Gillard, H. Legay, J. M. Floc'h, and J. Citerne, "Rigorous modeling of receiving active microstrip antenna,'' *Electron. Lett.,* vol. 27, no. 25, pp. 2357-2359, Dec. 1991.
[35] B. Robert, T. Razban, and A. Papiemik, "Compact amplifier integration in square patch antenna,'' *Electron. Lett.,* vol. 28, no. 19, pp. 1808-1810, Sept. 1992.
[36] P. S. Hall, "Analysis of radiation from active microstrip antennas," *Electron. Lett.,* vol. 29, pp. 127-129, Jan. 1993.
[37] C-Y. Yu and G. M. Reheiz, "A quasi-optical amplifier," *IEEE Microwuve Guided Wave Lett.,* vol. 3, pp. 164-166, June 1993.
[38] H. An, B. Nauwelaers, and A. Van de Capelle, "On the characterization of active receiving microstrip antennas," *IEEE Trans. Antennas Propagate.,* to appear.
[39] R. Ken, P. H. Siegel and R. J. Mattauch, "A simple quasi-optical mixer for 100-120 GHz," *IEEEM7TSInt. Microwave Symp. Dig.,* 1977, pp. 96-98.
[40] K. D. Stephen, N. Camilleri, and T. Itoh, "A quasi-optical polarizationduplexed balanced mixer for millimeter-wave applications," *IEEE Trans. Microwave Theory*
[41] *Tech.,* vol. MTT-31, pp. 164-170, Feb. 1983
[42] K. D. Stephen and T. Itoh, "A planar quasi-optical subharmonically pumped mixer characterized by isotropic conversion loss," *IEEE Trans. Microwave Theory Tech.,* vol. MTT-32, pp. 97-102, Jan. 1984.
[43] V. D. Hwang and T. Itoh, "Quasi-optical HEMT and MESFET self oscillating mixers,'' *IEEE Trans. Microwave Theonl Tech.,* vol. MTT-36, pp. 1701-1705, Dec. 1988.
[44] J. Zmuidzinas and H. G. LeDuc, "Quasi-optical slot antenna SIS mixers," *IEEE Trans. Microwave Theory Tech.,* vol. 40, pp. 1797-1804, Sept. 1992.
[45] P. A. Stimson, R. J. Dengler, H. G. LeDuc, S. R. Cypher, and P. H. Siegel, "A planar quasi-optical SIS receiver," *IEEE Trans. Microwuve Theory Tech.,* vol. 41, pp. 609-615, April 1993.
[46] S. V. Robertson, N. 1. Dib, G. Yang, and L. P. B. Katehi, "A folded slot antenna for planar quasi-optical mixer applications," in 1993 *IEEE AP-S In?. Symp. Dig.,* vol. 2, pp. 600-603, June 1993.
[47] K. Cha, S. Kawasaki, and T. Itoh, "Transponder using self-oscillating mixer and active antenna," to be appeared in the 1904 *IEEE MTT-S In?. Microwave Symp.,* San Diego, CA, May 23-27, 1994.
[48] G. O. Young, "Synthetic structure of industrial plastics (Book style with paper title and editor)," in *Plastics*, 2nd ed. vol. 3, J. Peters, Ed. New York McGraw-Hill, 1964, pp. 15–64.
[49] Tomohiro Seki, Kazuhiro Uehara and Kenichi Kagoshima, "A Three-Dimensional Active Antenna for a High-speed Wireless Communication Application ," *IEEE MTT-S Digest*, pp. 975-978, (1997).
[50] Peter Hagn, Andreas Przadka, Volker Gebhardt, Ulrich Bauernschmitt, "Ceramics: The Platform for Duplexers and Frontend-Modules," *IEEE Ultrasonics Symposium,* pp. 1-10, (2002).
[51] N. Khandelwal and R. W. Jackson, "An X-band SIP Active Antenna Module," *IEEE*, pp. 1019-1022. (2005).
[52] Hyungrak Kim, Ick-Jae Yoon, and Young Joong Yoon, "A Novel Fully Integrated Transmitter Front-End With High PAE," *IEEE Trans. Microwave Theory Tech.*, vol.53, no. 10, pp. 3206-3214, (2005).

[53] Yin Qin, Steven Gao and Alistair Sambell, "Broadband High-Efficiency Circularly Polarized Active Antenna and Array for RF Front-End Application," *IEEE Trans. Microwave Theory Tech.*, vol. 54, no. 7,pp. 2910-2916 ,(2006).

[54] Hyungrak Kim and Young Joong Yoon, "Wideband Design of the Fully Integrated Transmitter Front-End With High PAE," *IEEE Trans. Microwave Theory Tech.*, vol. 55, no. 5,pp. 916-924 ,(2007).

[55] Norberto Lopez, Cheng-Jung Lee, Ajay Gummalla and Maha Achour, "Compact Metamaterial Antenna Array for Long Term Evolution (LTE).

[56] Handset Application," *IEEE,* (2009).

[57] R. Malmqvist, C. Samuelsson, T. Nilsson, I. Ferrer, J. Svedin, and A. Dalera ," SiGe BiCMOS Wide-Band and Tunable Bandpass LNAs for Adaptive Multi-Functional Phased Arrays," *IEEE*, (2005).

[58] Kwang-Jin Koh and Gabriel M. Rebeiz," A 0.13-μm CMOS Digital Phase Shifter for K-band Phased Arrays," *IEEE*, pp. 383-386 , (2007).

[59] M. Gamal El Din , Koen Mertens , Bernd Geck and H. Eul," UWB Antenna iplexer with Mode Conversion using High Quality Passive and Active Technology,"2008 IEEE Radio Frequency Integrated Circuits Symposium , pp. 597-600 , (2008).

[60] D. L. Margerum, "Self-phased arrays," in *Microwave Scanning Antennas*, R. C. Hansen, Ed. Los Altos, CA: Peninsula, 1985, vol. 3, ch.5.

[61] B. S. Hewitt, "The evolution of radar technology into commercial systems," in *IEEE MTT-S Int. Microwave Symp. Dig.*, San Diego, CA, June 1994, pp. 1271–1274.

[62] S. L. Karode and V. F. Fusco, "Self-tracking duplex communication link using planar retrodirective antennas," *IEEE Trans. Antennas Propagate.*, pp. 993–1000, June 1999.

[63] M. I. Skolnik and D. D. King, "Self-phasing array antennas," *IEEE Trans. Antennas Propagate.*, vol. AP-12, pp. 142–149, March. 1964.

[64] E. D. Sharp and M. A. Diab, "Van Atta reflector array," *IRE Trans. Antennas Propagate.*, vol. AP-8, pp. 436–438, July 1960.

[65] C. Y. Pon, "Retrodirective array using the heterodyne technique," *IEEE Trans. Antennas Propagate.*, vol. AP-12, pp. 176–180, Mar. 1964.

[66] C. W. Pobanz and T. Itoh, "A conformal retrodirective array for radar applications uses a heterodyne phased scattering element," in *IEEE MTT-S Int. Microwave Symp. Dig.*, Orlando, FL, May 1995, pp. 905–908.

[67] S. L. Karode and V. F. Fusco, "Novel retrodirective beam forming techniques," in *Proc. 27th Eur. Microwave Conf.*, Jerusalem, Israel, Sept. 1997, pp. 81–86.

[68] Y. Chang, H. R. Fetterman, I. Newberg, and S. K. Panaretos, "Microwave phase conjugation using antenna arrays," *IEEE Trans. Microwave Theory Tech.*, vol. 46, pp. 1910–1919, Nov. 1998.

[69] R. Y. Miyamoto, Y. Qian, and T. Itoh, "Active retrodirective array for remote tagging and wireless sensor applications," in *IEEE MTT-S Int. Microwave Symp. Dig.*, Boston, MA, June 2000, pp. 1431–1434.

[70] Kai Chang, et al, "Active integrated antennas," *IEEE Trans. Microwave Theory Tech.*, vol. 50, pp. 937-944, Mar. 2002

[71] T. Brauner, "A Differential Active Patch Antenna Element for Array Applications," *IEEE Microwave and Wireless Components Letters*, vol. 13, pp. 161-163, Apr. 2003.

[72] Randall W. Clark, "An Integrated Active Microstrip Reflectarray Element With An Internal Amplifier," *IEEE Trans. Antennas Propagate.*, vol. 51, pp. 993–999, May 2003.

[73] Heng-Chia Chang, "Analysis of Coupled Phase-Locked Loops With Independent Oscillators for Beam Control Active Phased Arrays," *IEEE Trans. Microwave Theory Tech.*, vol. 52, pp. 1059–1066, Mar. 2004.

[74] Rodrigo Carrillo-Ramirez, "A Highly Integrated Millimeter- Wave Active Antenna Array Using BCB and Silicon Substrate," *IEEE Trans. Microwave Theory Tech.*, vol. 52, pp. 1648–1653, June 2004.

[75] Tomohiro Seki, "A 60-GHz Multilayer Parasitic Microstrip Array Antenna on LTCC Substrate for System-on- Package," *IEEE Microwave and Wireless Components Letters*, vol. 15, pp. 339-341, May 2005.

[76] Amane Miura, "*S*-Band Active Phased Array Antenna With Analog Phase Shifters Using Double-Balanced Mixers for Mobile SATCOM Vehicles," *IEEE Trans. Antennas Propagate.* vol. 53, pp. 2533–2541, Aug. 2005.

[77] Tomohiro Seki, "Millimeter-Wave High-Efficiency Multilayer Parasitic Microstrip Antenna Array on Teflon Substrate," *IEEE Trans. Microwave Theory Tech.*, vol. 53, pp. 2101-2106, June 2005.

[78] Mohsen Salehi, "A novel low mutual coupling microstrip antenna array design using defected ground structure," *Int. J. Electron. Commun. (AEÜ) 60*, pp. 718-723, 2006.

[79] Vassilis Douvalis, "A Monolithic Active Conical Horn Antenna Array for Millimeter and Sub millimeter Wave Applications," *IEEE Trans. Antennas Propagate.*, vol. 54, pp. 1393-1398, May 2006.

[80] Tiku Yu, "A 24 GHz 6-Bit CMOS Phased-Array Receiver," *IEEE Microwave and Wireless Components Letters*, vol. 18, pp. 422-424, June 2008.

[81] Ning Yang, "Fixed-Beam Frequency-Tunable Phase- Reversal Coplanar Stripline Antenna Array," *IEEE Trans. Antennas Propagate.*, vol. 57, pp. 671-681, Mar. 2009.

[82] Lakshminarayana S. Modur, "Transient Response of Injection-Locked Active Antenna Arrays," *IEEE Microwave and Wireless Components Letters*, vol. 9, pp. 546-549, June 2010.

[83] Dong-Woo Kang, "A Ku-Band Two-Antenna Four- Simultaneous Beams SiGe BiCMOS Phased Array Receiver," *IEEE Trans. Microwave Theory Tech.*, vol. 58, pp. 771-780, Apr. 2010.

[84] J. Lin and T. Itoh, "Active integrated antennas," *IEEE Trans. Microwave Theory Tech.*, vol. 42, pp. 2186–2194, (1994).

[85] James T. Aberle et al. , "Reconfigurable Antennas for Portable Wireless Devices," *IEEE* Antennas and Propagali On Magazine, Vol. 45, No. 6, December 2003.

[86] G. J. Simonis and K. G. Purchase," Optical Generation, Distribution, and Control of Microwaves Using Laser Heterodyne," *IEEE Trans. Microwave Theory Tech.*, vol. 38,no. 5, pp. 667-669, (1990).

[87] K. Hettak, G. Delisle, and M. Boulmalf ," A Novel Integrated Antenna for Millimeter-Wave Personal Communications Systems," *IEEE Trans. Antennas Propagate.,* vol. 46, no. 11, pp. 1757-1758, (1998).

[88] Huajun Yang, Yu Hu, Chenghong Li, Kang Xie, Jin Fu, Hu Wei," Optimum Design for Optical Antenna of Space Laser Communication System,"*IEEE ,* pp.2016-2019, (2006).

[89] P. Liao and R. A. York," A Varactor-Tuned Patch Oscillator for Active Arrays,"*IEEE Microwave and Guided Wave Letters*, vol. 4, no. 10, pp. 335-337,(1994).
[90] H. *S.* Tsai, M. J. W. Rodwell and R. A. York," Planar Amplifier Array With Improved Bandwidth Using Folded-Slots," *IEEE Microwave and Guided Wave Letters*, vol. 4, no. 4, pp. 112-114,(1994).
[91] Kwang-Chun Ho, Yung-Kwon Kim and Seung-Hong,"Design Characterstics of Photonic Antennas For Wireless Optical Communication," *IEEE,* pp.232-237, (2000).
[92] Robert A. Flynt, Lu Fan, Julio A. Navarro and Kai Chang," Low Cost and Compact Active Integrated Antenna Transceiver for System Applications," *IEEE Trans. Microwave Theory Tech.*, vol. 44, no. 10, pp. 1642-1649, (1996).
[93] Chris W. Hicks, Huan-Sheng Hwang, Michael B. Steer, James W. Mink and James F. Harvey," Spatial Power Combining for Two-Dimensional Structures," *IEEE Trans. Microwave Theory Tech.*, vol. 46, no. 6, pp. 784-791, (1998).
[94] Kohji Horikawa, Yoshinori Nakasuga and Hiroyo Ogawa*,"* Self-Heterodyning Optical Waveguide Beam Forming and Steering Network Integrated on Lithium Niobate Substrate," *IEEE Trans. Microwave Theory Tech.*, vol. 43, no. 9, pp. 2395-2401, (1995).
[95] H. S. Tsai and R. A. York," Multi-Slot 50-ΩAntennas for Quasi-Optical Circuits," *IEEE Microwave and Guided Wave Letters*, vol. 5, no. 6, pp. 180-182,(1995).

In: Knowledge Management and Communication … ISBN: 978-1-62808-250-0
Editor: Rafiqul Islam

Chapter 24

Comparative Performance of Rule Learners in Automated Web Page Classification

***M. Indra Devi*[1]*, *K. Selvakuberan*[2] *and R. Rajaram*[3]**
[1]Department of Information Technology,
Thiagarajar College of Engineering, Madurai, India
[2]Innovation Labs (Web 2.0)
TATA Consultancy Services,
ETL Infrastructures Services Ltd., SEZ,
Chennai, India
[3]Computer Science and Information Technology
Thiagarajar College of Engineering, Madurai, India

Abstract

Web page classification is a current research topic. The automated categorization (or classification) of web pages into topical categories has a long history, dating back at least in the early 1960. The Internet makes it possible to share and manipulate a vast quantity of information efficiently and effectively, but the rapid and chaotic growth experienced by the Net has generated a poorly organized environment that hinders the sharing and mining of useful data. The need for meaningful web-page classification techniques is therefore becoming an urgent issue. The main goal of this paper is to find whether Rules classifiers are helpful in web page classification and also we make a comparative study on the performance of these classifiers. In this paper we use Prism, ZeroR and OneR rule classifiers for our classification purposes.

Keywords: Feature Selection, web page classification, Rule classifiers

---

* E-mail id : midit@tce.edu.

INTRODUCTION

Over the past decade we have witnessed an explosive growth on the Internet, with millions of web pages on every topic easily accessible through the Web. The Internet is a powerful medium for communication between computers and for accessing online documents all over the world but it is not a tool for locating or organizing the mass of information. Tools like search engines assist users in locating information on the Internet. They perform excellently in locating but provide limited ability in organizing the web pages. Internet users are now confronted with thousands of web pages returned by a search engine using simple keyword search. Searching through those web pages is in itself becoming impossible for users. Thus it has been of more interest in tools that can help make a relevant and quick selection of information that we are seeking. The number of web pages indexed by Google and Yahoo search engines is in billions. In 2005, Yahoo claimed that its index covered more than 20 billion web resources, the largest search engine.[6] It is believed that the actual size of the Web is at least several times bigger than what search engines currently cover Describing and organizing this vast amount of content is essential for realizing the web's full potential as an information resource.

Automatic document categorization had arisen a lot of interest also for its possible Internet applications. One of these is automatically categorizing Web pages, or sites, into one or several of the categories that make up commercial hierarchical catalogues such as those embodied in Yahoo!, Infoseek, etc. When Web documents are catalogued in this way, rather than addressing a generic query to a general-purpose Web search engine, a searcher may find it easier to first navigate in the hierarchy of categories and then issue his search from (i.e. restrict his search to) a particular category of interest. Automatic categorization of web pages has obvious advantages, since the manual categorization of a large enough subset of the Web is problematic. Until the late '80s, the dominant approach to the problem involved knowledge engineering automatic categorizers, i.e. manually building a set of rules encoding expert knowledge on how to classify documents. A newer paradigm based called machine learning approach comes into picture. Within this paradigm, a general inductive process automatically builds a classifier by "learning", from a set of previously classified documents,[9] the characteristics of one or more categories; the advantages are a very good effectiveness, a considerable savings in terms of expert manpower, and domain independence.

The paper is organized as follows. Section 2 deals with the Related work. Section 3 focuses on Our Proposed approach. Section 4 describes the experimental set up and their results. Section 5 ends with the conclusion. Section 6 describes the references.

2. RELATED WORK

Susan Dumais, Hao Chen explores the use of hierarchical structure for classifying web Pages using Support Vector Machine Classifiers. The hierarchical structure is initially used to train different second-level classifiers. In the hierarchical case, a model is learned to distinguish a second-level category from other categories within the same top level.[1]

Chee-Hong Chan, Aixin Sun, Ee-Peng Lim propose a new system called categorizor that performs automated on-line personalized news classification. They use SVM classifier to classify news articles into general categories, or personalized categories.[2]

Aixin Sun,Ee-Peng Lim and Wee-Keong Ng propose the use of Support Vector Machine (SVM) classifiers to classify web pages using both their text and context feature sets.[3]

Offer Drori propose a fully automated software tool to analyze the texts of the required documents. The output of the tool is a set of significant words from the document, which effectively constitute the subject of the document by presenting it topics. The software tool uses statistical analysis to determine the most frequent words in the document. The text is analyzed after words that are meaningless for the subject are eliminated using a Stop List, and after limited processing of linguistic elements. [4]

3. Proposed Approach

We propose the use of Rule learning classifiers to solve the web page classification problem. The steps in our approach are:

1. Preprocessing phase
2. Feature selection phase
3. Classification phase
4. Comparative evaluation

Preprocessing phase

Preprocessing phase is necessary to remove the noise and inconsistent data since web pages are unstructured in nature. It includes removal of HTML tags from the web page, apply stemming algorithm to find the root word, removal of stop words, punctuation symbols.

Feature selection phase

The features are selected using the term frequency (wherein the term frequency (tf) is a number of times a term appears in a document) measure tf.

where tf can be defined as $tf\ (t_i\ ,d_j)=\#\ (t_i\ ,d_j)$

where $\#\ (t_i\ ,d_j)$ denotes the number of times the term t_i occurs in document d_j.

Given a training set, find the number of occurrence of all the words in the training set. The words which occur more frequently (above a certain threshold) are taken as representative features after stemming and stop word removal

Classification phase

Train the machine learning classifiers (OneR, ZeroR and Prism) on this feature set.

Test the learned classifiers with test set.

Comparative evaluation

Evaluate the classification performance of these classifiers and make a comparative study

3.1. Preprocessing Phase

Preprocessing of web pages is the first step for the given problem. Web pages cannot be processed as such because of the size, content and nature of the web pages. Dimensionality reduction is an essential data preprocessing technique for large-scale and streaming data classification tasks. It can be used to improve both the efficiency and the effectiveness of classifiers. In this phase we remove all the HTML tags, stop words, infrequent words, punctuation symbols and common words from the web pages. This reduces the dimensionality of the web pages. Processing time and memory requirement are reduced and this is the significant advantage for large database like Web.

3.2. Feature Selection Phase

Traditional dimensionality reduction approaches fall into two categories: Feature Extraction and Feature Selection [5][8]. Techniques in the feature extraction category are typically more effective than those in feature selection category. Feature extraction refers to the extraction from the various features of the web page.

Feature selection is a process that selects a subset of original features. The optimality of a feature subset is measured by an evaluation criterion [7]. A typical feature selection process consists of four basic steps namely, subset generation, subset evaluation, stopping criterion, and result validation. Subset generation produces candidate feature subsets for evaluation based on a certain search strategy. Each candidate subset is evaluated and compared with the previous best one according to a certain evaluation criterion. If the new subset turns out to be better, it replaces the previous best subset.The process of subset generation and evaluation is repeated until a given stopping criterion is satisfied. The process of feature selection is shown in figure 1.

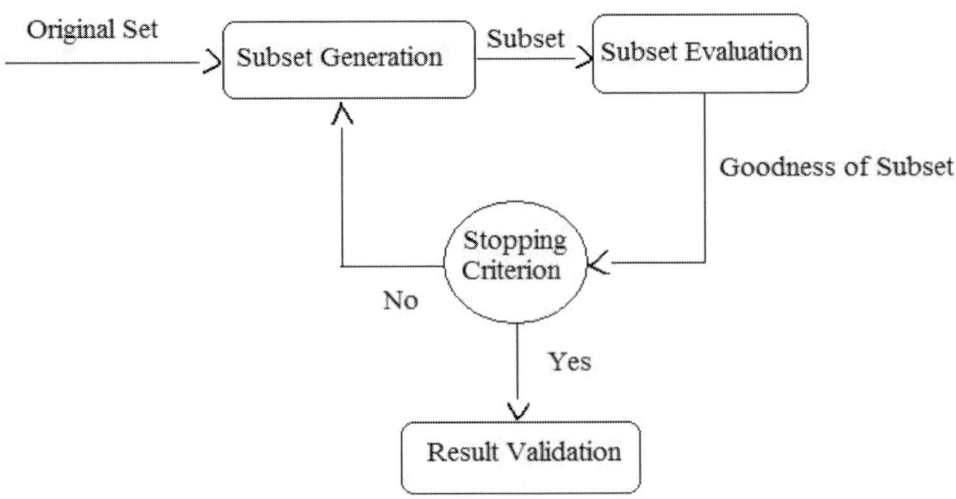

Figure 1. Feature Selection process.

3.3. Classification Phase

The classification is done by the Rule classifiers (OneR, ZeroR and Prism)

4. EXPERIMENTAL SETUP

For our experiment the database used is WEBKB data set and is downloaded from the UCI repository. It is a benchmarking dataset for machine learning problems. This is the university database having seven categories of web pages: Course, Project, Student, Faculty, Department, Staff and others.

We select all the pages in the course category (930 pages) and non course category (66 pages). We select 400 pages from the project category and 80 pages from the non project category. We select 1341 pages from the student category and 229 pages from the non student category. The experimental set up is shown in table 1. The set of features for the course, project and Student categories are listed in tables 2, 3 and 4 respectively.

Table 1. Experimental Setup

S.No	Name of the category	No. of positive examples	No. of negative examples
1.	Course	930	66
2.	Project	400	80
3.	Student	1341	229

Table 2. Set of features for "course" category

Course, class, syllabus, handout, homework, cs, lecture, notes, slides, solution, problem, program, instructor, information, project, paper, guide, study, prelim, professional, activities, resume, publications, language, research, teaching, contact, projects, professor, interests, department, personal, office, advisor, home, page, links, phone, iscourse (yes,no)

Table 3. Set of features for "project" category

University, research, computer, edu, project, department, laboratory, group, applications, information, work, computing, performance, lab, software, problems, faculty, tools, techniques, database, learning, communication, isproject (yes,no)

Table 4. Set of features for "student" category

Computer, university, department, page, research, information, student, systems, engineering, computing, work, institute, technology, graduate, conference, email, interest, tech, learning, technical, resume, group, advisor, report, isstudent (yes,no)

4.1. Results and Analysis

In table 5 we have presented the results for OneR, ZeroR and Prism for all the three categories. The figures 2, 3 and 4 show the comparative analysis of three learners in Course, Project and Student category respectively.

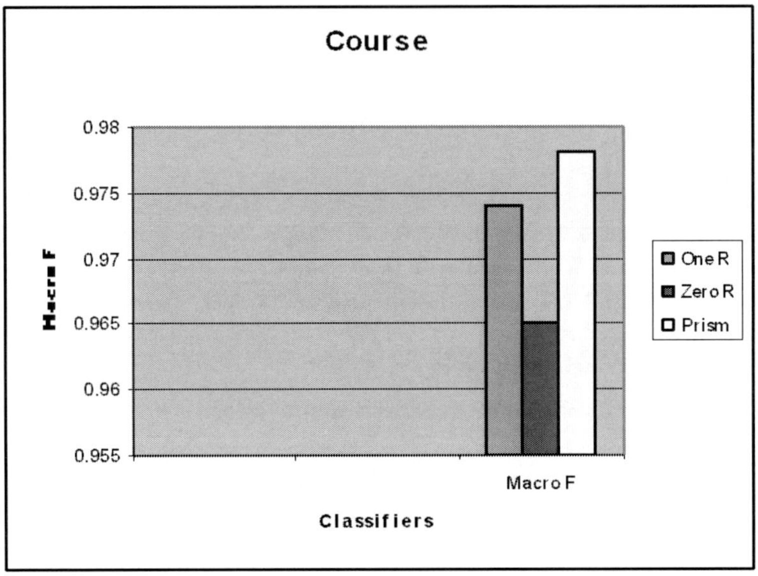

Figure 2. Comparative analysis for "Course" category.

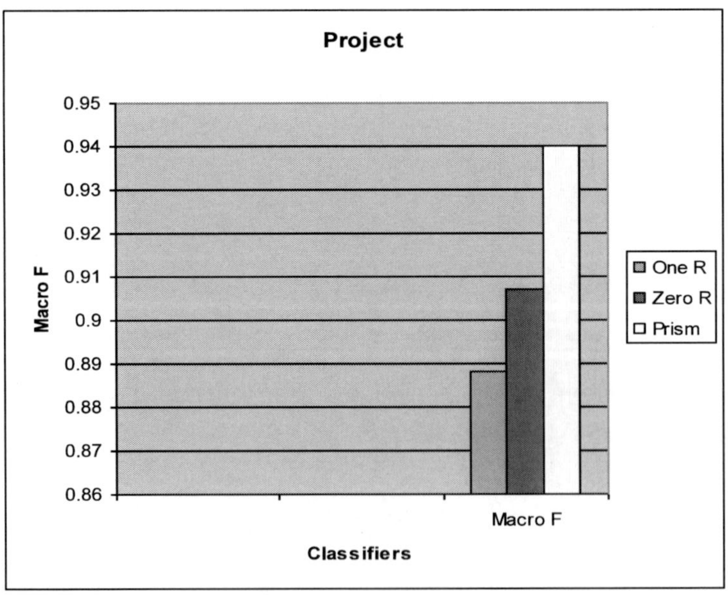

Figure 3. Comparative analysis for "Project" category.

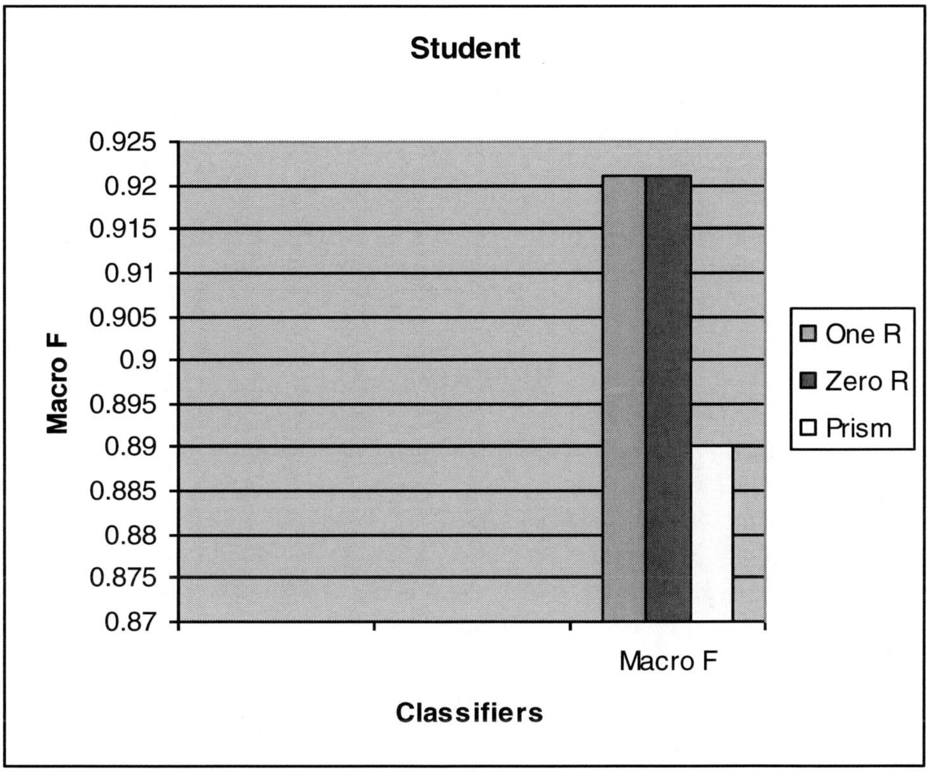

Figure 4. Comparative analysis for "Student" category.

Table 5. Experimental Results and analysis

S. No	Name of the category	Name of the Rule Learner	Precision	Recall	CCI instances	Macro F
1.	Course (Totally 996 instances)	One R	0.957	0.991	950	0.974
		Zero R	0.932	1	932	0.965
		Prism	0.963	0.994	956	0.978
2	Project (Totally 480 instances)	OneR	0.867	0.91	392	0.888
		ZeroR	0.831	1	402	0.907
		Prism	0.921	0.962	419	0.94
3	Student (Totally 1570 instances)	OneR	0.853	1	1343	0.921
		ZeroR	0.853	1	1343	0.921
		Prism	0.862	0.92	1211	0.89

Conclusion

In this paper we solve the web page classification problem using Rule learners namely OneR, ZeroR and Prism. Our experimental results show that both Rule learners classify the web pages with acceptable accuracy. From our results we analyzed that Prism classifier classifies the web pages little bit more efficiently than the other two learners and are the better tool for solving web page classification problem. As a future work we try to combine Rule learners with other machine learning techniques to solve the problem more efficiently.

References

[1] Susan Dumais, Hao Chen, (2000) "Hierarchical Classification of Web Content", *SIGIR* 2000 , ACM, pp 256 –263.

[2] Chee-Hong Chan, Aixin Sun and Ee-Peng Lim(2001) "Automated Online News classification with Personalization", *Proceedings of the 4th international conference of Asian Digital Library (ICADL 2001),* pages 320-329, Bangalore, India.

[3] Aixin Sun,Ee-Peng Lim and Wee-Keong Ng(2002) "Web Classification Using Support Vector Machine", *Proceedings of the 4th international workshop on Web Information and Data Management held in conj.* With CIKM, USA,Nov 2002.

[4] Offer Drori,(2003) "Using frequently occurring words to identify the subject of a document", *Proceedings of Informing Science.*

[5] Rudy Setino and Huan Liu. (1997) "Feature Selection via Discretization", *IEEE Transactions on Knowledge and Data Engineering,* Vol 9,Issue 4, July 1997,642-645.

[6] Terdiman, D. (2005). Yahoo passes Google in search index capacity. *News.com*, August 8, 2005.

[7] Vafaie, H., and De Jong, K. (1993), "Robust feature selection algorithms", *Proceedings on Fifth International Conference on Tools with Artificial Intelligence*, Boston, MA: IEEE Computer Society Press, pp. 356--363.

[8] Yiming Yang and John O. Pederson (1997), "Comparative study of feature selection in text categorization", *Proceedings on Fourteenth International Conference on Machine Learning (ICML'97)*, pp. 412-420.

[9] Tom M. Mitchell (1999), "The role of unlabeled data in supervised learning", *Proceedings of the Sixth International Colloquium on Cognitive Science,* San Sebastian, Spain.

[10] Salzberg, S. (1992) "Improving classification methods via feature selection", *Proceedings of ACM.*

INDEX

A

B

D

E

F

G

M

N

O

P

Q

R

S

T

U

V

W

X

Y